Mass Spectrometric Proteomics 2.0

Mass Spectrometric Proteomics 2.0

Editor

Paolo Iadarola

MDPI

Basel • Beijing • Wuhan • Barcelona • Belgrade • Novi Sad • Cluj • Manchester

Editor
Paolo Iadarola
Universita degli Studi di Pavia
Pavia
Italy

Editorial Office
MDPI
St. Alban-Anlage 66
4052 Basel, Switzerland

This is a reprint of articles from the Special Issue published online in the open access journal *International Journal of Molecular Sciences* (ISSN 1422-0067) (available at: https://www.mdpi.com/journal/ijms/special_issues/Spectrometric_Proteomics_2021).

For citation purposes, cite each article independently as indicated on the article page online and as indicated below:

Lastname, A.A.; Lastname, B.B. Article Title. *Journal Name* **Year**, *Volume Number*, Page Range.

ISBN 978-3-7258-0635-5 (Hbk)
ISBN 978-3-7258-0636-2 (PDF)
doi.org/10.3390/books978-3-7258-0636-2

Cover image courtesy of Paolo Iadarola

Contents

About the Editor

Paolo Iadarola

Paolo Iadarola graduated in Chemistry at the University of Pavia in 1972, and is currently an Adjunct Professor of the Department of Biology and Biotechnologies "L.Spallanzani" at the same university. Throughout his career, he has been involved in several projects including the extraction, purification, and structural characterization of proteins from different sources, including proteomic studies and biomarker discoveries for the identification of potential markers of chronic obstructive pulmonary disease (COPD) and other pulmonary disorders (COVID-19) in the biological fluids (urine, plasma, sputum, exhaled breath condensate) of patients.

Preface

The intention of the Special Issue dedicated to mass spectrometric proteomics is to highlight the results obtained when applying sophisticated methods that have the ability to identify proteins from biological fluids/cells/tissues, as well as to address an array of biological questions (including measuring dynamic changes in protein expressions, modifications, and interactions). The final purpose is to investigate the molecular mechanisms behind different human disorders. All researchers who study protein modifications involved in a specific pathology will benefit from the information obtained through the application of this strategy.

Paolo Iadarola
Editor

International Journal of
Molecular Sciences

MDPI

ditorial

Mass Spectrometric Proteomics 2.0

aolo Iadarola [1,*] and Simona Viglio [2,3]

1 Department of Biology and Biotechnologies "L.Spallanzani", University of Pavia, 27100 Pavia, Italy
2 Department of Molecular Medicine, Biochemistry Unit, University of Pavia, 27100 Pavia, Italy; simona.viglio@unipv.it
3 Lung Transplantation Unit, IRCCS Policlinico San Matteo Foundation, 27100 Pavia, Italy
* Correspondence: paolo.iadarola@unipv.it

Citation: Iadarola, P.; Viglio, S. Mass Spectrometric Proteomics 2.0. *Int. J. Mol. Sci.* **2024**, *25*, 2960. https://doi.org/10.3390/ijms25052960

Received: 17 February 2024
Accepted: 29 February 2024
Published: 4 March 2024

This Special Issue, "Mass Spectrometric Proteomics 2.0", presents the second volume of a series dedicated to the dissemination of results obtained from the application of MS-based proteomics across different areas. Although not all problems in MS proteomics have been solved, this is currently a mature technique that has prompted the publication of thousands of articles in the biochemical sector to date [1]. That being said, the question arises as to whether the release of a "new" series on proteomics would make sense. A careful reading of this Special Issue suggests that it can rightly find its own place in the international literature among the journals dedicated to proteomics. The contributions that this volume has attracted have dealt with such diverse fields of life science research that they confirm, if required, the leadership of MS-based proteomics in handling the complexity of biological challenges that other methods cannot manage. Flexibility, reliability, and speed of execution are the three fundamental features of proteomics that have emerged from these papers. Interestingly, most of these reports are not purely focused on cataloging the total proteins present in any given samples but also provide new insights into the role played by some proteins in specific human disorders. This provides proof that proteomics has outgrown its infancy and entered the era of "systems biology". Given its unique characteristics, proteomics, in synergy with other complementary methods (i.e., transcriptomics and metabolomics), makes it possible to obtain a global and integrated view of biological questions [2]. Another aspect that has not been neglected in this volume is the pivotal role played by proteomics in the area of biomarker discovery for the early diagnosis of human disorders. On the assumption that modifications occurring in an organism in response to different stimuli can be mirrored (to some degree) by changes in protein profiles, investigating how proteins are modulated between different conditions may shed light on the biological mechanisms involved in these processes [3,4]. The area focused on the investigation by MS of the proteome dynamics over time has witnessed enormous progress in recent years [5,6]. A crucial role in MS proteomics is also played by the availability of sophisticated methodological strategies to explore the molecular mechanisms behind different human disorders on a deeper level. In this context, an article that delves into the development of a new software tool for the visualization and validation of protein turnover rates must be acknowledged [7].

Given these premises, we hope that this new volume, "Mass Spectrometric Proteomics 2.0", will soon be published.

Author Contributions: Writing—original draft preparation, P.I.; writing—review and editing, S.V. All authors have read and agreed to the published version of the manuscript.

Conflicts of Interest: The authors declare no conflict of interest.

References

1. Chandrasekhar, K.; Dileep, A.; Ester Lebonah, D.; Pramoda Kumari, J. A short review on proteomics and its applications. *Int. Let* *Nat. Sci.* **2014**, *12*, 77–84. [CrossRef]
2. Patterson, S.D.; Aebersold, R.H. Proteomics: The first decade and beyond. *Nat. Genet.* **2003**, *33*, 311–323. [CrossRef]
3. Drabovich, A.P.; Pavlou, M.P.; Batruch, I.; Diamandis, E.P. Proteomic and mass spectrometry technologies for biomarker discover In *Proteomic and Metabolomic Approaches to Biomarker Discovery*, 2nd ed.; Issaq, H.J., Veenstra, T.D., Eds.; Academic Press (Elsevier Waltham, MA, USA, 2013; pp. 17–37.
4. Geyer, P.E.; Holdt, L.M.; Teupser, D.; Mann, M. Revisiting biomarker discovery by plasma proteomics. *Mol. Syst. Biol.* **201** *13*, 942. [CrossRef]
5. Al-Amrani, S.; Al-Jabri, Z.; Al-Zaabi, A.; Alshekaili, J.; Al-Khabori, M. Proteomics: Concepts and applications in human medicin *World J. Biol. Chem.* **2021**, *12*, 57–69. [CrossRef] [PubMed]
6. Mann, M.; Kumar, K.; Zeng, W.F.; Strauss, M.T. Artificial intelligence for proteomics and biomarker discovery. *Cell Syst.* **2021**, *1* 759–770. [CrossRef]
7. Deberneh, H.M.; Sadygov, R.G. Software tool for visualization and validation of protein turnover rates using heavy wate metabolic labeling and LC-MS. *Int. J. Mol. Sci.* **2022**, *23*, 14620. [CrossRef]

International Journal of
Molecular Sciences

MDPI

Article

Comparative Analysis of Different Proteins and Metabolites in the Liver and Ovary of Local Breeds of Chicken and Commercial Chickens in the Later Laying Period

Yuan Tang †, Lingqian Yin †, Li Liu, Qian Chen, Zhongzhen Lin, Donghao Zhang, Yan Wang and Yiping Liu *

Farm Animal Genetic Resources Exploration and Innovation Key Laboratory of Sichuan Province, Sichuan Agricultural University, Chengdu 611130, China; 2021202030@stu.sicau.edu.cn (Y.T.); yinlingqian@stu.sicau.edu.cn (L.Y.); liuli55@stu.sicau.edu.cn (L.L.); 2021302176@stu.sicau.edu.cn (Q.C.); zzlin599@163.com (Z.L.); 2019202006@stu.sicau.edu.cn (D.Z.); wangyan519@sicau.edu.cn (Y.W.)

* Correspondence: 11982@sicau.edu.cn

† These authors contributed equally to this work.

Abstract: The liver and ovary perform a vital role in egg production in hens. In the later laying period, the egg-laying capacity of female hens, particularly that of local breeds, declines significantly. Hence, it is essential to study the features and conditions of the ovary and liver during this period. In this research, we characterized the proteins and metabolites in the liver and ovary of 55-week-old Guangyuan gray chickens (Group G) and Hy-Line gray chickens (Group H) by using liquid chromatography chip/electrospray ionization quadruple time-of-flight/mass spectroscopy (LC-MS/MS). In total, 139 differentially expressed proteins (DEPs) and 186 differential metabolites (DMs) were identified in the liver, and 139 DEPs and 36 DMs were identified in the ovary. The upregulated DEPs and DMs in both the liver and ovary of Group G were primarily enriched in pathways involved in amino acid and carbohydrate metabolism. This suggests that energy metabolism was highly active in the Guangyuan gray chickens. In contrast, the upregulated DEPs and DMs in Group H were mainly enriched in pathways associated with lipid metabolism, which may explain the higher egg production and the higher fatty liver rate in Hy-Line gray hens in the later laying period. Additionally, it was found that the unique protein s-(hydroxymethyl) glutathione dehydrogenase (ADH4) in Group G was implicated in functions such as fatty acid degradation, glycolysis, and pyruvate metabolism, whereas the unique proteins, steroid sulfatase (STS), glucosylceramidase (LOC107050229), and phospholipase A2 Group XV (PLA2G15), in Group H were involved in the metabolism of steroid hormones and glycerol phosphate. In conclusion, variations in how carbohydrates, lipids, and amino acids are processed in the liver and ovary of local breeds of chicken and commercial hens towards the end of their laying period could explain the disparities in their egg production abilities.

Keywords: aged hens; proteomics; metabolomics; UHPLC-MS/MS; reproductive performance

Citation: Tang, Y.; Yin, L.; Liu, L.; Chen, Q.; Lin, Z.; Zhang, D.; Wang, Y.; Liu, Y. Comparative Analysis of Different Proteins and Metabolites in the Liver and Ovary of Local Breeds of Chicken and Commercial Chickens in the Later Laying Period. *Int. J. Mol. Sci.* **2023**, *24*, 14394. https://doi.org/10.3390/ijms241814394

Academic Editor: Paolo Iadarola

Received: 6 August 2023
Revised: 19 September 2023
Accepted: 20 September 2023
Published: 21 September 2023

1. Introduction

Poultry play a vital role in providing food for humans, and the reproductive performance of poultry directly impacts the development and benefits of the poultry industry. The utilization cycle of commercial chickens has been extended from 72 to 80 weeks as a result of enhancements in egg production capabilities [1], and Hy-Line chickens still maintain an egg production rate of about 80% at 53 weeks of age [2]. However, the egg production performance of laying hens declines significantly in the later laying period, which has a considerable impact on economic profitability. For the purpose of reducing breeding costs and increasing the utilization effectiveness, extending the egg-laying cycle with better reproductive performance of hens during the later laying period is a crucial task.

Previous research on animals' reproductive performance has primarily concentrated on studying the ovary and liver. In the later laying period, the ovary and liver of chickens undergo changes in their structure and declines in their function, which subsequently impact their reproductive capacity [3]. The proper development of follicles in the ovary is crucial for egg production, and this process is regulated by the hypothalamic–pituitary–gonad (HPG) axis [4,5]. As chickens reach the later laying period, the release of sex hormones decreases, the number of pre-grade follicles decreases, and yolk synthesis and accumulation reduces, ultimately resulting in a decline in ovarian function [6]. The liver has multiple functions in poultry. It plays a major part in fatty acid metabolism and is responsible for producing the building blocks needed for yolk formation [7]. Estrogen (E2) controls this process and ensures that the follicles receive the necessary nutrients [8]. However, laying hens often accumulate excessive amounts of fat due to continuous egg production and high-energy diets during the later stages of laying [3]. This often leads to liver failure [9] and lipid metabolism disorders [10,11] in laying hens. Consequently, the overall health and egg production performance of the poultry during the later stages of laying can be adversely affected.

Previously, researchers have focused on improving the reproductive performance of later laying hens through feed additives to improve the synthesis of hormones [12,13], eggshell quality [14], and antioxidant capacity [15–17]. With the development of technologies such as the transcriptomics, many genes and transcription factors have been identified that affect the reproductive performance of poultry during the later laying period. For example, *ACSF2*, a gene involved in lipid metabolism, was highly expressed in the liver of high-producing chickens [18]. However, transcriptomic information does not fully explain the complexity and dynamics of the regulation of reproduction in poultry. Proteomics and metabolomics, as new technologies in the post-genomic era, provide new insights into and means to understanding the molecular changes in reproduction in poultry. Proteomics and metabolomics are now widely used in avian breeding, such as in screening for key growth factors by evaluating differently expressed proteins (DEPs) and differential metabolites (DMs) in the pectoral muscle and intramuscular fat of chickens with contrasting growth rates [19], investigating the mechanisms of the functional regulation of granulosa cells in poultry by identifying DEPs and DMs in the ovary of laying chickens before and after sexual maturity [20], and screening for disease biomarkers of laying hens with fatty liver hemorrhagic syndrome (FLHS) by proteomics [21].

Until now, few studies have been reported on key regulatory proteins and metabolites in the liver and ovary of later laying chickens. The Guangyuan gray chicken is a high-quality meat and egg chicken in China. However, as it is a local breed, its egg production rate is only 80% during the peak laying period, and it mostly stops laying eggs at 55 weeks of age. In this study, to investigate the DEPs and DMs in the liver and ovary of local breeds of chicken and commercial laying hens during the later laying period, liver and ovary tissues from Guangyuan gray chickens (Group G) and Hy-Line gray chickens (Group H) were collected for proteomic and metabolomic analyses to distinguish the different proteins and metabolites in the liver and ovary of different chicken breeds during the later laying period. This research will lay the foundation for improving the performance of Guangyuan gray chickens in the later laying period.

2. Results

2.1. Comparison of the Morphological and Histological Characteristics of the Liver and Ovary between Guangyuan Gray Chickens and Hy-Line Gray Chickens

Body weight, liver weight, ovarian weight, oviduct length, and follicle number were collected from Guangyuan gray chickens (Group G) and Hy-Line gray chickens (Group H). As shown in Table 1, compared to Group H, Group G had a markedly lower body weight, liver weight, number of yellow follicles (YF, 5–10 mm), number of F1–F6 follicles, and oviduct length, but no apparent difference in counts of white follicles (WF, <5 mm).

According to the morphological and histological observations, the livers of Group G were smaller in size and reddish in color (Figure 1A), and HE staining showed that the cytoplasm of the hepatocytes in Group G was mildly stained, but the nuclei of the cells were darkly stained (Figure 1B). Oil Red O staining showed that the cellular interstitial space was narrow, and the distribution of lipid droplets was uniform (Figure 1C). In contrast, the livers from Group H were larger and yellowish in color (Figure 1D). The hepatocytes were neatly and clearly arranged, with a pink cytoplasm and light blue nuclei (Figure 1E), and the cytoplasm was sparse with large lipid droplets (Figure 1F). Observations of the ovaries showed that the ovaries in Group G contained only a small number of primordial follicles and pre-grade follicles (Figure 1G,H), and the ovaries were low in lipid droplets and had a low rate of yolk deposition in the follicles (Figure 1I). In contrast, Group H had ovaries containing more than four graded follicles (>12 mm), follicles filled with yolk and containing more than two layers of granulosa cells (Figure 1J,K), and ovaries with a high content of lipid droplets and a high rate of yolk deposition in the follicles (Figure 1L).

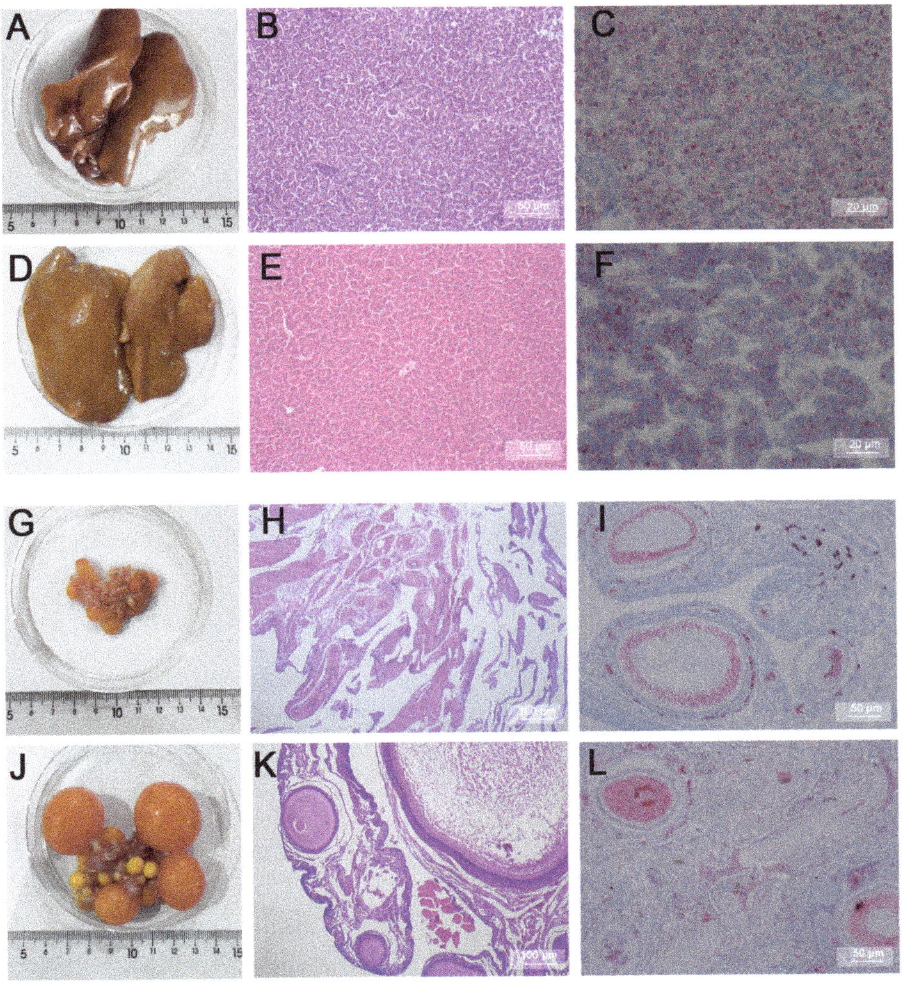

Figure 1. Histological analysis of the liver and ovary in Groups G and H. (**A**) appearance of the liver from Group G. (**B**,**C**) HE result and Oil Red O result of liver staining in Group G. (**D**) appearance of the liver from Group H. (**E**,**F**) HE result and Oil Red O result of liver staining in Group H. (**G**) appearance of the ovaries from Group G. (**H**,**I**) HE result and Oil Red O result of ovary staining in Group G. (**J**) appearance of the ovaries from Group H. (**K**,**L**) HE result and Oil Red O result of ovary staining in Group H. n = 3.

Table 1. Measurements of liver and ovary indices at 55 weeks in Groups G and H.

Groups	Body Weight (g)	Liver Weight (g)	Number of F1–F6 Follicles	Number of YF	Number of WF	Oviduct Length (mm)
G	1611.00 ± 194.44	29.89 ± 3.06	1.50 ± 2.12	12.80 ± 3.39	17.70 ± 5.73	43.25 ± 10.59
H	1858.00 ± 185.34 **	36.21 ± 1.13 **	5.70 ± 0.48 **	16.80 ± 3.15 *	14.20 ± 2.85	76.32 ± 6.68 **

YF, yellow follicles, diameter 5–10 mm; WF, white follicles, diameter < 5 mm. All results are presented as the mean ± SD. n = 10. * $p < 0.05$, ** $p < 0.01$.

2.2. Identification of Differentially Expressed Proteins in the Liver and Ovary

Prior to the LC-MS/MS analysis, the quality of the samples was examined using BCA and SDS-PAGE, which showed that no degradation of the proteins occurred and that the amount of protein was sufficient for the subsequent experiments (Tables S1 and S2 Figure S1). The results revealed that in the liver, the number of spectra, peptides, and proteins identified in the non-redundant (NR) database and the Uniprot database were 140,517, 30,026, and 4167, respectively (Figure S2A). The mass distribution of the proteins showed that the minority of proteins had a mass of 180–190 kDa, while the majority of proteins were distributed in the 20–30 kDa range (Figure S2B). The results for the distribution of protein coverage showed that the majority of proteins had a peptide sequence coverage of 5–10% and the minority of proteins had a coverage of 90–95% (Figure S2C). In the ovary, the number of spectra, peptides, and proteins identified was 159,983, 34,497, and 4964, respectively (Figure S2D). The mass distribution and peptide sequence coverage of the proteins were the same as in the liver (Figure S2E,F). A principal component analysis (PCA) of the proteins revealed a high degree of intra-group aggregation and significant inter-group separation, proving the reliability of the model (Figure 2A,B). The results of the DEP analysis showed that 139 DEPs were identified in the liver between Groups G and H of which 41 were downregulated and 98 were upregulated (Figure 2C). In total, 139 DEPs were identified in the ovary, of which 73 were downregulated and 66 were upregulated (Figure 2D). A hierarchical clustering analysis showed that the significant DMs obtained from the previous analysis could effectively separate the two groups (Figure 2E,F). Notably, the volcano plots and hierarchical clustering analyses did not include unique proteins for both groups, whereas all subsequent functional analyses included unique proteins.

2.3. Functional Enrichment Analysis of Differentially Expressed Proteins in the Liver and Ovary

A GO analysis showed that DEPs in the liver were involved in several biological processes (BP), with more DEPs included in functions such as cellular processes, metabolic processes, and bio-regulation. Among the cellular components (CC) in which the DEPs were involved, more DEPs were included in functions such as cells, cell fractions, and organelles, while among the molecular functions (MF) in which the DEPs were involved, more DEPs were included in functions such as binding and catalytic activity (Figure 3A). The results of the GO analysis of DEPs in the ovary were consistent with those of the liver (Figure 3B).

The results of the KEGG analysis showed that the DEPs identified in the liver were enriched in 93 pathways. Among them, the upregulated DEPs were enriched in 46 and 13 unique pathways in the livers of Groups G and H, respectively. Among the unique pathways in Group G, four pathways were associated with amino acid metabolism, four pathways were associated with carbohydrate metabolism, and three pathways were involved in lipid metabolism. Among them, ALDH18A1, which is involved in lipid metabolism, and ADH4, which is involved in a variety of metabolic pathways, were unique proteins in Group G. In Group H, DEPs were enriched in two pathways related to lipid metabolism and in the pathway related to amino acid metabolism. All three DEPs were involved in tryptophan metabolism, with CYP1A4 being a unique protein in Group H (Tables 2 and S3). The DEPs upregulated in Groups G and H in the ovary were enriched in 36 and 33 unique pathways, respectively. In Group G, seven pathways were involved in amino acid metabolism and four pathways were involved in carbohydrate metabolism. Notably, ADH4 was also expressed in the ovary of Group G. It is involved in the fatty acid degradation and tyrosine metabolism pathways in the liver, and in the retinol metabolism, methane metabolism, pyruvate metabolism, and tyrosine metabolism pathways in the ovary. In Group H, in addition to the pathways related to amino acid metabolism and carbohydrate metabolism, three pathways associated with the lipid metabolism and oocyte meiosis pathways were also identified (Tables 3 and S4). In addition, STS, LOC107050229, and PLA2G15 were enriched in the lipid metabolic pathway in the ovary, and were also unique proteins in Group H.

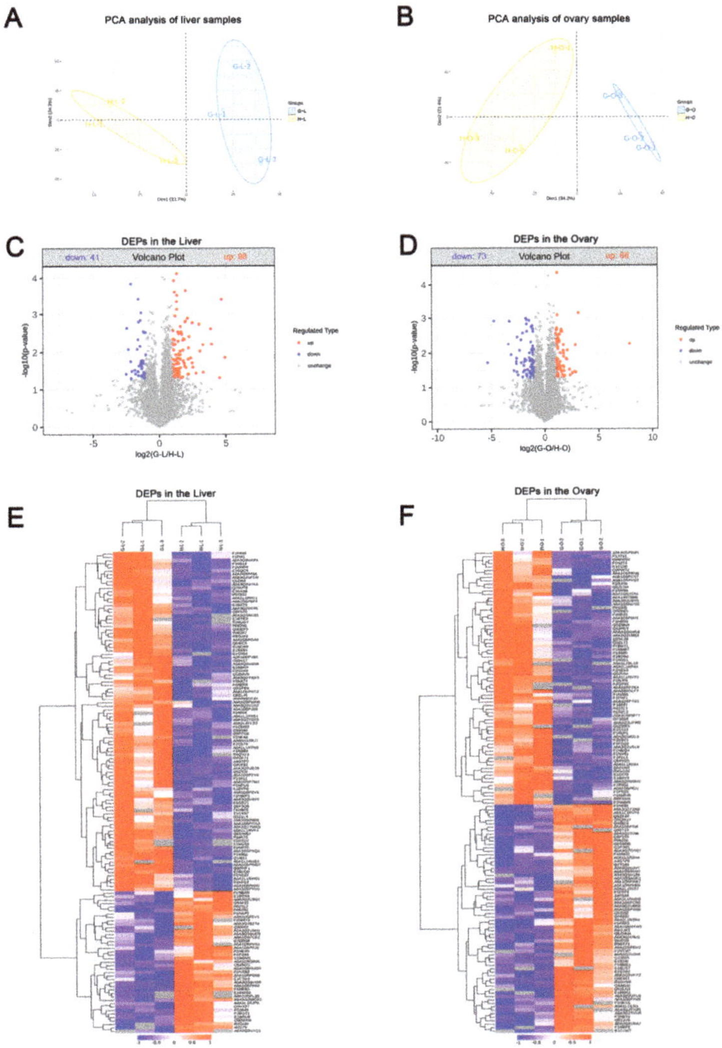

Figure 2. Differential expression analysis of proteins in the liver and ovary between Groups G and H. (**A**,**B**) principal component analysis (PCA) of proteins in the liver and ovary. (**C**,**D**) volcano map of differentially expressed proteins (DEPs) in the liver and ovary. Upregulated DEPs were indicated by red dots, downregulated DEPs by blue dots, and proteins with no significant changes by gray dots. (**E**,**F**) hierarchical clustering analysis of DEPs in the liver and ovary. Red and blue regions indicate significantly upregulated or downregulated proteins, respectively; gray regions indicate no quantitation information.

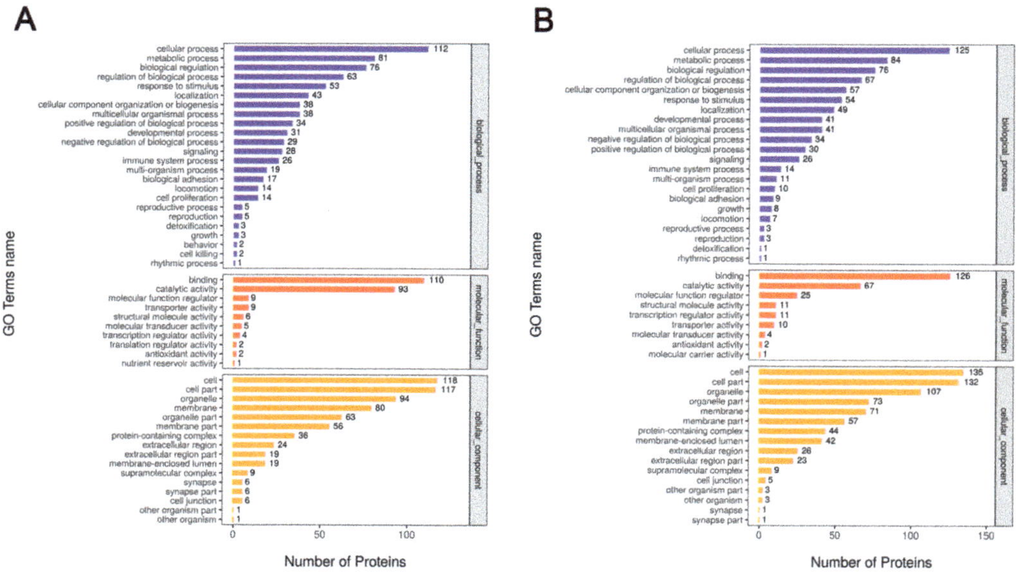

Figure 3. Bar graphs of the GO annotations of the DEPs. (**A**) GO functional annotations of DEPs in the liver including biological processes (BP), cellular components (CC), and molecular functions (MF). The *y*-axis indicates the number of proteins. (**B**) GO functional annotations of DEPs in the ovary.

Table 2. Pathways of upregulated differentially expressed proteins (DEPs) enriched in the liver between Groups G and H.

Group	Classification	KEGG Pathway	Protein ID	Name
G	Carbohydrate metabolism	Pyruvate metabolism	sp I P07032 I ACYP1_CHICK	ACYP1
		Pyruvate metabolism	tr I E1BZT9 I E1BZT9_CHICK	ACSS1B
		Pyruvate metabolism	tr I F1NI89 I F1NI89_CHICK	ADH4
		Propanoate metabolism	tr I E1BZT9 I E1BZT9_CHICK	ACSS1B
		Glycolysis, gluconeogenesis	tr I E1BZT9 I E1BZT9_CHICK	ACSS1B
		Glycolysis, gluconeogenesis	tr I F1NI89 I F1NI89_CHICK	ADH4
		Glycolysis, gluconeogenesis	tr I A0A3Q2UE29 I A0A3Q2UE29_CHICK	HK2
		Fructose and mannose metabolism	tr I A0A3Q2UE29 I A0A3Q2UE29_CHICK	HK2
	Amino acid metabolism	Valine, leucine, and isoleucine degradation	tr I A0A1D5P6E8 I A0A1D5P6E8_CHICK	CDH15
		Tyrosine metabolism	tr I F1NI89 I F1NI89_CHICK	ADH4
		Histidine metabolism	tr I E1BVP5 I E1BVP5_CHICK	ASPA
		Arginine and proline metabolism	tr I F1N914 I F1N914_CHICK	PYCR1
		Arginine and proline metabolism	tr I E1BSN3 I E1BSN3_CHICK	L3HYPDH
		Arginine and proline metabolism	tr I A0A1L1RR71 I A0A1L1RR71_CHICK	ALDH18A1
	Lipid metabolism	Fatty acid degradation	tr I F1NI89 I F1NI89_CHICK	ADH4
		Fatty acid biosynthesis	tr I A0A1L1RR71 I A0A1L1RR71_CHICK	ALDH18A1
		Fatty acid biosynthesis	sp I P21611 I B2MG_CHICK	B2M
		Arachidonic acid metabolism	tr I Q5F370 I Q5F370_CHICK	PTGR2
H	Lipid metabolism	Cutin, suberin, and wax biosynthesis	sp I Q5ZM72 I FACR1_CHICK	FAR1
		Biosynthesis of unsaturated fatty acids	tr I E7EDS8 I E7EDS8_CHICK	FADS1
	Amino acid metabolism	Tryptophan metabolism	sp I P79760 I CP1A4_CHICK	CYP1A4
		Tryptophan metabolism	tr I F1NEK9 I F1NEK9_CHICK	ACMSD
		Tryptophan metabolism	tr I A0A1D5P0J6 I A0A1D5P0J6_CHICK	TDO2

Table 3. Pathways of upregulated DEPs enriched in the ovary between Groups G and H.

Group	Classification	KEGG Pathway	Protein ID	Name
G	Amino acid metabolism	Tyrosine metabolism	tr I F1NTM7 I F1NTM7_CHICK	GOT1
		Phenylalanine, tyrosine, and tryptophan biosynthesis	tr I F1NTM7 I F1NTM7_CHICK	GOT1
		Cysteine and methionine metabolism	tr I F1NTM7 I F1NTM7_CHICK	GOT1
		Arginine biosynthesis	tr I F1NTM7 I F1NTM7_CHICK	GOT1
		Arginine and proline metabolism	tr I F1NTM7 I F1NTM7_CHICK	GOT1
		Alanine, aspartate, and glutamate metabolism	tr I F1NTM7 I F1NTM7_CHICK	GOT1
		Glycolysis, tyrosine metabolism	tr I F1NI89 I F1NI89_CHICK	ADH4
	Carbohydrate metabolism	Starch and sucrose metabolism	tr I A0A3Q2U8J2 I A0A3Q2U8J2_CHICK	G6PC3
		Pyruvate metabolism	tr I F1NI89 I F1NI89_CHICK	ADH4
		Glycolysis, gluconeogenesis	tr I A0A3Q2U8J2 I A0A3Q2U8J2_CHICK	G6PC3
		Glycolysis, gluconeogenesis	tr I F1NI89 I F1NI89_CHICK	ADH4
		Galactose metabolism	tr I A0A3Q2U8J2 I A0A3Q2U8J2_CHICK	G6PC3
H	Amino acid metabolism	Valine, leucine, and isoleucine degradation	tr I A0A1L1RZ64 I A0A1L1RZ64_CHICK	HMGCS1
		Tryptophan metabolism	tr I A0A3Q3AW56 I A0A3Q3AW56_CHICK	GCDH
		Lysine degradation	tr I A0A3Q3AW56 I A0A3Q3AW56_CHICK	GCDH
		Histidine metabolism	tr I E1C378 I E1C378_CHICK	HNMT
	Carbohydrate metabolism	Propanoate metabolism	tr I F1P5S5 I F1P5S5_CHICK	AC0SS3
		Pentose phosphate pathway	tr I A0A1L1RP44 I A0A1L1RP44_CHICK	RBKS
		Citrate cycle (TCA cycle)	tr I Q5F3V2 I Q5F3V2_CHICK	RCJMB04_6f14
		Butanoate metabolism	tr I A0A1L1RZ64 I A0A1L1RZ64_CHICK	HMGCS1
	Lipid metabolism	Steroid hormone biosynthesis	tr I A0A1D5PEM6 I A0A1D5PEM6_CHICK	STS
		Sphingolipid metabolism	tr I A0A1D5NWE6 I A0A1D5NWE6_CHICK	LOC107050229
		Glycerophospholipid metabolism	tr I A0A1D5PU31 I A0A1D5PU31_CHICK	PLA2G15
	Cell growth and death	Oocyte meiosis	tr I A0A3Q2U5L9 I A0A3Q2U5L9_CHICK	RPS6KA1

2.4. Identification of Differential Metabolites in the Liver and Ovary

A comparison of the spectral overlap of the total ion chromatograms (TIC) of quality control (QC) samples was carried out and the results showed that the response strengths and holding times of the peaks were basically superimposed, suggesting that there was little alteration due to instrumental errors during the whole experiment (Figure S3). A partial least squares discriminant analysis (PLS-DA) was conducted on all metabolites identified in the positive and negative ion modes to assess the association between the metabolites' expression and the samples' class. In addition, an orthogonal PLS-DA (OPLS-DA) was used to modify the results of the PLS-DA (Figure S4).

In the liver, the results of the PLS-DA showed more significant intra-group aggregation and inter-group segregation of the samples, indicating that the model was reliable (Figure 4A,B). By using UHPLC-MS/MS, 1619 differential metabolites (DMs) with identifying information were found in Groups G and H in this study ($p < 0.05$). Chemical classification showed that the largest proportion of these metabolites was made up of organic acids and derivatives (28.474%), followed by lipids and lipid-like molecules (17.542%) (Figure 4C). The results of the analysis of DMs showed that 978 metabolites were upregulated and 1514 metabolites were downregulated in the liver between two groups (Fold change (FC) > 1.5 or FC < 0.67) (Figure 4D,E). According to the results of analyzing the significant DMs, 186 significant DMs with annotation information were identified in the liver between two groups, of which 63 DMs were upregulated and 77 DMs were downregulated. A correlation analysis showed that 119 DMs were identified as relevant in the positive ion mode and 67 DMs were identified in the negative ion mode (Figure 4F,G). In the ovary, the PLS-DA analysis also showed the high reliability of the model (Figure 5A,B). In the positive and negative ion modes, 1191 DMs with annotation information were identified, most of which belonged to lipids and lipid-like molecules (28.67%), followed by organic acids and derivatives (21.662%) (Figure 5C). The analysis of the DMs showed that 223 DMs were upregulated and 225 DMs were downregulated (Figure 5D,E). In total, 36 significant

DMs with annotation information were screened, of which 17 DMs were upregulated and 12 DMs were downregulated. The correlation analysis showed that 14 DMs were identified to be correlated in the positive ion mode, while 22 DMs were identified to be correlated in the negative ion mode (Figure 5F,G).

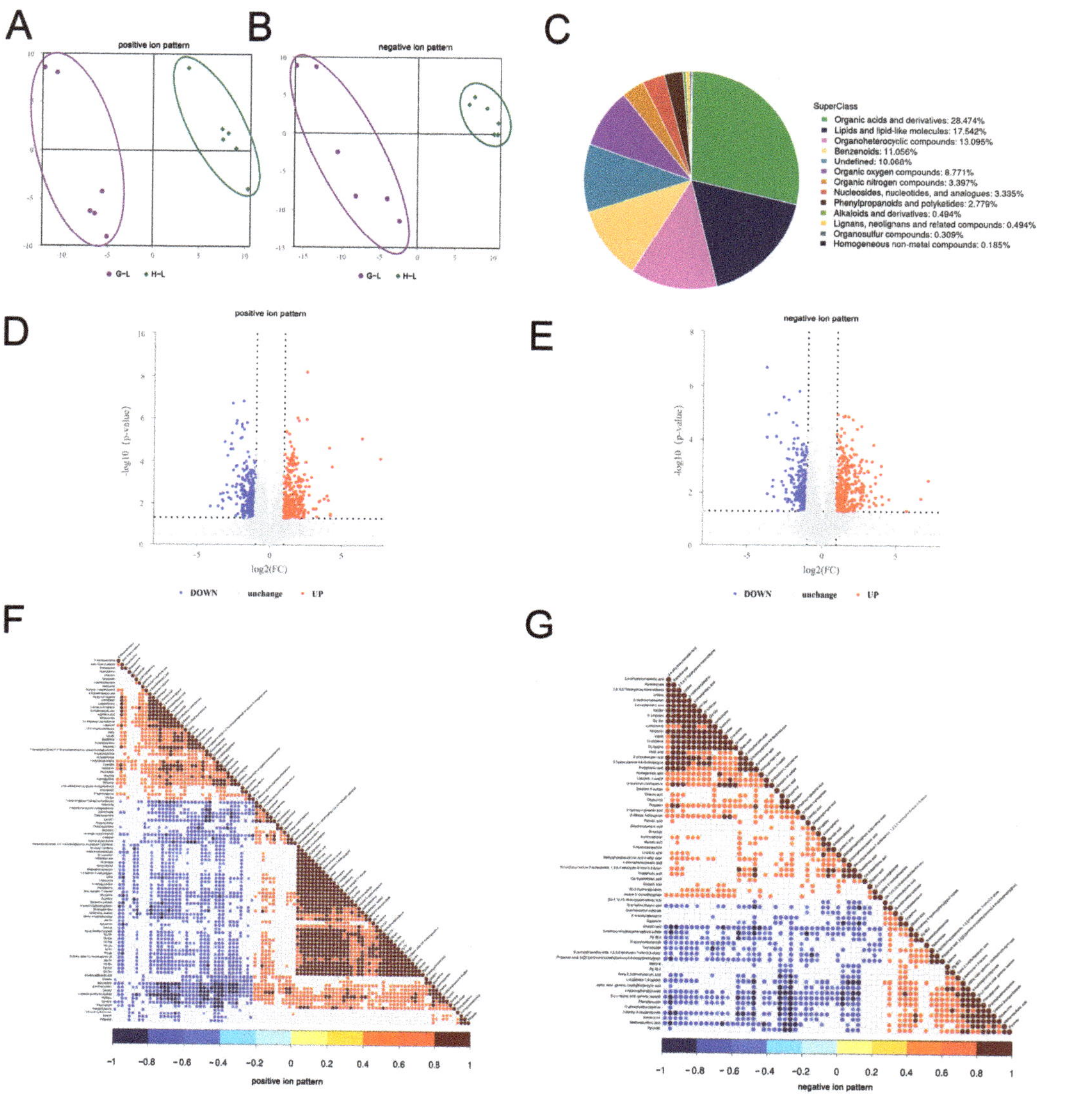

Figure 4. The patterns of differential metabolites (DMs) in the liver between Groups G and H were examined in the positive and negative ion modes. (**A**,**B**) PLS-DA distribution of 12 samples in the positive and negative ion modes. (**C**) the chemical classification of all liver metabolites was determined. (**D**,**E**) volcano plots displayed the DMs identified in the positive and negative ion modes. Upregulated DMs were indicated by red dots, downregulated DMs by blue dots, and metabolites with no significant changes by gray dots. (**F**,**G**) heat maps depicted correlations in the positive and negative ion modes, where red represented positive associations, blue represented negative associations, and white represented non-significant associations.

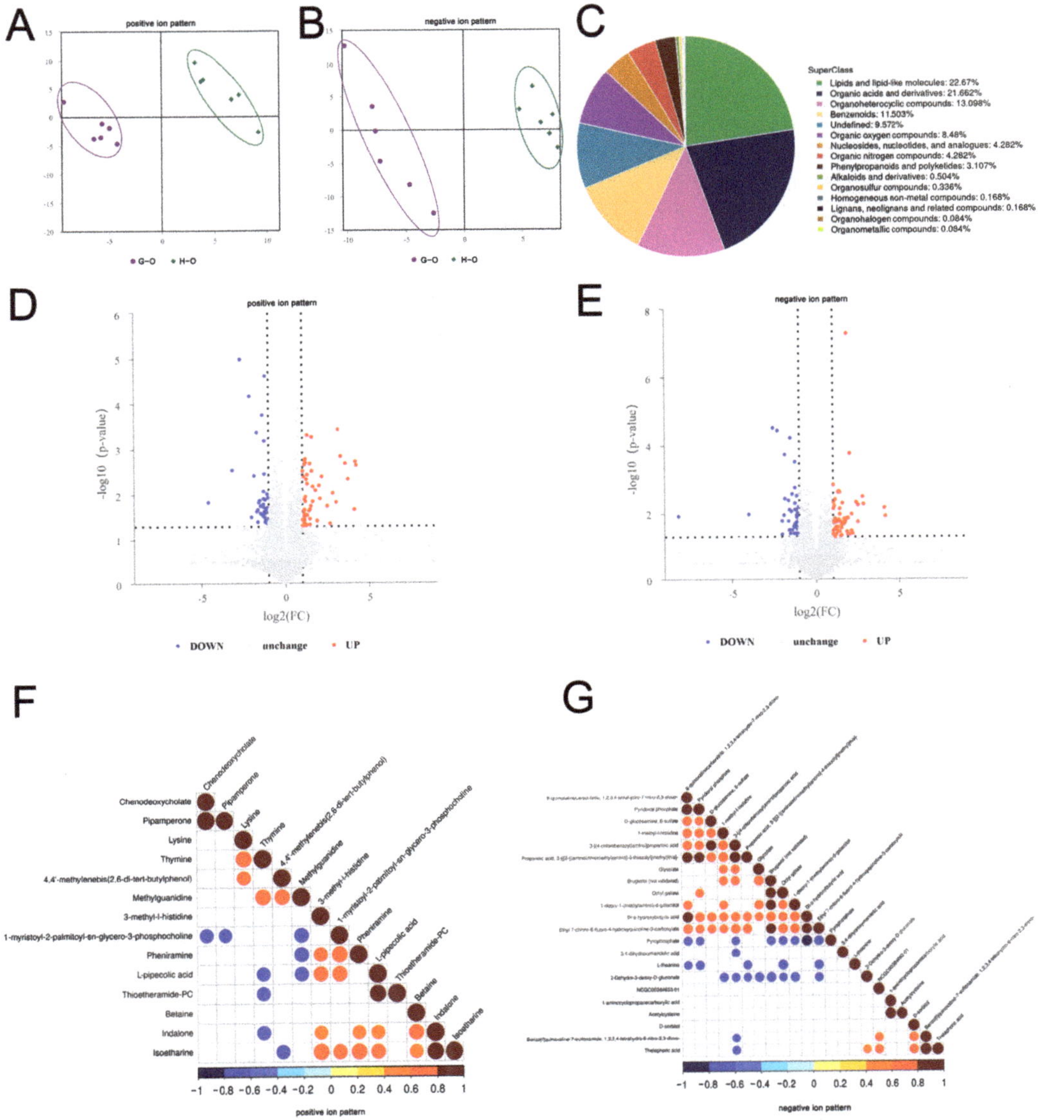

Figure 5. The patterns of DMs in the ovary between Groups G and H were examined in the positive and negative ion modes. (**A**,**B**) PLS-DA distribution of 12 samples in the positive and negative ion modes. (**C**) the chemical classification of all ovary metabolites was determined. (**D**,**E**) volcano plots displayed the DMs identified in the positive and negative ion modes. Upregulated DMs were indicated by red dots, downregulated DMs by blue dots, and metabolites with no significant changes by gray dots. (**F**,**G**) heat maps depicted correlations in the positive and negative ion modes, where red represented positive associations, blue represented negative associations, and white represented non-significant associations.

2.5. Functional Enrichment Analysis of Differential Metabolites in the Liver and Ovary

The significant DMs identified in this study were functionally annotated with reference to the KEGG and HMDB databases, and DMs without annotation information were excluded. The results showed that the upregulated DMs were enriched in 42 and 14 unique pathways in the liver of Groups G and H, respectively. In Group G, six pathways were associated with amino acid metabolism and six pathways were associated with carbohydrate metabolism. In addition, three pathways were linked to the endocrine system and two pathways were linked to lipid metabolism. In contrast, in Group H, four pathways were associated with lipid metabolism, and DMs were also involved in amino acid metabolism and translation. Among them, the pathways related to amino acid metabolism in Group G mainly included valine, leucine, histidine, tyrosine, and phenylalanine, while the DMs related to amino acid metabolism in Group H were all involved in the tryptophan metabolism pathway (Tables 4 and S5). In addition, DMs upregulated in the ovaries of Groups G and H were enriched in six and eight unique pathways, respectively. In Group G, two pathways were linked to amino acid metabolism and two pathways were associated with carbohydrate metabolism. In Group H, four pathways were involved in carbohydrate metabolism and two pathways were involved in amino acid metabolism. Furthermore, the DMs were enriched in ABC transporters and biosynthesis of primary bile acid (Tables 5 and S6).

Table 4. Pathways of upregulated differential metabolites (DMs) enriched in the liver between Groups G and H.

Group	Classification	KEGG Pathway	Metabolite
Group G	Amino acid metabolism	Valine, leucine, and isoleucine degradation	Ketoleucine
		Valine, leucine, and isoleucine biosynthesis	Ketoleucine, pyruvate
		Histidine metabolism	4-Imidazoleacetic acid, anserine, histamine, N-acetylhistamine
		Alanine, aspartate, and glutamate metabolism	Pyruvate
		Phenylalanine, tyrosine, and tryptophan biosynthesis	4-Hydroxyphenylpyruvate, phenylpyruvate
		Phenylalanine metabolism	Phenylpyruvate, pyruvate
	Carbohydrate metabolism	Pyruvate metabolism	Pyruvate
		Propanoate metabolism	Dl-a-hydroxybutyric acid
		Pentose and glucuronate interconversions	Pyruvate
		Glycolysis, gluconeogenesis	Pyruvate
		Citrate cycle (TCA cycle)	Pyruvate
		Ascorbate and aldarate metabolism	L-arabinono-1,4-lactone, pyruvate
	Endocrine system	Progesterone-mediated oocyte maturation	Camp
		Insulin signaling pathway	Camp
		GnRH signaling pathway	Camp
	Lipid metabolism	Sphingolipid metabolism	O-phosphoethanolamine
		Primary bile acid biosynthesis	Taurocholate
Group H	Lipid metabolism	Glycerolipid metabolism	Glyceric acid
		Fatty acid elongation	Palmitic acid
		Fatty acid biosynthesis	Cis-9-palmitoleic acid, palmitic acid
		Alpha-linolenic acid metabolism	Linolenic acid
	Amino acid metabolism	Tryptophan metabolism	Serotonin, 3-(2-hydroxyethyl) indole, 3-hydroxyanthranilic acid, N-acetyl-5-hydroxytryptamine

Table 5. Pathways of upregulated DMs enriched in the ovary between Groups G and H.

Group	Classification	KEGG Pathway	Metabolite
Group G	Amino acid metabolism	Lysine degradation	L-pipecolic acid
		Histidine metabolism	1-Methyl-l-histidine, 3-methyl-l-histidine
	Carbohydrate metabolism	Propanoate metabolism	Dl-a-hydroxybutyric acid
		Glyoxylate and dicarboxylate metabolism	Glycolate
Group H	Carbohydrate metabolism	Pentose phosphate pathway	2-Dehydro-3-deoxy-D-gluconate
		Pentose and glucuronate interconversions	2-Dehydro-3-deoxy-D-gluconate
		Galactose metabolism	D-sorbitol
		Fructose and mannose metabolism	D-sorbitol
	Amino acid metabolism	Tyrosine metabolism	3,4-Dihydroxymandelic acid
		Cysteine and methionine metabolism	1-Aminocyclopropanecarboxylic acid
	Membrane transport	ABC transporters	D-sorbitol
	Lipid metabolism	Primary bile acid biosynthesis	Chenodeoxycholate

2.6. Comprehensive Analysis of DEPs and DMs in the Liver and Ovary

By comparing the unique pathways of DEPs and DMs enriched in the liver and ovary of Group G and H, the overlapping enrichment pathways of DEPs and DMs were screened. The analysis showed that in the liver of Group G, upregulated DEPs and DMs were co-enriched in the MAPK signaling pathway, glutathione metabolism, ferroptosis, pyruvate metabolism, glycolysis/gluconeogenesis, propanoate metabolism, histidine metabolism, and the degradation of valine, leucine, and isoleucine (Figure 6A). In contrast, in Group H, upregulated DEPs and DMs were co-enriched in the tryptophan metabolism pathway (Figure 6B). In the ovary of Group G, upregulated DEPs and DMs were co-enriched in the thiamine metabolism pathway (Figure 6C), while in Group H, upregulated DEPs and DMs were co-enriched in ABC transporters and the pentose phosphate pathway (Figure 6D).

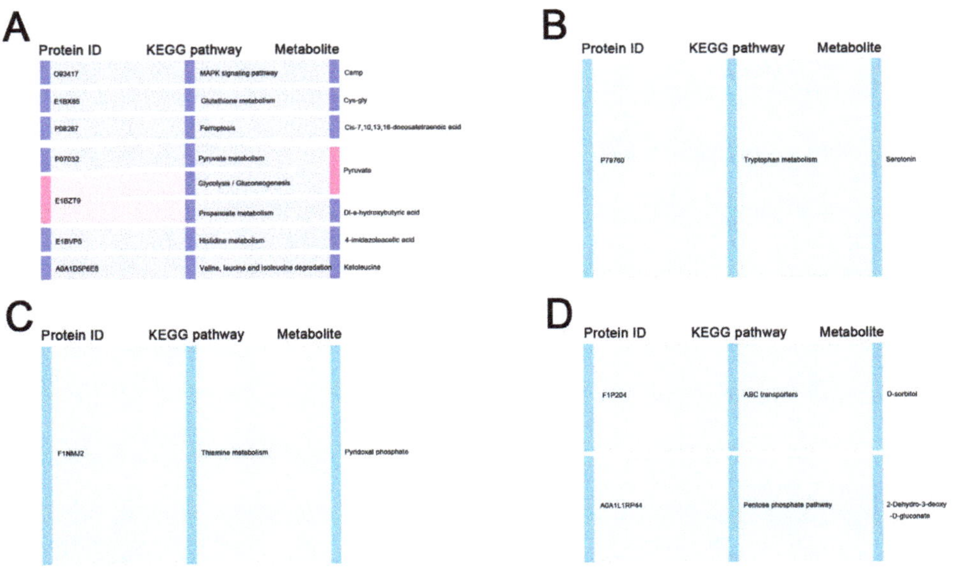

Figure 6. Comprehensive analysis of the KEGG pathways of DEPs and DMs in the liver and ovary between Groups G and H. (**A**,**B**) Sankey diagram of the KEGG pathways of DEPs and DMs in the liver of Groups G and H. (**C**,**D**) Sankey diagram of the KEGG pathways of DEPs and DMs in the ovary of Groups G and H.

2.7. Validation of DEPs Associated with Carbohydrate Metabolism and Lipid Metabolism Pathway in the Liver and Ovary

The biochemical indices in the serum of the two groups were determined. The result showed that the level of FSH in Group G was significantly lower than in Group H. The levels of the follicle stimulating hormone (FSH) (Figure 7A), luteinizing hormone (LH) (Figure 7B), lecithin (LEC) (Figure 7C), vitellogenin (VTG) (Figure 7D), very-low-density lipoprotein (VLDLy) (Figure 7E), total cholesterol (TC) (Figure 7F), and triglyceride (TG) (Figure 7G) were significantly lower than those in Group H. We also validated the protein expression levels of the DEPs. In the liver, the expression of ADH4 was significantly higher in Group G than in Group H, and it was hardly expressed in Group H. And the protein level of FADS1 was significantly higher in Group H than in Group G (Figure 7H,I). In the ovary, the protein levels of STS and HMGCS1 were significantly higher in Group H than in Group G, and STS was basically not expressed in Group G (Figure 7J,K). The protein expression trends of these DEPs were consistent with the results of proteomic analysis.

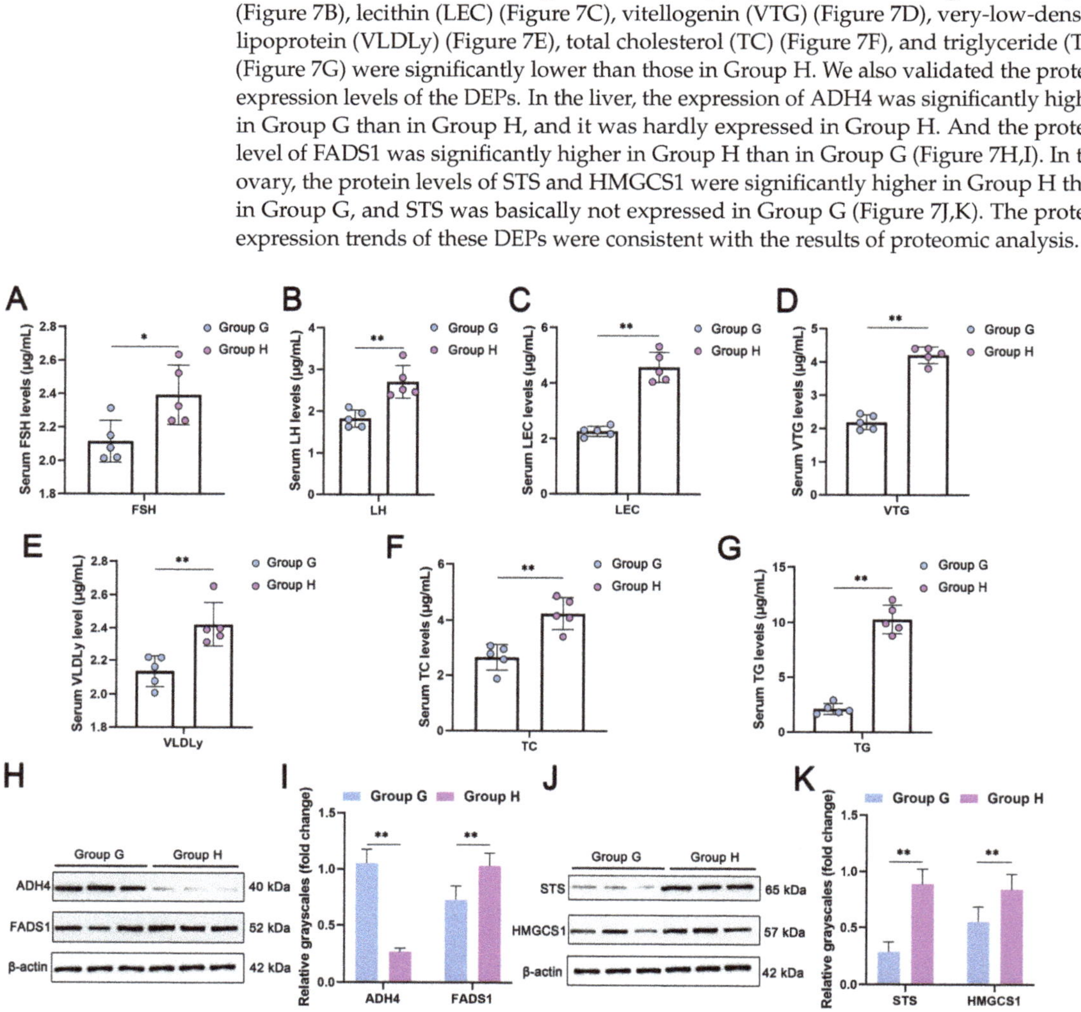

Figure 7. Validation of blood biochemical indicators and protein expression of DEPs found in Groups G and H. (**A**) follicle stimulating hormone (FSH) levels in serum. (**B**) luteinizing hormone (LH) levels in serum. (**C**) lecithin (LEC) levels in serum. (**D**) vitellogenin (VTG) levels in serum. (**E**) very-low-density lipoprotein (VLDLy) levels in serum. (**F**) total cholesterol (TC) levels in serum. (**G**) triglyceride (TG) levels in serum. (**H**,**I**) protein expression of DEPs (ADH4 and FADS1) enriched in carbohydrate and lipid metabolic pathways in the liver. (**J**,**K**) protein levels of DEPs (STS and HMGCS1) enriched in carbohydrate and lipid metabolic pathways in the ovary. All results are presented as the mean ± SD. n = 3. * $p < 0.05$; ** $p < 0.01$.

3. Discussion

The laying performance of local breeds of chickens differs considerably from that of commercial laying hens and remains so during the late laying period [22]. The liver is the major organ of energy metabolism and fatty acid metabolism [21], and yolk deposition in the follicle depends on the transport of lipids and proteins from the liver to the ovary [23]. In this study, the proteomic and metabolomic profiles were compared between the liver and ovary of 55-week-old Guangyuan gray chickens and Hy-Line gray chickens. Morphological and histological studies showed that the livers of Group H were yellowish in color, with more loosely arranged liver cells, and were richer in lipid droplets than those of Group G under the same feeding conditions. The altered morphology of liver cells in Group H may be attributed to the long-term absorption of large amounts of nutrients, such as TC and TG, by the liver in order to maintain a high egg production [23]. Eventually, the fat absorbed by the liver exceeds the transport capacity of the apolipoproteins, leading to fatty liver syndrome (FLS) [24]. The ovaries of Group G were basically atrophied, whereas grade follicles filled with yolk still existed in the ovaries of Group H. In addition, the serum levels of FSH and LH, which are related to follicular development [25]; the yolk precursors LEC, VTG, and VLDLy [26]; and TC and TG, which are related to steroid hormone synthesis and lipid deposition [27], were significantly higher in Group H. All the results indicated that the liver and ovarian functions of commercial laying hens are still active in the later laying period, which is the reason for their superior laying performance.

In general, carbohydrate metabolism is more vigorous in the liver of broiler chickens, whereas lipid metabolism is more vigorous in laying hens [28] because broiler chickens need more energy to maintain rapid growth but laying hens need lipids to maintain egg production. Proteomics studies have shown that upregulated DEPs in the liver and ovary of Group G were mainly enriched in pathways related to carbohydrate metabolism, which may be because Guangyuan gray chickens, as a local breed, tend to be used for both meat and eggs. DEPs such as acetyl-coenzyme A synthetase (ACSS1B), s-(hydroxymethyl) glutathione dehydrogenase (ADH4), and hexokinase (HK2), which are involved in carbohydrate metabolism, were identified in the liver of Group G, among which acylphosphatase-1 (*ACYP1*) and *ACSS1B* were also shown to be involved in the regulation of heat stress in chickens [29,30], which may be because Guangyuan gray chickens have a more energetic metabolism and produce more heat, and therefore rely on ACYP1 and ACSS1B to reduce the apoptosis of liver cells in hot environments [29]. In the ovary, glucose-6-phosphatase (G6PC3) [31] and ADH4 have been identified as being involved in carbohydrate metabolism. Notably, ADH4, a unique protein of Group G involved in the catabolism of a variety of substrates such as ethanol, retinol, steroids, and lipids [32,33], has been demonstrated to be a key gene for energy metabolism in cancer cells [34] and can be critical in energy metabolism in the liver and ovary in Group G. Among the DEPs identified in the liver of Group H, only one DEP was enriched in a unique pathway related to carbohydrate metabolism. However, propionate-CoA ligase (ACSS3), ribokinase (RBKS), ATP-citrate synthase (RCJMB04_6f14), and hydroxymethylglutaryl-CoA synthase (HMGCS1) were identified to be involved in carbohydrate metabolism and were found in the ovary of Group H, of which *ACSS3* was also found to be involved in the regulation of heat stress [35] and *HMGCS1* is also involved in the degradation of amino acids and cholesterol [36], which may, to some extent, alleviate fatty liver syndrome in Hy-Line gray chickens.

Notably, in addition to carbohydrate and lipid metabolism, DEPs were enriched in pathways related to amino acid metabolism in Groups G and H. Among them, DEPs such as pyrroline-5-carboxylate reductase (PYCR1), ALDH18A1, aspartate aminotransferase, and cytoplasmic (GOT1) were identified to be significantly enriched in the arginine biosynthesis pathway in the liver of Group G, which is able to inhibit fatty acid synthesis and promote fatty acid β-oxidation [37], which may interfere with the synthesis of substances such as TG and TC in the liver of Guangyuan gray chicken.

On the other hand, most of the upregulated DEPs in the liver and ovary of Group H were enriched in pathways associated with lipid metabolism. No DEPs were identified in the ovary of Group G that were enriched in unique pathways related to lipid metabolism, but ADH4, beta-2-microglobulin (B2M), delta-1-pyrroline-5-carboxylate synthase (ALDH18A1), and 15-oxoprostaglandin 13-reductase (PTGR2), which are involved in lipid metabolism, were identified in the liver. Among them, ADH4 is involved in fatty acid degradation, whereas the overexpression of *PTGR2* suppressed *PPARG*'s transcriptional activity and inhibited lipid synthesis. ALDH18A1 was also a unique protein found in Group G. The loss of function of ALDH18A1 was found to be associated with larger lipid droplets and was positively correlated with leanness in chickens [38], which was consistent with our findings. On the other hand, DEPs such as fatty acid desaturase 1 (FADS1), cytochrome P450 1A4 (CYP1A4), and tryptophan 2,3-dioxygenase (TDO2) were identified in the liver of Group H. These are involved in lipid metabolism, of which CYP1A4 is a protein unique to Group H involved in the metabolism of a variety of compounds, including substances such as steroids and fatty acids [39]. It was identified that steroid sulfatase (STS), glucosylceramidase (LOC107050229), and phospholipase A2 Group XV (PLA2G15) were all involved in the metabolism of functional lipids in the ovary of group H. Among them, *STS* was identified as a key gene for the secretion of reproductive hormones [40]. These three DEPs were unique proteins of group H and may be associated with the high egg production of Hy-Line gray hens at the end of the laying period.

In metabolomic studies, the functional enrichment of DMs was consistent with that of the DEPs. The most abundant metabolites identified in the liver and ovary were organic acids and their derivatives, and lipids and lipid-like molecules, respectively. The involvement of pyruvate, Dl-a-hydroxybutyric acid, L-arabinono-1, 4-lactone, and glycolate in carbohydrate metabolism was identified in the liver and ovary of Group G. Among them, 4-lactone and glycolate have been have been discovered to be critical in fighting oxidative stress [41]. 2-Dehydro-3-deoxy-D-gluconate and D-sorbitol were identified in the ovary of Group H and are involved in carbohydrate metabolism. Sorbitol catalyzes the conversion of fructose in the liver. Excess fructose increases liver stress and contributes to the accumulation of fat [42]. Additionally, the more abundant lipids and lipid-like molecules in the liver of Group G were o-phosphoethanolamine and taurocholate; taurocholate is involved in the biosynthesis of primary bile acid, and bile acid can promote the digestion and absorption of fat [43].

The more abundant lipid and lipid-like molecules in the liver of Group H were glyceric acid, palmitic acid, cis-9-palmitoleic acid, and linolenic acid; the most abundant in the ovary was chenodeoxycholate. A former study found that the liver of laying hens with fatty liver hemorrhage syndrome (FLHS) contained high levels of beta-hydroxybutyric acid, oleic acid, palmitoleic acid, glutamic acid, and other metabolites, which could be used as biomarkers for diagnosing disease [44]. These results were consistent with our findings. Therefore, although Hy-Line gray hens performed better in terms of egg production than Guangyuan gray chickens, prolonged overlaying also caused liver lesions in Hy-Line gray hens. It is also noteworthy that DMs identified in the liver and ovary of Group H were also significantly enriched in the ABC transporter pathway, suggesting the higher transmembrane transport capacity for amino acids, sugars, and other nutrients in Hy-Line gray hens [45]. Moreover, 1-aminocyclopropanecarboxylic acid was identified in the ovary of Group H, which is involved in the synthesis of methionine, a major component of apolipoproteins that significantly increases serum TG levels in laying hens and is important for yolk deposition [46]. In summary, the results of this study suggest that DEPs and DMs involved in the metabolism of carbohydrates, lipids, and amino acids in the liver and ovary of local breeds of chicken and commercial chickens may be responsible for the differences in egg production performance between them in the later laying period.

4. Materials and Methods

4.1. Birds and Sample Collection

All animal experiments were carried out in accordance with the relevant regulations formulated by the Experimental Animal Operation Standards and Welfare Management Committee of Sichuan Agricultural University (approval no. DKY2021202030).

The chickens used in this experiment were reared by Sichuan Tianguan Agricultural and Animal Husbandry Co., Ltd. (Guangyuan, China). One hundred 1-day-old female chicks were reared, with fifty each of Guangyuan gray chicks and Hy-Line gray chicks. At six weeks of age, each breed was divided into 10 groups, with replicates in each group, and reared in a single cage. At 55 weeks of age, 10 healthy chickens with a similar bodyweight were randomly selected from each of the two breeds (Group G and Group H). Blood was obtained from the pterygoid vein, followed by weighing and euthanizing the chickens by neck dislocation. Liver and ovarian tissues were collected, the livers were weighed, and the pre-grade follicles and grade follicles were counted. The liver and ovary tissues were separated into two sections. One section was immersed in a 4% paraformaldehyde fixative (Sigma-Aldrich, St. Louis, MO, USA) for histological observations, and the other was rapidly refrigerated in liquid nitrogen and kept at −80 °C for later analysis.

4.2. Protein Extraction, Digestion and Quality Control

Six liver and ovary samples were collected from each breed for protein profiling. Three replicate groups were established for each of the different tissues of each breed, and each group consisted of a mixed sample of tissues from two chickens of the same breed. An SDT buffer (4% SDS, 1 mM DTT) was used for extracting protein under a high pH. The protein quantification was conducted with the BCA Kit (Bio-Rad, Hercules, CA, USA). Protein digestion followed the Filter-Aided Sample Preparation (FASP) procedure [47]. Each sample initially contained 200 μg of protein, which was mixed with 30 μL of the SDT buffer comprising 4% SDS, 100 mM DTT, and 150 mM Tris-HCl at pH 8.0. We then removed small molecules by treating the samples with a UA buffer containing 8 M urea. Then, 100 μL of iodoacetamide was added, and the samples were incubated in the dark for 30 min. The filters were washed sequentially with the UA buffer and a 25 mM NH_4HCO_3 buffer. Next, 4 μg of trypsin (Promega, Madison, WI, USA) was added to a protein/enzyme ratio of 50:1, and the protein suspension was digested at 37 °C overnight, then, the peptides in the filtrate were collected. The digested peptides from each sample were desalted on C18 cartridges (Empore™ SPE Cartridges C18 (standard density) on a bed with an inner diameter of 7 mm and a volume of 3 mL, Sigma-Aldrich, St. Louis, MO, USA), then, they were concentrated by vacuum centrifugation and reconstituted in 40 μL of 0.1% formic acid. Finally, the peptide content was calculated at 280 nm.

A quality control of the protein extraction process was performed using SDS-PAGE. Briefly, 20 μg of protein from each sample was mixed with the loading buffer diluted fivefold and heated for 5 min. The proteins were then separated on a 12.5% SDS-PAGE gel. Finally, we visualized the protein bands by staining with Coomassie Blue R-250.

4.3. LC-MS/MS Analysis of Proteins

In this study, a 4D label-free approach and shotgun approach were used for the proteomics experiments, i.e., ion mobility separation was added to the original three dimensions of proteomic separation (retention time, m/z, and intensity) [48]. The samples were separated using the HPLC liquid phase system Easy nLC, loaded with 400 ng of peptides per run. The peptides were loaded onto a reverse-phase trap column (Acclaim PepMap100, 100 μm × 2 cm, nanoViper C18, Thermo Fisher Scientific, Waltham, MA, USA) connected to the C18 reverse-phase analytical column (Easy Column; 10 cm long, 75 μm inner diameter, 3 μm resin, Thermo Fisher Scientific, Waltham, MA, USA) in Buffer A (0.1% formic acid). The separation of peptides was achieved using a linear gradient of Buffer B (84% acetonitrile and 0.1% formic acid) at a flow rate of 300 nL/min. Then, we conducted the LC-MS/MS analysis using a timsTOF Pro mass spectrometer (Bruker Daltonics, Billerica, MA, USA) coupled with Nanoelute equipment (Bruker Daltonics, Billerica, MA, USA). The mass spectrometer in this study was operated in positive ion mode and the data acquisition mode was the DDA-PASEF mode. The mass range of ion mobility for mass spectrometry acquisition was 100–1700 (m/z) and 0.6–1.6 (1/k0). Additionally, 10 cycles of PASEF MS/MS with a target intensity of 1.5 k and a threshold of 2500 were performed. To prevent repeated analysis, active exclusion with a release time of 15 s was enabled.

4.4. Bioinformatic Analysis

The NR databases (GenBank, Refseq, SwissProt, etc.) and the UniProt databases (SwissProt, TrEmbl) were used for identifying proteins in this study. MaxQuant software 1.6.1 was used for quantitative analysis and the QC of the protein mass spectrometry data [49 In the analysis, the raw data from the mass spectrometry analysis were entered and the relevant parameters were set to check the database for identification and quantitative analysis. The following configurations were used: a maximum tolerance of 2 missed sites was allowed, with primary and secondary ion mass tolerances set at ±6 ppm and 20 ppm Moreover, carbamidomethyl (C) and oxidation (M) were designated as fixed and variable modifications, respectively. The database pattern used to calculate the false discovery rate (FDR) was reversed, with a peptide FDR of ≤0.01 and a protein FDR of ≤0.01 as the screening criteria. A quantitative comparison of the proteins between groups was performed using LFQ, and the main algorithm was applied after pairwise correction of the peptide and protein multiples. FC > 2 or FC < 0.50, with $p < 0.05$, was used to screen DEPs that were significantly up- and downregulated between the groups, and volcano plots were generated using the volcano 3D R package (R 4.2.2). We utilized the Complexheatmap R 4.2.2 (R Foundation for Statistical Computing, Vienna, Austria), Cluster 3.0 (Michelson Laboratories Tokyo, Japan), and Java Treeview software 3.0 (Eisen Lab at Stanford University, Stanford CA, USA) for sample classification and protein expression level analysis. The classification was performed using the Euclidean distance algorithm and average linkage for clustering Hierarchical clustering heatmaps were generated based on significant changes in protein expression levels (FC > 2, $p < 0.05$). The DEPs were then functionally annotated with Gene Ontology (GO) terms using Blast2Go software (https://www.blast2go.com, accessed on 20 April 2023, BioBam Bioinformatics, Valencia, Spain), and Kyoto Encyclopedia of Genes and Genomes (KEGG) annotation was performed on the DEPs using KAAS software (https://www.genome.jp/tools/kaas, accessed on 20 April 2023, Kanehisa Laboratories at Kyoto University, Kyoto, Japan). In addition, unique proteins were obtained by using the screening criterion that half or more than half of the samples in one group were not null and all data in the other group were null.

4.5. Metabolite Extraction

Six liver and six ovarian samples from each of the two breeds were selected for metabolic profile analysis between the two breeds. After slowly thawing the samples, 5 mg of each sample was added to a pre-cooled solution of methanol, acetonitrile, and water in a ratio of 2:2:1. Then, we vortexed the mixture and treated it ultrasonically at a low temperature for 30 min, followed by ultrasonic treatment at −20 °C for 10 min, and we finally centrifuged the sample at 14,000× *g* at 4 °C for 20 min. The supernatant was then dried and 100 μL of an aqueous solution of acetonitrile was added. We centrifuged the mixture at 14,000× *g* for 15 min and collected the supernatant for ultra-high pressure liquid chromatograph (UHPLC)-MS/MS analysis. The temperature was maintained at 4 °C throughout the experiment.

4.6. UHPLC-MS/MS Analysis and Data Analysis

The samples were separated on a Vanquish LC UHPLC (Thermo Fisher Scientific, Waltham, MA, USA) with a HILIC column (ACQUITY UPLC BEH Amide, 1.7 μm, 2.1 mm × 100 mm, Waters, Milford, MA, USA). The detection conditions were as follows: column temperature, 25 °C; flow rate, 300 μL/min. The composition of Mobile Phase A was composed of water, 25 mM acetic acid, and 25 mM ammonia; Phase B was acetonitrile. By gradient elution, within 17 min, B was linearly changed from 98% to 2%, then to 98%, and finally returned to and maintained at 98%. QC samples were inserted in the sample queue to ensure the system stability and data reliability. Metabolite spectra were obtained using a Q Exactive Orbitrap mass spectrometer (Thermo Fisher Scientific, Waltham, MA, USA). The positive and negative ion modes of electric spray ionization (ESI) were used for primary and secondary mass spectra acquisition, with the following ESI settings: Gas1 and Gas2: 60, CUR: 30 psi, ion source temperature: 600 °C, ISVF: ±5500 V. Mass spectra were acquired in DDA mode with secondary spectra obtained by segmented acquisition. Primary and secondary spectra were acquired with the following settings: mass-to-charge ratio detection ranges of 80–1200 Da and 70–1200 Da, resolutions of 60,000 and 30,000, cumulative scan times of 100 ms and 50 ms, asnd a dynamic exclusion time of 4 s.

The raw data format was converted to mzXML using ProteoWizard. Metabolite data were extracted using XCMS software 3.5.1 (Scripps Research Institute, La Jolla, CA, USA), and a substance match value of 0.7 was obtained by considering retention time, molecular mass (molecular mass error: <25 ppm), and mass spectral match (m/z < 10 ppm, peak width = c (10, 60), prefilter = c (10, 100)). The obtained data were then identified using an in-house database (Shanghai Applied Protein Technology, Shanghai, China) and verified twice. Based on the standards for metabolite identification, the metabolites in this study were classified as Level 2 or higher [50]. PLS-DA was performed on the metabolites that differed between the groups, and the PLS-DA was corrected by OPLS-DA. FC analysis was performed on all metabolites, including those without identification information, to screen for up- or downregulated DMs on the basis of FC > 1.5 or FC < 0.67, and $p < 0.05$. Significant DMs were screened among DMs with identifying information using OPLS-DA, with variable importance in projection (VIP) > 1 as a criterion, and subsequent analysis was performed. Pearson's correlation analysis was used to identify the relationship between two breeds. In addition, we used Blast2GO and KAAS to perform GO and KEGG pathway annotation of the target metabolite collection. R package (ropls) was used to visualize analyses such as volcano maps and correlation heat maps.

4.7. Detection of Serum Biochemical Parameters

The separated serum was centrifuged at 4000 r/min to remove impurities. Then, the hormone levels were tested with an ELISA kit (MyBioSource, Wuhan, China) for chicken VTG, LH, FSH, LEC, and VLDLy according to the manufacturer's instructions. In addition, TC and TG test kits (Solarbio, Beijing, China) were used to detect the TC and TG levels. The assay was repeated 14 times for three QC samples (1.23 ng/mL, 4.25 ng/mL, and 19.45 ng/mL, respectively) to test the reproducibility of the kit between wells in the assay plate, and the results showed that the CV value of each sample was less than 10% (4.23%, 7.52%, and 7.67%, respectively), indicating that the kit had high precision. The results on the serum hormone levels were visualized using GraphPad Prism 9.0.0 (GraphPad Software, La Jolla, CA, USA).

4.8. Oil Red O Staining

Frozen samples were sectioned to 8 μm, fixed in 10% formalin (Solarbio, Beijing, China) for 10 min, and washed. The sections were immersed in 60% isopropanol (Sinopharm, Beijing, China) for 2 min. The sections were stained with an Oil Red O solution (Sangon Biotech, Shanghai, China) for 15 min while protected from the light. The sections were again immersed in 60% isopropanol for 5 s to remove the staining solution and washed again in ice-cold distilled water. The nuclei were stained with Mayer's hematoxylin (Sangon Biotech, Shanghai, China) for 5 min, washed, dried, and embedded in glycerol gelatin (Xilong Scientific, Shantou, China). The sections were observed under a BX53F inverted microscope (Olympus, Tokyo, Japan) and photographed.

4.9. Hematoxylin–Eosin (HE) Staining

The samples were first immobilized in 4% paraformaldehyde (Solarbio, Beijing, China) for 24 h, then immersed in 70%, 80%, 90%, 95%, and 100% ethanol solutions for 30 min to dehydrate the samples. They were then placed in xylene for 2 h to make the samples transparent and embedded in paraffin wax for 3 h. The embedded samples were sectioned into 5 μm pieces and immersed in xylene for 20 min to dewax the samples. The sections were then immersed in a series of ethanol solutions from high to low concentrations and finally in distilled water. The sections were stained with a hematoxylin solution (Beyotime,

Haimen, China) for 4 min, fractionated in hydrochloric acid and ethanol for 3 s each, rinsed in running water for 1 h, immersed in distilled water for 10 min, and dehydrated in 70% and 90% ethanol solutions for 10 min each, followed by staining with the eosin staining solution for 3 min. The stained sections were dehydrated by immersion in an ethanol solution and then immersed in xylene to make the sections transparent, and they were finally sealed and stained with gum. The sections were sealed with resin, observed under a BX53F inverted microscope (Olympus, Tokyo, Japan), and photographed.

4.10. Western Blot Validation of DMs Expression

After tissue collection, proteins were extracted using the Total Tissue Protein Extraction Kit (Servicebio, Wuhan, China) according to the provided instructions. Protein concentrations were determined using the BCA kit (Servicebio, Wuhan, China) and the samples were made consistent. Protein samples (5 μL) were loaded onto SDS-PAGE gels (separation gel: 10%, concentration gel: 5%) along with a 4:1 ratio of protein to sample loading buffer for electrophoresis. The proteins were then moved to the PVDF membranes with the wet transfer device and blocked for 1 h with blocking solution (Beyotime, Haimen, China). The primary antibodies (Table S7) were co-incubated with target proteins overnight at 4 °C, followed by removal of unbound antibodies using TBST (Servicebio, Wuhan, China). The membranes were then treated with appropriate secondary antibodies, excess secondary antibody was washed off, and the protein bands were visualized using the Ultra Hypersensitive ECL Chemiluminescence Kit (Servicebio, Wuhan, China). Quantification was performed using ChemiScope Analysis software 6200 (CLINX, Shanghai, China).

4.11. Statistical Analysis

This study presented all the findings in the form of the mean ± standard deviation (SD). To determine the significance, we conducted a statistical analysis using either one-way ANOVA or an unpaired Student's *t*-test with the help of SPSS 26.0 software (IBM Corporation, Armonk, NY, USA). The significance levels were defined as $^{*} p < 0.05$, $^{**} p < 0.01$, and $^{ns} p \geq 0.05$.

5. Conclusions

In conclusion, in this study, morphological and histological observations of the liver and ovary of local breeds of chicken and commercial chickens in the later laying period revealed that the serum levels of reproductive hormones in Guangyuan gray chickens were significantly lower than those in Hy-Line gray chickens, there were fewer lipid droplets in the livers of the Guangyuan gray chickens, and their ovaries were atrophied. In contrast, the Hy-Line gray chickens had abundant lipid droplets in the liver and showed fatty liver-like lesions, but the ovaries were still active with follicles in all stages. The results of the proteomic and metabolomic analyses showed that the DEPs and DMs in the liver and ovary were mainly involved in carbohydrate metabolism, lipid metabolism, and amino acid metabolism. These findings can be used to improve the egg production of local breeds of chicken in the later laying period.

Supplementary Materials: The following supporting information can be downloaded at: https://www.mdpi.com/article/10.3390/ijms241814394/s1.

Author Contributions: Conceptualization, Y.T. and L.Y.; Data curation, Y.T., Q.C., L.L. and Z.L.; Funding acquisition, Y.L.; Investigation, D.Z. and Q.C.; Methodology, Y.T., L.Y. and L.L.; Supervision, Z.L.; Writing—original draft, Y.T. and L.Y.; Writing—review and editing, Y.W. and L.Y. All authors have read and agreed to the published version of the manuscript.

Funding: This research was funded by the National Key Research and Development Project (grant no. 2022YFD1601608), the Sichuan Province Key Research and Development Project (grant no. 2023YFQ0035), and the Sichuan Science and Technology Program (grant no. 2021YFYZ0031).

Institutional Review Board Statement: All animal experiments were carried out in accordance with the relevant regulations formulated by the Experimental Animal Operation Standards and Welfare Management Committee of Sichuan Agricultural University (approval no. DKY2021202030).

Informed Consent Statement: Not applicable.

Data Availability Statement: The mass spectrometry proteomics data have been deposited to the ProteomeXchange Consortium (http://proteomecentral.proteomexchange.org, accessed on 5 September 2023) via the iProX partner repository with the dataset identifier PXD045098. Metabolomics raw data have been deposited in Metabolights with the dataset identifier MTBLS8517.

Conflicts of Interest: The authors declare no conflict of interest.

Abbreviations

LC-MS/MS	liquid chromatography chip/electrospray ionization quadruple time of flight/mass spectroscopy
DEPs	differentially expressed protein
DMs	differential metabolites
HPG	hypothalamic-pituitary-gonad
E2	estrogen
FLHS	fatty liver hemorrhagic syndrome
YF	yellow follicles
WF	white follicles
NR	non-redundant
PCA	principal component analysis
BPs	biological processes
CCs	cellular components
MFs	molecular functions
TIC	total ion chromatograms
QC	quality control
PLS-DA	partial least squares discriminant analysis
OPLS-DA	orthogonal partial least squares discriminant analysis
FC	fold change
FSH	follicle stimulating hormone
LH	luteinizing hormone
LEC	lecithin
VTG	vitellogenin
VLDLy	very-low-density lipoprotein
TC	total cholesterol
TG	triglyceride
FLS	fatty liver syndrome
FASP	Filter-Aided Sample Preparation
FDR	false discovery rate
GO	Gene Ontology
KEGG	Kyoto Encyclopedia of Genes and Genomes
UHPLC	ultra-high pressure liquid chromatograph
ESI	electric spray ionization
VIP	variable importance for the projection
HE	Hematoxylin–Eosin

References

1. Liu, Z.; Sun, C.; Yan, Y.; Li, G.; Shi, F.; Wu, G.; Liu, A.; Yang, N. Genetic variations for egg quality of chickens at late laying period revealed by genome-wide association study. *Sci. Rep.* **2018**, *8*, 10832. [CrossRef]
2. Wang, J.; Zhang, H.; Xu, L.; Long, C.; Samuel, K.; Yue, H.; Sun, L.; Wu, S.; Qi, G. Dietary supplementation of pyrroloquinoline quinone disodium protects against oxidative stress and liver damage in laying hens fed an oxidized sunflower oil-added diet. *Animal* **2016**, *10*, 1129–1136. [CrossRef]
3. Song, J.; Huang, M.; Shi, X.; Li, X.; Chen, X.; He, Z.; Li, J.; Xu, G.; Zheng, J. T329S Mutation in the FMO3 Gene Alleviates Lipid Metabolic Diseases in Chickens in the Late Laying Period. *Animals* **2021**, *12*, 48. [CrossRef]

4. Hanlon, C.; Takeshima, K.; Bedecarrats, G.Y. Changes in the Control of the Hypothalamic-Pituitary Gonadal Axis Across Thre Differentially Selected Strains of Laying Hens (*Gallus gallus* domesticus). *Front. Physiol.* **2021**, *12*, 651491. [CrossRef]
5. Liu, H.; Wang, J.; Li, L.; Han, C.; He, H.; Xu, H. Transcriptome analysis revealed the possible regulatory pathways initiatin female geese broodiness within the hypothalamic-pituitary-gonadal axis. *PLoS ONE* **2018**, *13*, e0191213. [CrossRef]
6. Lillpers, K.; Wilhelmson, M. Age-dependent changes in oviposition pattern and egg production traits in the domestic hen. *Poul Sci.* **1993**, *72*, 2005–2011. [CrossRef]
7. Bai, L.; Li, H. Innate immune regulatory networks in hepatic lipid metabolism. *J. Mol. Med.* **2019**, *97*, 593–604. [CrossRef]
8. Walzem, R.; Davis, P.; Hansen, R. Overfeeding increases very low density lipoprotein diameter and causes the appearance of unique lipoprotein particle in association with failed yolk deposition. *J. Lipid Res.* **1994**, *35*, 1354–1366. [CrossRef]
9. Huang, Z.; Lv, Z.; Dai, H.; Li, S.; Jiang, J.; Ye, N.; Zhu, S.; Wei, Q.; Shi, F. Dietary mulberry-leaf flavonoids supplementatio improves liver lipid metabolism and ovarian function of aged breeder hens. *J. Anim. Physiol. Anim. Nutr.* **2022**, *106*, 1321–133 [CrossRef]
10. Shini, A.; Shini, S.; Bryden, W.L. Fatty liver haemorrhagic syndrome occurrence in laying hens: Impact of production syster *Avian Pathol.* **2019**, *48*, 25–34. [CrossRef]
11. Yao, H.; Hu, Y.; Wang, Q.; Zhang, Y.; Rao, K.; Shi, S. Effects of dietary dimethylglycine supplementation on laying performanc egg quality, and tissue index of hens during late laying period. *Poult. Sci.* **2022**, *101*, 101610. [CrossRef] [PubMed]
12. Jiang, J.; Qi, L.; Dai, H.; Hu, C.; Lv, Z.; Wei, Q.; Shi, F. Dietary stevioside supplementation improves laying performance an eggshell quality through increasing estrogen synthesis, calcium level and antioxidant capacity of reproductive organs in age breeder hens. *Anim. Feed Sci. Technol.* **2020**, *269*, 114682. [CrossRef]
13. Liu, H.-K.; Long, D.; Bacon, W. Interval between preovulatory surges of luteinizing hormone increases late in the reproductiv period in turkey hens. *Biol. Reprod.* **2002**, *66*, 1068–1075. [CrossRef] [PubMed]
14. Feng, J.; Zhang, H.J.; Wu, S.G.; Qi, G.H.; Wang, J. Uterine transcriptome analysis reveals mRNA expression changes associate with the ultrastructure differences of eggshell in young and aged laying hens. *BMC Genom.* **2020**, *21*, 770. [CrossRef] [PubMed]
15. He, W.; Wang, H.; Tang, C.; Zhao, Q.; Zhang, J. Dietary supplementation with astaxanthin alleviates ovarian aging in aged layin hens by enhancing antioxidant capacity and increasing reproductive hormones. *Poult. Sci.* **2023**, *102*, 102258. [CrossRef]
16. Li, H.; Hou, Y.; Chen, J.; Wu, H.; Huang, L.; Hu, J.; Zhang, Z.; Lu, Y.; Liu, X. Dietary naringin supplementation on layin performance and antioxidant capacity of Three-Yellow breeder hens during the late laying period. *Poult. Sci.* **2022**, *101*, 10202 [CrossRef]
17. Wu, H.; Yuan, J.; Yin, H.; Jing, B.; Sun, C.; Tsopmejio, I.S.N.; Jin, Z.; Song, H. Flammulina velutipes stem regulates oxidativ damage and synthesis of yolk precursors in aging laying hens by regulating the liver–blood–ovary axis. *Poult. Sci.* **2023**, *102* 102261. [CrossRef]
18. Tian, W.; Zheng, H.; Yang, L.; Li, H.; Tian, Y.; Wang, Y.; Lyu, S.; Brockmann, G.A.; Kang, X.; Liu, X. Dynamic expression profile regulatory mechanism and correlation with egg-laying performance of ACSF gene family in chicken (*Gallus gallus*). *Sci. Rep.* **2018** *8*, 8457. [CrossRef]
19. Liu, R.; Wang, H.; Liu, J.; Wang, J.; Zheng, M.; Tan, X.; Xing, S.; Cui, H.; Li, Q.; Zhao, G. Uncovering the embryonic development related proteome and metabolome signatures in breast muscle and intramuscular fat of fast-and slow-growing chickens. *BMC Genom.* **2017**, *18*, 816. [CrossRef]
20. Zhang, L.; Wang, E.; Peng, G.; Wang, Y.; Huang, F. Comprehensive Proteome and Acetyl-Proteome Atlas Reveals Hepatic Lipid Metabolism in Layer Hens with Fatty Liver Hemorrhagic Syndrome. *Int. J. Mol. Sci.* **2023**, *24*, 8491. [CrossRef]
21. Pearce, J. Carbohydrate metabolism in the domestic fowl. *Proc. Nutr. Soc.* **1971**, *30*, 254–259. [CrossRef] [PubMed]
22. Du, Y.; Liu, L.; He, Y.; Dou, T.; Jia, J.; Ge, C. Endocrine and genetic factors affecting egg laying performance in chickens: A review. *Br. Poult. Sci.* **2020**, *61*, 538–549. [CrossRef] [PubMed]
23. Lv, Z.; Xing, K.; Li, G.; Liu, D.; Guo, Y. Dietary genistein alleviates lipid metabolism disorder and inflammatory response in laying hens with fatty liver syndrome. *Front. Physiol.* **2018**, *9*, 1493. [CrossRef]
24. Trott, K.; Giannitti, F.; Rimoldi, G.; Hill, A.; Woods, L.; Barr, B.; Anderson, M.; Mete, A. Fatty liver hemorrhagic syndrome in the backyard chicken: A retrospective histopathologic case series. *Vet. Pathol.* **2014**, *51*, 787–795. [CrossRef]
25. Johnson, A.L. Regulation of follicle differentiation by gonadotropins and growth factors. *Poult. Sci.* **1993**, *72*, 867–873. [CrossRef] [PubMed]
26. Kashap, A.; Dalvi, R.A.S.; Kapale, P. Study of Serum Biochemical Metabolites during Late Laying Phase of Layer Chicken. *Indian Res. J. Ext. Educ.* **2017**, 5–9.
27. Dedousi, A.; Kritsa, M.-Z.; Đukić Stojčić, M.; Sfetsas, T.; Sentas, A.; Sossidou, E. Production performance, egg quality characteristics, fatty acid profile and health lipid indices of produced eggs, blood biochemical parameters and welfare indicators of laying hens fed dried olive pulp. *Sustainability* **2022**, *14*, 3157. [CrossRef]
28. Zheng, A.; Chang, W.; Liu, G.; Yue, Y.; Li, J.; Zhang, S.; Cai, H.; Yang, A.; Chen, Z. Molecular differences in hepatic metabolism between AA broiler and big bone chickens: A proteomic study. *PLoS ONE* **2016**, *11*, e0164702. [CrossRef]
29. Liu, Z.; Liu, Y.; Xing, T.; Li, J.; Zhang, L.; Jiang, Y.; Gao, F. Transcriptome analysis reveals the mechanism of chronic heat stress on meat quality of broilers. *J. Anim. Sci. Biotechnol.* **2022**, *13*, 110. [CrossRef]

30. Srikanth, K.; Kumar, H.; Park, W.; Byun, M.; Lim, D.; Kemp, S.; Te Pas, M.F.; Kim, J.-M.; Park, J.-E. Cardiac and skeletal muscle transcriptome response to heat stress in Kenyan chicken ecotypes adapted to low and high altitudes reveal differences in thermal tolerance and stress response. *Front. Genet.* **2019**, *10*, 993. [CrossRef]
31. Van Every, H.A.; Schmidt, C.J. Transcriptomic and metabolomic characterization of post-hatch metabolic reprogramming during hepatic development in the chicken. *BMC Genom.* **2021**, *22*, 380. [CrossRef] [PubMed]
32. Duester, G. Genetic dissection of retinoid dehydrogenases Gregg Duester. *Chem.-Biol. Interact.* **2001**, *130*, 469–480. [CrossRef] [PubMed]
33. Ambolet-Camoit, A.; Ottolenghi, C.; Leblanc, A.; Kim, M.J.; Letourneur, F.; Jacques, S.; Cagnard, N.; Guguen-Guillouzo, C.; Barouki, R.; Aggerbeck, M. Two persistent organic pollutants which act through different xenosensors (alpha-endosulfan and 2, 3, 7, 8 tetrachlorodibenzo-p-dioxin) interact in a mixture and downregulate multiple genes involved in human hepatocyte lipid and glucose metabolism. *Biochimie* **2015**, *116*, 79–91. [CrossRef]
34. Xu, B.; Peng, Z.; An, Y.; Yan, G.; Yao, X.; Guan, L.; Sun, M. Identification of energy metabolism-related gene signatures from scRNA-Seq data to predict the prognosis of liver cancer patients. *Front. Cell Dev. Biol.* **2022**, *10*, 858336. [CrossRef] [PubMed]
35. Zhang, Q.; Luo, Y.K.; Zhang, B.H.; Chan, Y.Z.; Huang, L.L.; Wang, Y.; Liang, J.M.; Zhang, X.Q. RNA-Seq study of hepatic response of yellow-feather chickens to acute heat stress. *Ann. Anim. Sci.* **2020**, *20*, 55–69. [CrossRef]
36. Shi, X.; Sun, X.; Xu, P.; Zhang, C.; Zhang, M.; Yuan, X.; Jiang, J.; Jin, K.; Chen, C.; Zuo, Q. HMGCS1 Promotes male differentiation of chicken embryos by regulating the generate of cholesterol. *All Life* **2021**, *14*, 577–587. [CrossRef]
37. Fouad, A.; El-Senousey, H.; Yang, X.; Yao, J. Dietary L-arginine supplementation reduces abdominal fat content by modulating lipid metabolism in broiler chickens. *Animal* **2013**, *7*, 1239–1245. [CrossRef]
38. Handley, M.T.; Mégarbané, A.; Meynert, A.M.; Brown, S.; Freyer, E.; Taylor, M.S.; Jackson, I.J.; Aligianis, I.A. Loss of ALDH 18A1 function is associated with a cellular lipid droplet phenotype suggesting a link between autosomal recessive cutis laxa type 3A and Warburg Micro syndrome. *Mol. Genet. Genom. Med.* **2014**, *2*, 319–325. [CrossRef]
39. Luptakova, L.; Dvorcakova, S.; Demcisakova, Z.; Belbahri, L.; Holovska, K.; Petrovova, E. Dimethyl Sulfoxide: Morphological, Histological, and Molecular View on Developing Chicken Liver. *Toxics* **2021**, *9*, 55. [CrossRef]
40. Li, J.; Li, C.; Li, Q.; Li, G.; Li, W.; Li, H.; Kang, X.; Tian, Y. Novel regulatory factors in the hypothalamic-pituitary-ovarian axis of hens at four developmental stages. *Front. Genet.* **2020**, *11*, 591672. [CrossRef]
41. Duque, P.; Vieira, C.P.; Vieira, J. Advances in novel animal vitamin C biosynthesis pathways and the role of prokaryote-based inferences to understand their origin. *Genes* **2022**, *13*, 1917. [CrossRef] [PubMed]
42. Kosmalski, M.; Śliwińska, A.; Drzewoski, J. Non-Alcoholic Fatty Liver Disease or Type 2 Diabetes Mellitus—The Chicken or the Egg Dilemma. *Biomedicines* **2023**, *11*, 1097. [CrossRef]
43. Hu, X.; Wang, Y.; Sheikhahmadi, A.; Li, X.; Buyse, J.; Lin, H.; Song, Z. Effects of glucocorticoids on lipid metabolism and AMPK in broiler chickens' liver. *Comp. Biochem. Physiol. Part B Biochem. Mol. Biol.* **2019**, *232*, 23–30. [CrossRef] [PubMed]
44. González-Muñoz, M.; Bastida, S.; Jiménez, O.; Vergara, G.; Sánchez-Muniz, F. The effect of dietary fat on the fatty acid composition and cholesterol content of Hy-line and Warren hen eggs. *Grasas Y Aceites* **2009**, *60*, 350–359. [CrossRef]
45. Li, J.; Yang, C.; Ran, J.; Yu, C.; Yin, L.; Li, Z.; Liu, Y. The age-dependent variations for fatty acid composition and sensory quality of chicken meat and associations between gene expression patterns and meat quality. *Livest. Sci.* **2021**, *254*, 104736. [CrossRef]
46. Nukreaw, R.; Bunchasak, C.; Markvichitr, K.; Choothesa, A.; Prasanpanich, S.; Loongyai, W. Effects of methionine supplementation in low-protein diets and subsequent re-feeding on growth performance, liver and serum lipid profile, body composition and carcass quality of broiler chickens at 42 days of age. *J. Poult. Sci.* **2011**, *48*, 229–238. [CrossRef]
47. Wiśniewski, J.R.; Zougman, A.; Nagaraj, N.; Mann, M. Universal sample preparation method for proteome analysis. *Nat. Methods* **2009**, *6*, 359–362. [CrossRef]
48. Cox, J.; Hein, M.Y.; Luber, C.A.; Paron, I.; Nagaraj, N.; Mann, M. Accurate proteome-wide label-free quantification by delayed normalization and maximal peptide ratio extraction, termed MaxLFQ. *Mol. Cell. Proteom.* **2014**, *13*, 2513–2526. [CrossRef]
49. Cox, J.; Mann, M. MaxQuant enables high peptide identification rates, individualized ppb-range mass accuracies and proteome-wide protein quantification. *Nat. Biotechnol.* **2008**, *26*, 1367–1372. [CrossRef]
50. Blaženović, I.; Kind, T.; Ji, J.; Fiehn, O. Software tools and approaches for compound identification of LC-MS/MS data in metabolomics. *Metabolites* **2018**, *8*, 31. [CrossRef]

International Journal of
Molecular Sciences

MDPI

Article

Less Severe Polymicrobial Sepsis in Conditional *mgmt*-Deleted Mice Using LysM-Cre System, Impacts of DNA Methylation and MGMT Inhibitor in Sepsis

Kritsanawan Sae-khow [1,2], Pornpimol Phuengmaung [2,3], Jiraphorn Issara-Amphorn [3], Jiradej Makjaroen [4], Peerapat Visitchanakun [2,3], Atsadang Boonmee [5], Salisa Benjaskulluecha [5], Tanapat Palaga [5,†] and Asada Leelahavanichkul [1,2,6,*,†]

1 Medical Microbiology, Interdisciplinary and International Program, Graduate School, Chulalongkorn University, Bangkok 10330, Thailand; kritsanawan_29@hotmail.com
2 Department of Microbiology, Faculty of Medicine, Chulalongkorn University, Bangkok 10330, Thailand; pphuengmaung@gmail.com (P.P.); peerapat.visitchanakun@gmail.com (P.V.)
3 Center of Excellence in Translational Research in Inflammation and Immunology (CETRII), Faculty of Medicine, Chulalongkorn University, Bangkok 10330, Thailand; jiraphorn298@gmail.com
4 Center of Excellence in Systems Biology, Research Affairs, Faculty of Medicine, Chulalongkorn University, Bangkok 10330, Thailand; jiradejmak@gmail.com
5 Department of Microbiology, Faculty of Science, Chulalongkorn University, Bangkok 10330, Thailand; atsadang88@gmail.com (A.B.); salisafaii@gmail.com (S.B.); tanapat.p@chula.ac.th (T.P.)
6 Division of Nephrology, Department of Medicine, Faculty of Medicine, Chulalongkorn University, Bangkok 10330, Thailand
* Correspondence: aleelahavanit@gmail.com; Tel.: +66-2-256-4132
† These authors contributed equally to this work.

Abstract: The O6-methylguanine-DNA methyltransferase (MGMT) is a DNA suicide repair enzyme that might be important during sepsis but has never been explored. Then, the proteomic analysis of lipopolysaccharide (LPS)-stimulated wild-type (WT) macrophages increased proteasome proteins and reduced oxidative phosphorylation proteins compared with control, possibly related to cell injury. With LPS stimulation, *mgmt* null (*mgmt*$^{flox/flox}$; LysM-Cre$^{cre/-}$) macrophages demonstrated less profound inflammation; supernatant cytokines (TNF-α, IL-6, and IL-10) and pro-inflammatory genes (*iNOS* and *IL-1β*), with higher DNA break (phosphohistone H2AX) and cell-free DNA, but not malondialdehyde (the oxidative stress), compared with the littermate control (*mgmt*$^{flox/flox}$; LysM-Cre$^{-/-}$). In parallel, *mgmt* null mice (MGMT loss only in the myeloid cells) demonstrated less severe sepsis in the cecal ligation and puncture (CLP) model (with antibiotics), as indicated by survival and other parameters compared with sepsis in the littermate control. The *mgmt* null protective effect was lost in CLP mice without antibiotics, highlighting the importance of microbial control during sepsis immune modulation. However, an MGMT inhibitor in CLP with antibiotics in WT mice attenuated serum cytokines but not mortality, requiring further studies. In conclusion, an absence of *mgmt* in macrophages resulted in less severe CLP sepsis, implying a possible influence of guanine DNA methylation and repair in macrophages during sepsis.

Keywords: sepsis; lipopolysaccharide; macrophages; epigenetics; *mgmt*

Citation: Sae-khow, K.; Phuengmaung, P.; Issara-Amphorn, J.; Makjaroen, J.; Visitchanakun, P.; Boonmee, A.; Benjaskulluecha, S.; Palaga, T.; Leelahavanichkul, A. Less Severe Polymicrobial Sepsis in Conditional *mgmt*-Deleted Mice Using LysM-Cre System, Impacts of DNA Methylation and MGMT Inhibitor in Sepsis. *Int. J. Mol. Sci.* **2023**, *24*, 10175. https://doi.org/10.3390/ijms241210175

Academic Editor: Paolo Iadarola

Received: 12 May 2023
Revised: 27 May 2023
Accepted: 29 May 2023
Published: 15 June 2023

1. Introduction

The imbalanced immune responses in patients with sepsis, a potentially life-threatening condition from severe infection regardless of the organisms, result in severe hyperinflammation, despite effective microbial control by antibiotics [1–3]. Although there is improved supportive care in sepsis [4], the inflammatory blockage during sepsis-hyperinflammation [5–12] may still help reduce sepsis mortality. Among several sepsis pro-inflammatory factors, the presence of LPS (a major cell wall component of Gram-negative bacteria) in the blood, referred to as endotoxemia [13–16], possibly due to Gram-negative bacteremia or translocation of LPS from the gut into the blood circulation (leaky gut) [17–19] is a well-known cause. Interestingly, monocytes (or macrophages) are important cells controlling microbial molecules (LPS and other substances) [20,21], and the responses to LPS lead to several cell activities, including epigenetic modifications, chromatin remodeling, and interferences on cell energy status [22–24]. Epigenetics, the phenotypic alterations without the changes in the DNA sequence [25,26] used for the switch-on and -off DNA transcription through the modifications on DNA and histone (methylation and acetylation) or with the noncoding RNA (microRNA) [27], are one of the interesting macrophage responses after LPS activation. Indeed, the modifications of DNA and histone rely on several enzymes to modify the physiological outcomes [28]. DNA methylation, among all modifications, for regulating DNA repair is a fascinating mechanism in LPS responses. Because any forms of oxidative stress from both regular procedures or pathogenic processes (such as LPS stimulation and exposure to alkylating agents) in the cells can induce DNA modifications, DNA methylation is a common process regularly found in every body cell [29]. Indeed, DNA methylation, especially the N-methylated purines and O6-methylguanine (O6MeG), are common DNA damage that can trigger point mutation with high mutagenicity and carcinogenicity [30]. The O6MeG, which the alkylating agent usually activates, is mutagenic because polymerase enzymes typically mismatch insert thymine instead of cytosine (O6MeG:T) due to the similar strength of the hydrogen bonds to cytosine and thymine [31]. Not only alkylating agents and environmental compounds but several endogenous factors, particularly oxidative stress, are also responsible for producing O6MeG during routine cell activities [32]. Base damage, single-strand breaks (SSB), double-strand breaks (DSB), and inter-strand cross-links are all types of DNA damage caused by normal metabolic processes (hydrolysis, deamination, alkylation, and oxidation) which occur about 50,000 lesions per cell per day (or about 30,000 nucleoside sites in DNA per cell) [33]. Although the number of modifications at the O6 position of guanine (O6MeG) may not be significantly high compared to the methylation in the total lesions on DNA [34], the amount of O6MeG may be increased in sepsis or LPS hyper-responsiveness due to high levels of oxidative stress as indirectly indicated by inflammation-induced cancer in some situations [35]. Interestingly, O6MeG may result in more severe DNA damage and cell death than methylation at the other DNA locations due to the easier point mutation and more potent DNA damage [36], which might also profoundly affect macrophages than other mechanisms of epigenetic changes.

To maintain genome stability, DNA repair is necessary, partly through the removal of methyl groups, on the DNA by base excision repair initiated by the alkyladenine-DNA glycosylase, the family of alkylation B (AlkB) homologs proteins, and the suicidal enzyme O6-methylguanine-DNA methyltransferase (MGMT) [30,37]. Following DNA repair, these enzymes' methylated and alkylated forms are rapidly broken down. Intact MGMT prevents the malignant transformation of various tissues. MGMT blocking is employed for adjuvant chemotherapy [30] in malignancies that depend on the rate of MGMT re-synthesis in the cancer cells [37]. The DNA methylation and damage in the activated macrophages are conceivable because of several activator signals and immune activation-induced oxidative stress [38], particularly after induction with LPS, a microbial molecule with an inflammatory-inducing solid property. Indeed, DNA methylation in macrophages is directly induced by LPS [39] and indirectly activated by reactive oxygen species (ROS) generated by LPS responses [40]. In patients with sepsis, higher levels of genomic DNA hypermethylation patterns are linked to pro-inflammatory pathways [41]. The effects of this methylation and MGMT on macrophages, the cells with a potentially high stress-induced DNA methylation [42], are still unknown, despite a strong conclusion on the involvement of O6MeG in malignant cells. Accordingly, sepsis and LPS cause DNA methylation in several sites, perhaps including Q6MeG [41], and the MGMT enzyme reduces O6MeG during the DNA repair that revitalizes the cell functions. Then, the MGMT inhibitors (such as Lomeguatrib) enhance the death of cancer cells through interference with DNA repair [43], and this interference on macrophages during sepsis may reduce macrophage hyper-inflammation [44]. Additionally, the screening of epigenetic inhibitors reveals that MGMT inhibitors change the expression of inflammatory cytokines in LPS activated macrophages [45], and the MGMT inhibitors not only neutralize O6MeG in DNA but also link to the repair of other pathways [46]. Despite several active research areas controlling macrophage responses through epigenetic manipulations is an exciting method for managing immune responses during sepsis [10,47].

Then, we hypothesized that failure of DNA repair causes a reduction in macrophage functions, especially cytokine production, that might attenuate sepsis-induced hyper inflammatory responses. Here, we explored the impact of *mgmt* on macrophage responses to LPS and cecal ligation and puncture (CLP) sepsis model using the conditional *mgmt* deletion mice with LysM-Cre system that selectively affected *mgmt* only in myeloid cells.

2. Results

2.1. Proteomic Analysis of Lipopolysaccharide (LPS)-Activated Wild-Type Macrophages and Impacts of mgmt Null and MGMT Inhibitor on LPS Stimulation

The alteration of wild-type (WT) macrophages after LPS stimulation was explored by proteomic analysis through the list of the genes that generated proteins, enrichment pathway, and KEGG analysis (Figures 1A–C and 2A–C). There were proteins from 119 and 206 up-and down-regulated genes in LPS-activated macrophages compared with the control (Figures 1A and 2A). Functions of the elevated proteins in LPS-stimulated cells compared with the control, as indicated by the enrichment analysis, involved in the proteasome, carbohydrate metabolism, antigen presentation, and several infections (Figure 1B). The KEGG analysis of these LPS-enhanced proteins in proteasome was demonstrated in Figure 1C, and the genes that appeared in the analysis were Rpn3, Rpn6, Rpn7, Rpn8, Rpn9 PA28α, and PA28β (Figure 1C; red color boxes). Notably, the RPN is a subunit of the 19S regulatory particle (RP) at the lid portion of the 26S proteasome [48], while PA28α and PA28β are proteasome activators forming a heteropolymer that binds to both ends of the 20S proteasome, referred to as "immunoproteasome", using for the processing of certain antigens [49]. In parallel, functions of the LPS-decreased proteins in macrophages by the enrichment analysis involved in cell energy (citrate cycle and oxidative phosphorylation), synthesis of protein and lipid, and several diseases (Figure 2B). Additionally, the KEGG analysis of these LPS-suppressed proteins in the oxidative phosphorylation, a significant mechanism for providing cell energy, was demonstrated in Figure 2C, including several Ndufs (NADH: Ubiquinone Oxidoreductase Core Subunits) of NADH (nicotinamide adenine dinucleotide + hydrogen) dehydrogenases [50], a component of cytochrome C reductase and oxidase, consisting of a few QCRs (quinol-cytochrome c oxidoreductases) and COX5B (cytochrome c oxidase subunit 5B) [51,52], with several parts of F-type ATPase, such as OSCP (oligomycin-sensitivity-conferring protein) [53]. The list and details of the genes generating the up and down- proteins from LPS-activated macrophages using the GO enrichment analysis were demonstrated in Supplementary Tables S1 and S2.

Because the reduced cell energy status (mitochondrial oxidative phosphorylation) possibly leads to abnormal molecules and proteins degraded by the proteasome [54,55], the decreased proteins in oxidative phosphorylation and increased proteins of proteasome might be due to LPS-induced cell injury [56–58], partly including the DNA methylation. Then, the impact of LPS activation was further explored in macrophages from mgmt null and littermate control (mgmt control) mice. Indeed, mgmt null macrophages demonstrated lower inflammatory responses as indicated by supernatant cytokines (TNF-α, IL-6, and IL-10) and expression of pro-inflammatory M1 macrophage polarization genes, including nitric oxide synthase (iNOS) and interleukin-1β (IL-1β) without an impact on M2 polarization genes; arginase-1 (Arg-1), transforming growth factor-β (TGF-β), and resistin-like molecule-1 (Fizz-1) (Figure 3A–H). Likewise, administration of the MGMT inhibitor in LPS-activated wild-type (WT) macrophages also demonstrated anti-inflammatory effects through these parameters (except for supernatant IL-6) (Figure 3I–P), suggesting a possible use of MGMT inhibitor to attenuate sepsis-induced hyperinflammation. These data also imply the reduced macrophage cytokine production, perhaps due to methylation of the DNA that is responsible for the production of these cytokines, due to the lack of MGMT to repair the DNA [42]. Because mgmt null macrophages might demonstrate higher cell injury than the control cells due to persistent DNA methylation from the loss of MGMT enzymes, several injury parameters were evaluated. As such, mgmt null macrophages demonstrated higher supernatant cell-free DNA and the DNA break (phosphohistone H2A.X) but not cell reactive oxygen species (evaluated by Malondialdehyde; MDA) (Figure 4A–D). Notably, phosphorylation of the histone variant H2AX is a critical factor for DNA damage response to assembly of the DNA repair proteins at the chromatin damaged sites [59], and MDA is the final peroxidation products of polyunsaturated fatty acids activated by several inducers, including LPS [60]. Perhaps, the loss of MGMT enzyme leads to more profound DNA damage without repair, as indicated by the DNA break indicator (Figure 4C,D), which results in more potent cell injury (cell-free DNA) (Figure 4A) but does not affect lipid peroxidation injury (MDA) (Figure 4B).

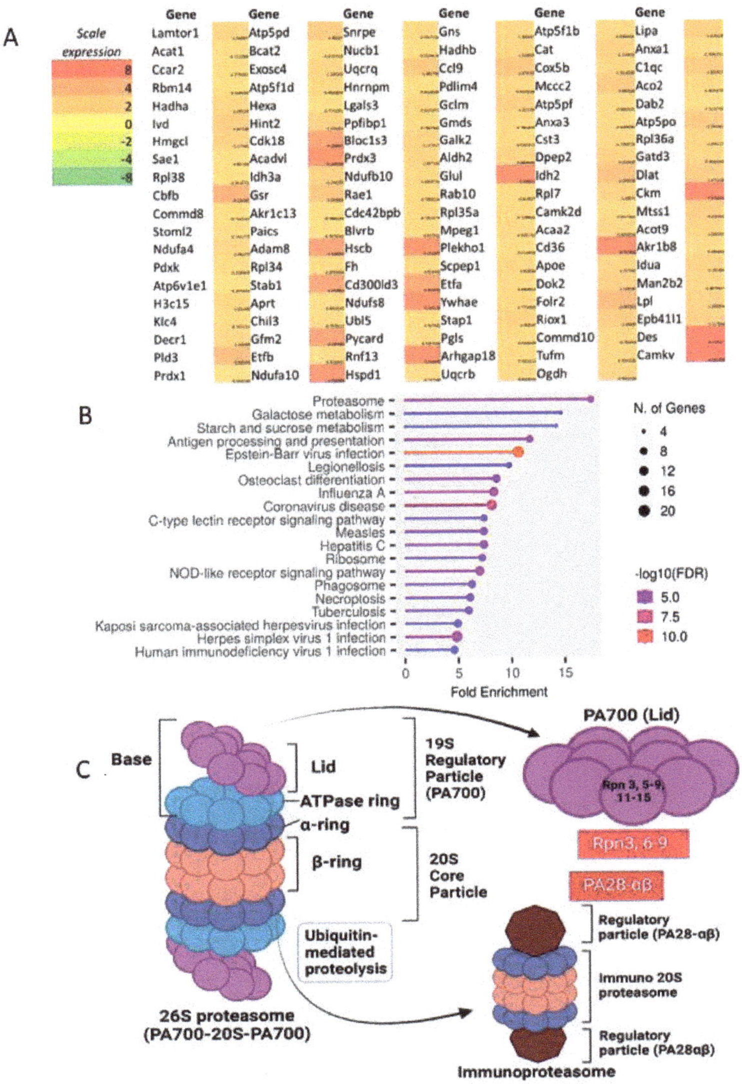

Figure 1. The proteome profiles of genes that generate peptides from wild-type bone marrow-derived

macrophages after 24 h lipopolysaccharide (LPS) activation compared with media control as indicated by heatmap analysis (average value from 3 samples) using log2 of the count per million (TPM) of the upregulated genes (**A**) fold enrichment pathway of this list (**B**). KEGG analysis of the path (proteasome) (**C**) is demonstrated. The red boxes with the white letters in the gene's name of KEGG pathway (**C**) indicate the analysis results. Macrophages were isolated from three different mice to perform triplicate experiments. Picture (**C**) is created by BioRender.com accessed on 15 March 2023.

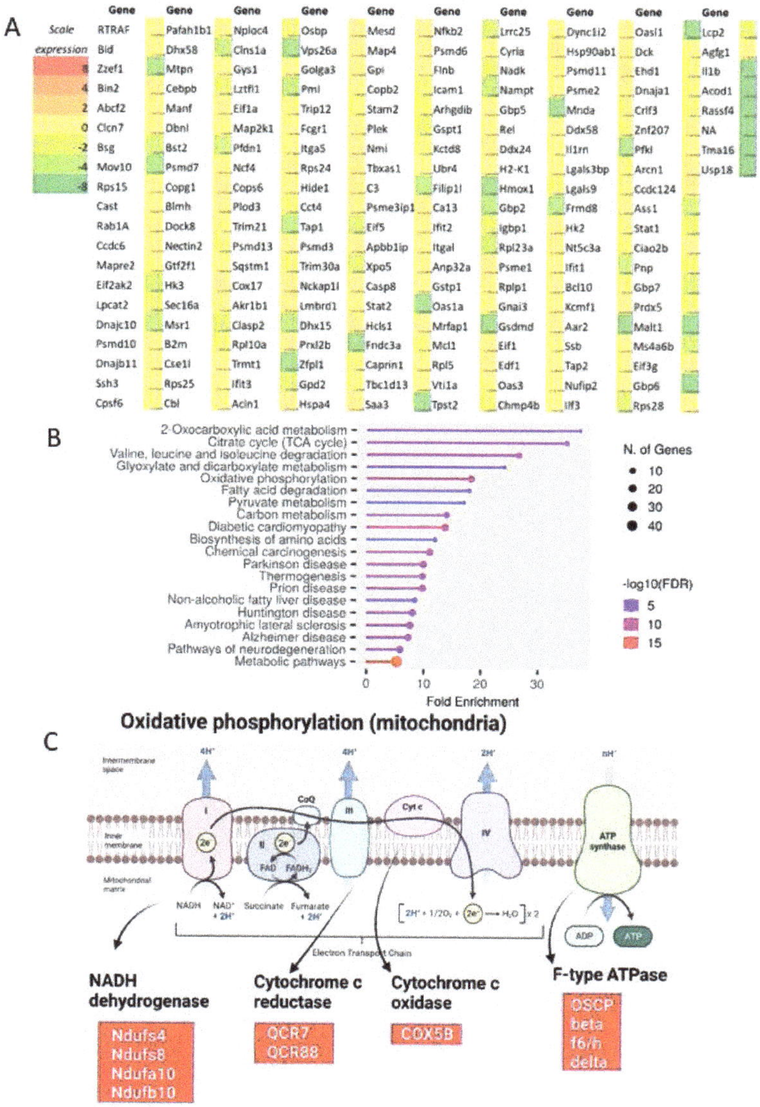

Figure 2. The proteome profiles of genes that generate peptides from wild-type bone marrow-derived macrophages after 24 h lipopolysaccharide (LPS) activation compared with media control as indicated by heatmap analysis (average value from 3 samples) using log2 of the count per million (TPM) of the downregulated genes (**A**) fold enrichment pathway of this list (**B**). In addition, KEGG analysis of the pathway (oxidative phosphorylation) (**C**) is demonstrated. The red boxes with the white letters in the gene's name of KEGG pathway (**C**) indicate the analysis results. Macrophages were isolated from three different mice to perform triplicate experiments. Picture (**C**) is created by BioRender.com accessed on 15 March 2023.

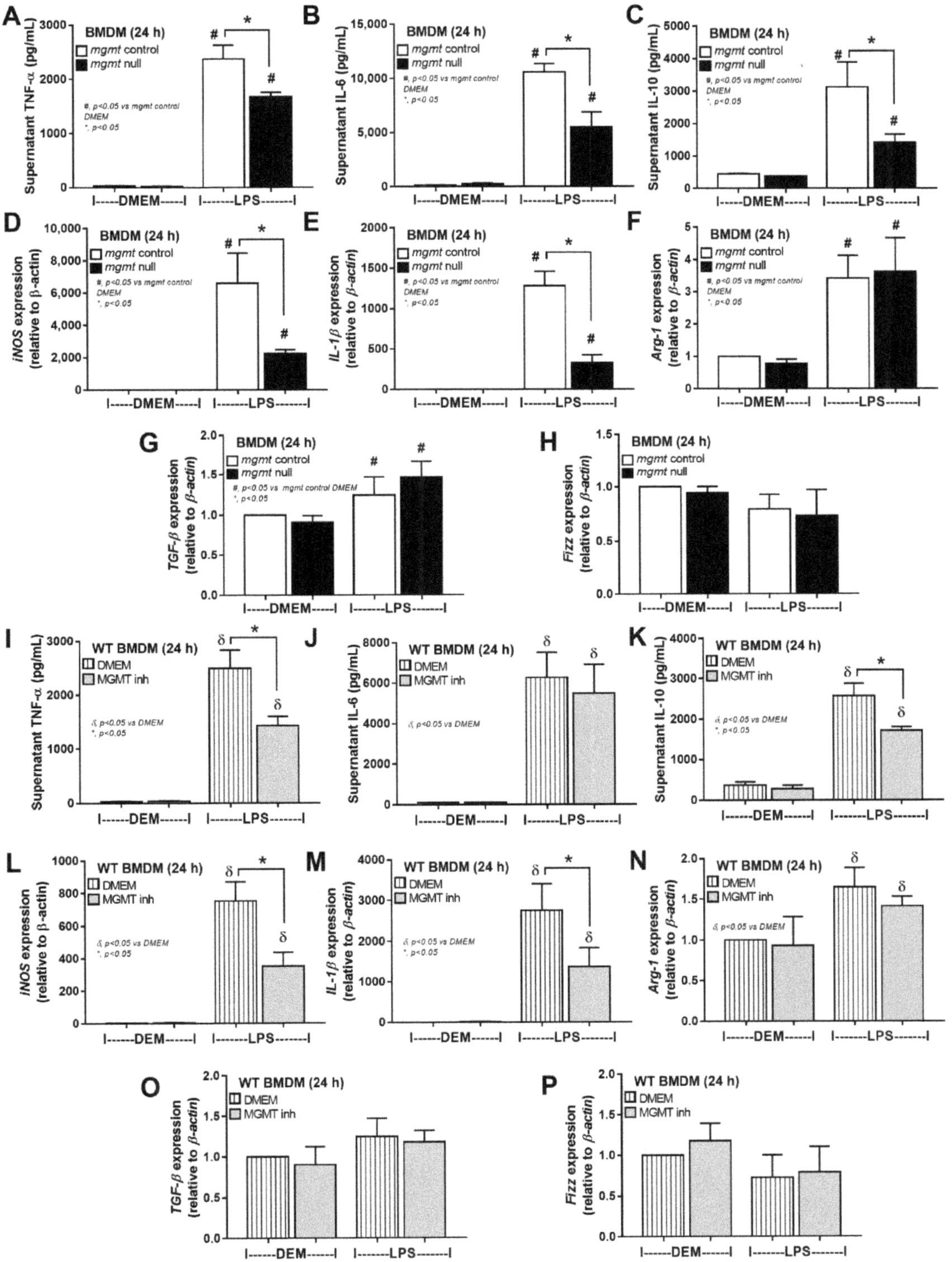

Figure 3. The characteristics of bone marrow-derived macrophages (BMDM) from mgmt control (mgmt$^{fl/fl}$; LysM-Cre$^{-/-}$) or mgmt null (mgmt$^{fl/fl}$; LysM-Cre$^{cre/-}$) mice at 24 h after activation by lipopo-

lysaccharide (LPS) as indicated by supernatant cytokines (TNF-α, IL-6, and IL-10) (**A–C**), expressio of pro-inflammatory genes of M1 polarization (iNOS and IL-1β) (**D,E**), and anti-inflammatory gene of M2 polarization (Arg-1, TGF-β, and Fizz-1) (**F–H**) are demonstrated. Parallelly, the characteristic of BMDM from wild-type mice at 24 h after activation by media (DMEM) or MGMT inhibito (Lomeguatrib) with or without LPS as indicated by these parameters (**I–P**) is also demonstrated Triplicated independent experiments were performed. Mean ± SEM with the one-way ANOV followed by Tukey's analysis was used. #, $p < 0.05$ vs. *mgmt* control DMEM; δ, $p < 0.05$ vs. contro DMEM; *, $p < 0.05$ between the indicated groups.

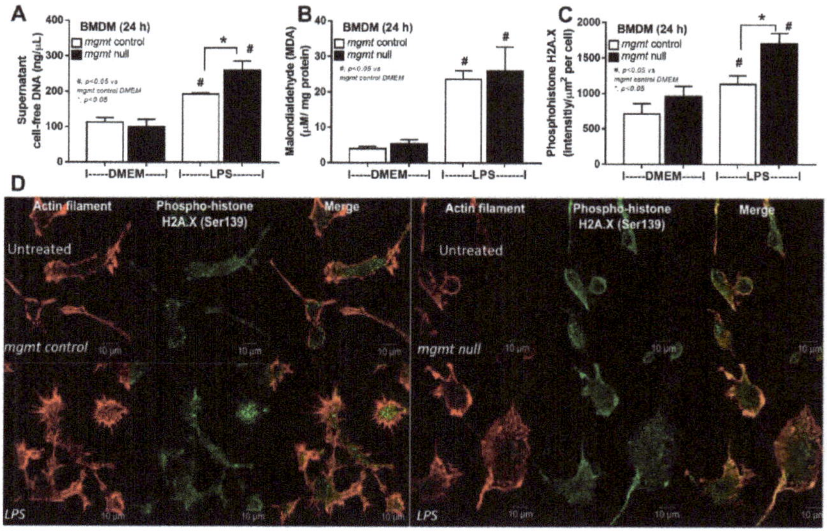

Figure 4. The characteristics of bone marrow-derived macrophages (BMDM) from mgmt control ($mgmt^{fl/fl}$; LysM-$Cre^{-/-}$) or mgmt null ($mgmt^{fl/fl}$; LysM-$Cre^{cre/-}$) mice at 24 h after activation by lipopolysaccharide (LPS) as indicated by supernatant cell-free DNA (**A**), malondialdehyde (MDA; a reactive stress molecule) (**B**), and immunofluorescent stained for DNA break (phospho-histone H2A.X; green color) and actin filament (red color) in intensity score and representative pictures (**C,D**) are demonstrated. Triplicated independent experiments were performed. Mean ± SEM with the one-way ANOVA followed by Tukey's analysis was used. #, $p < 0.05$ vs mgmt control DMEM; *, $p < 0.05$ between the indicated groups.

2.2. *The Mgmt Null Mice Demonstrated Less Severe Cecal Ligation and Puncture (CLP) Sepsis than the Littermate Control*

Because of the impacts of mgmt-manipulated macrophages (Figures 3 and 4) and the importance of antibiotics in sepsis [61], further exploration in mgmt null mice ($mgmt^{fl/fl}$; LysM-$Cre^{cre/-}$) and mgmt littermate control ($mgmt^{fl/fl}$; LysM-$Cre^{-/-}$) using CLP surgery with and without antibiotics were performed. With antibiotic use, mgmt null mice demonstrated less severe sepsis than sepsis in littermate control, as indicated by survival analysis, kidney injury (serum creatinine and renal histology score), liver damage (serum alanine transaminase and liver histological score), spleen apoptosis, cell-free DNA, endotoxemia, bacteremia, and serum cytokines (TNF-α, IL-6, and IL-10) (Figures 5A–G and 6A–G). However, the protective effect against sepsis of mgmt null mice was lost in CLP without antibiotics as indicated by survival analysis, serum creatinine, alanine transaminase, and serum cytokines (TNF-α, IL-6, and IL-10) (Figure 7A–F) supporting the necessity of microbial control during immune modulation in sepsis [62]. For clinical translation purposes, Lomeguatrib (an MGMT inhibitor) was further tested in WT mice with an antibiotic-administered CLP model. Although the inhibitor could not attenuate mortality and organ injury (kidney and liver), the serum cytokines of the treated mice were lower than the control mice (Figure 7G–L). Notably, there was no kidney and liver injury in sham mice with the inhibitor, implying less toxicity to the kidney and liver of the inhibitor.

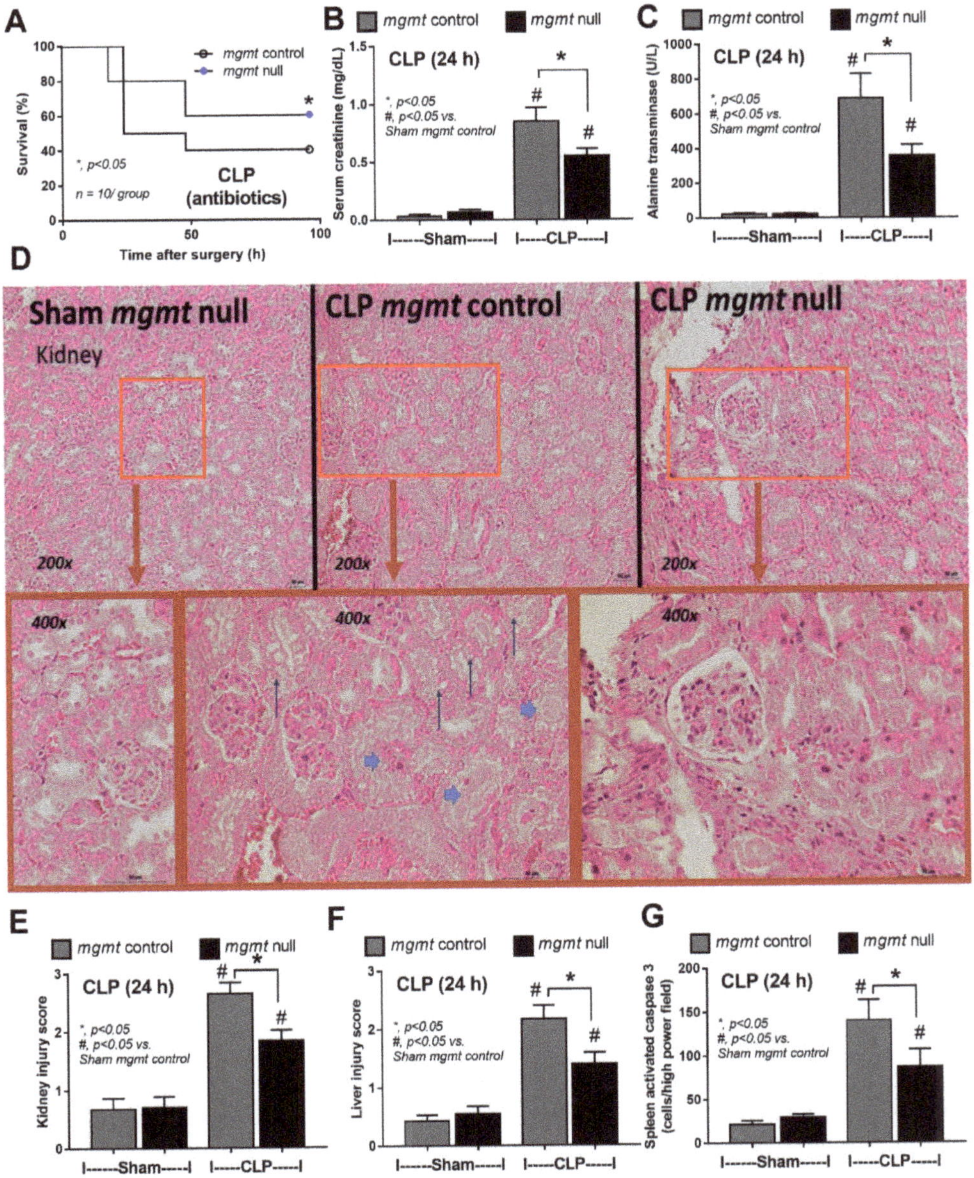

Figure 5. The characteristics of mice in mgmt control (mgmt$^{fl/fl}$; LysM-Cre$^{-/-}$) or mgmt null (mgmt$^{fl/fl}$; LysM-Cre$^{cre/-}$) group after cecal ligation and puncture (CLP) sepsis or sham control (Sham) surgery as indicated by survival analysis (**A**), and the parameters at 24 h-post surgery, including serum creatinine (**B**), alanine transaminase (**C**), renal histological score with representative pictures (**D**,**E**), liver injury score (**F**), and spleen apoptosis (activated caspase three immunohistochemistry) (**G**), is demonstrated (n = 10/group for A and n = 5–7/group for (**B**–**G**)). Mean ± SEM with the one-way ANOVA followed by Tukey's analysis was used. #, $p < 0.05$ vs. Sham mgmt control; *, $p < 0.05$ between the indicated groups. Inset pictures, increased magnification; narrow arrows, tubular cell injury; broad arrows, renal tubular casts.

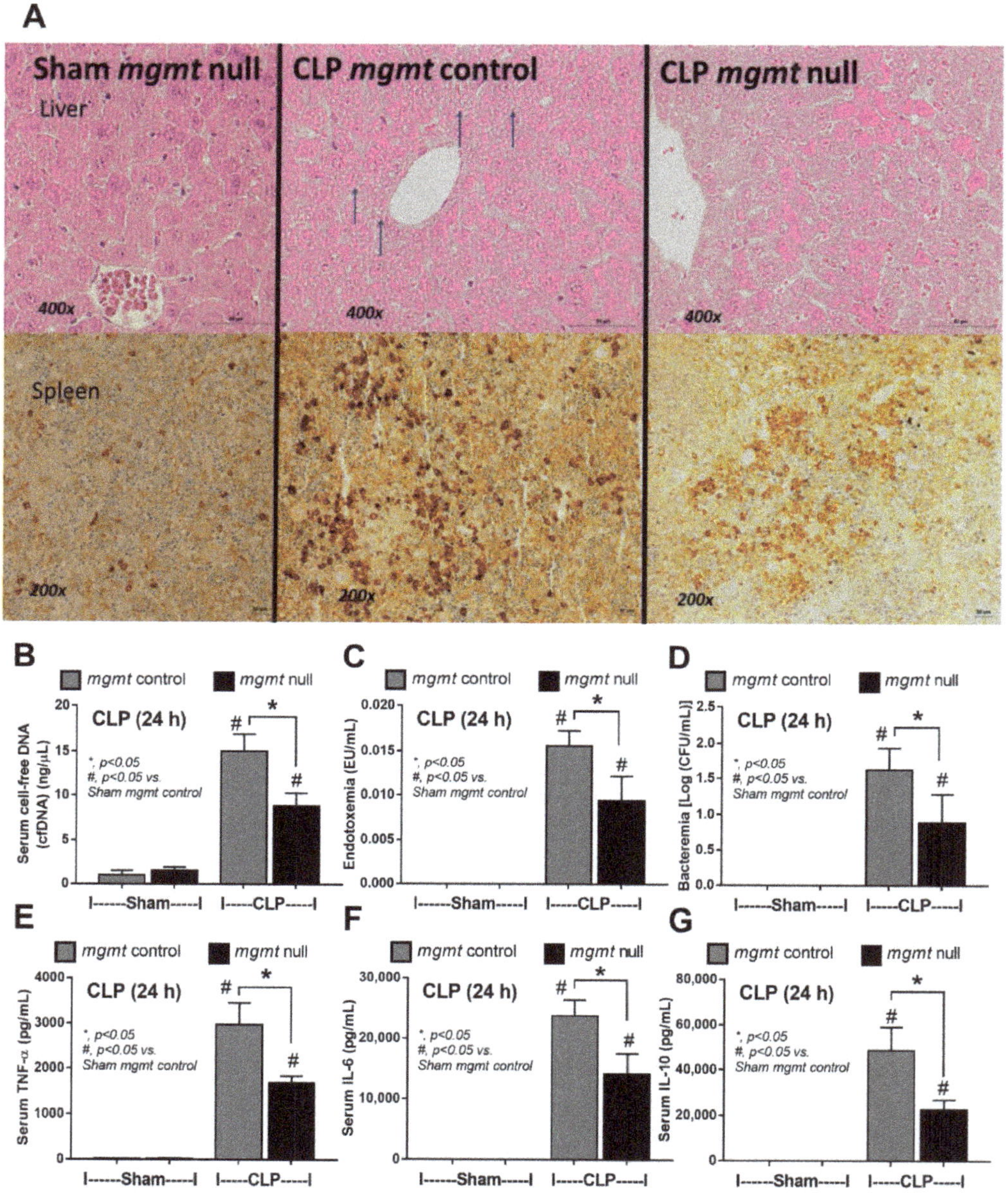

Figure 6. The characteristics of mice in mgmt control (mgmt$^{fl/fl}$; LysM-Cre$^{-/-}$) or mgmt null (mgmt$^{fl/fl}$; LysM-Cre$^{cre/-}$) group after 24 h of cecal ligation and puncture (CLP) sepsis (with antibiotic use) or sham control (Sham) surgery as indicated by representative pictures of the liver in Hematoxylin & eosin (H&E) staining and spleen activated caspase 3 (**A**), serum cell-free DNA (cf-DNA) (**B**), endotoxemia (**C**), bacteremia (**D**), and serum cytokines (TNF-α, IL-6, and IL-10) (**E**–**G**) are demonstrated (n = 5–7/group). Mean ± SEM with the one-way ANOVA followed by Tukey's analysis was used. #, $p < 0.05$ vs. Sham mgmt control; *, $p < 0.05$ between the indicated groups. Narrow arrows, hepatocyte apoptosis.

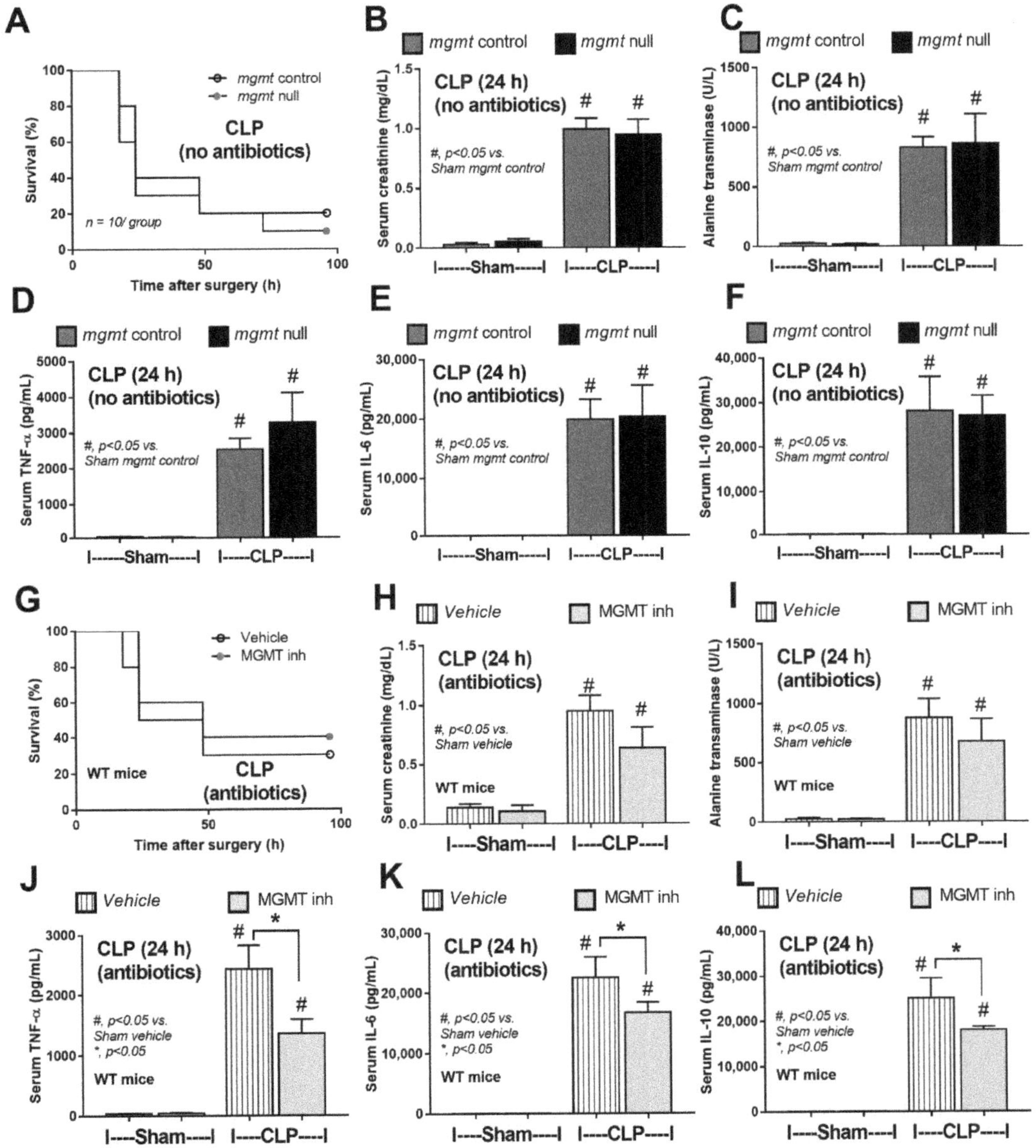

Figure 7. The characteristics of mice in mgmt control (mgmt$^{fl/fl}$; LysM-Cre$^{-/-}$) or mgmt null (mgmt$^{fl/fl}$; LysM-Cre$^{cre/-}$) group after 24 h of cecal ligation and puncture (CLP) sepsis (without antibiotic use) or sham control (Sham) surgery as indicated by survival (**A**), and the parameters at 24 h-post surgery, including serum creatinine (**B**), alanine transaminase (**C**), and serum cytokines (TNF-α, IL-6, and IL-10) (**D**–**F**) are demonstrated. In parallel, the characteristics of wild-type (WT) mice after CLP (with antibiotics) or Sham surgery with or without MGMT inhibitor (Lomeguatrib) as indicated by survival analysis (**G**), and the parameters at 24 h-post surgery, including serum creatinine (**H**), alanine transaminase (**I**), and serum cytokines (TNF-α, IL-6, and IL-10) (**J**–**L**), are demonstrated (n = 10/ group for survival studies and n = 5–7/group for others). Mean ± SEM with the one-way ANOVA followed by Tukey's analysis was used. #, $p < 0.05$ vs. Sham *mgmt* control; *, $p < 0.05$ between the indicated groups.

3. Discussion

3.1. Possible LPS-Induced Macrophage Injury from the Proteomic Analysis in the Wild-Type Cells

The presence of lipopolysaccharide (LPS) in blood circulation (endotoxemia) can be found in several conditions [63–65], predominantly due to gut barrier damage [3,18,20,66 and Gram-negative bacteremia [2,3,67]. As such, LPS (one of the pathogen-associated molecular patterns; PAPMs) stimulates all cells in the body, including immune cells, and macrophages are the major cells responsible for recognising and controlling most foreign molecules, including LPS [68,69]. Although only the high abundance proteins are detectable by proteomic analysis, evaluating these proteins might crudely identify the direction of cell response. Here we demonstrated a higher number of down-regulated proteins than the up-regulated groups in LPS-stimulated macrophages, as listed in Supplementary Tables S1 and S2. With enrichment pathway analysis, the upregulated proteins correlated with the proteasome, the protein complexes used for degrading the damaged proteins by proteolysis with several protease enzymes [70,71]. Despite several functions of the proteasome in cell homeostasis the 26S proteasome regulates DNA repair by degrading the proteins or acting as a molecular chaperone to promote the disassembly of the repair complex [54]. Increased proteasome in LPS-activated macrophages indirectly indicates an increased abundance of unneeded proteins after LPS stimulation supporting previous publications [72,73].

On the other hand, enrichment pathways of the down-regulated proteins were correlated with cell metabolisms and energy status, including 2-oxocarboxylic acids (association with pyruvate pathway), citrate cycle (glycolysis), and oxidative phosphorylation (mitochondrial function). Perhaps, the increased cell energy utilization during macrophage responses to LPS (cytokine production, inflammatory signaling synthesis, etc.) uses up these proteins that partly be associated with mitochondrial injury and increased reactive oxygen species [74]. Although proteins for the modulation of DNA and histone were not found in the proteome results, possibly due to too little abundance of these proteins, the proteomic analysis indirectly supported a possible cell injury, including LPS-induced DNA methylation as previously published [57].

The DNA methylation is more frequently demonstrated at the cytosine site than the guanine position, especially at the cytosine-phosphate-guanine (CpG), as more than 70–80% of CpG sites in humans are modified, despite the lower abundance of CpG in humans compared with prokaryotic cells [75,76]. Both DNA and histone alterations are critical regulators of gene expression through the chromatin structures [77], using several key enzymes to control the chromatin accessibility that is well-known in cancer [78] but has fewer data in sepsis [79]. For DNA methylation, it is the transfer of a methyl group, frequently to the C-5 position of the cytosine ring of DNA-by-DNA methyltransferase (DNMT) in any cytosines of the genome, especially at the CpG regions [80]. In previous publications, the enhanced DNMT activity increases DNA methylation and aggravates pro-inflammatory macrophages and the *dnmt1* deletion of enhanced anti-inflammation [42] and attenuates macrophage inflammatory responses [81]. Despite the common cytosine methylation at CpG sites [82], methylation of guanine at the O-6 positions (O6MeG) that are controlled by O6-methylguanine-DNA methyltransferase (MGMT) caused genotoxicity [83,84] as the insufficient MGMT worsens cell injury [85].

Interestingly, DNA methylation impairs DNA transcription and induces programmed cell death, especially apoptosis [86]. Because (i) previous studies of DNA methylation in sepsis are mentioned [41,87–89], (ii) the possibility that increased O6MeG (due to the loss of *mgmt* for DNA repair) might enhance cell injury [85], (iii) the availability of *mgmt* inhibitor for anti-cancer [90] that possibly be helpful for sepsis [91], and (iv) epigenetic changes and in vitro tests of *mgmt* inhibitors in LPS-activated macrophages [45,92], further tests on mice with the depletion of MGMT enzyme only in the myeloid cells are interesting. Theoretically, MGMT deficiency should interfere with the macrophage activities leading to an anti-inflammatory direction which might be beneficial for treating sepsis hyper-inflammation [93–95]. Then, further tests on *mgmt* null macrophages and mice were performed.

3.2. Impact of *mgmt* and DNA Repair in Sepsis, In Vitro and In Vivo

To test the impacts of MGMT enzyme in sepsis, bone marrow-derived macrophages from *mgmt* null mice (*mgmt*$^{fl/fl}$; *LysM-Cre*$^{cre/-}$) and an MGMT inhibitor were used. As such, *mgmt* null macrophages and MGMT inhibitors induce anti-inflammatory effects as indicated by reduced supernatant cytokines and the down-regulated genes of M1 pro-inflammatory macrophages (*iNOS* and *IL-1β*) compared with the control. Due to the influence of MGMT enzyme on the DNA repairs through the removal of O6MeG (methyl group at the 6th oxygen molecule on guanine) of DNA, the deletion of MGMT in macrophages might be responsible for the enhanced cell injury as indicated here, by the higher abundance of DNA break and supernatant cell-free DNA in *mgmt* null macrophages than the control cells after LPS activation. Accordingly, phosphor-H2AX (γ-H2AX) is a marker of DNA double-strand break (DSB), a disruption of both DNA strands compromising genomic stability [96], which regularly occurs in any eukaryotic cells in the order of 10 to 50 per cell per day, depending on the cell cycles and cell types [97], due to the exogenous and endogenous inducers, including radiation, chemicals, LPS, ROS, DNA replication, and repair [98,99]. Interestingly, LPS increased ROS, and both factors (LPS and ROS) are well-known to induce DNA methylation [57,100], especially the methylation at essential sites on DNA, which is mostly mentioned as DNA break in the mutated cells [101]. In non-immune cells, several alkylating agents induce DNA methylation, especially at the guanine sites more than the cytosine positions, causing DNA mismatch binding that is toxic to the cells resulting, at least in part, in DNA break, cell mutation, or cell death [102,103]. In macrophages, LPS-induced DNA damage through LPS-mediated ROS is well-known as the damage is detectable in more than 95% of macrophages (in vitro) within only 30 min of LPS incubation, and the damage is completely protected by anti-oxidants [99]. The increased ROS and cell death in macrophages after LPS activation are also mentioned [104,105]. While DNA damage may enhance macrophage inflammation through several DNA sensors responding to the damage [106], the damage that cannot be repaired might negatively affect macrophages. Then, the interfered equilibrium between the generation of DNA break and DNA repair might be another important intervention on macrophages [40]. Here, more prominent DNA break and cell-free DNA in *mgmt* null macrophages over the control cells indirectly indicate the possible presence of DNA methylation at the guanine (O6MeG), despite the technical limitation on the direct O6MeG detection [107]. Notably, the detectable DNA damage in control untreated macrophages supports DNA injury from normal cellular processes, possibly due to stress-induced ROS, which might be more prominent in macrophages than other cell types [105,108,109]. The DNA methylation at guanine seems important in LPS-activated macrophages, perhaps due to the maintenance of cell viability and attenuation of cell injury, despite the uncommon DNA methylation at the guanine site compared with cytosine sites [80]. The extended use of an MGT inhibitor (Lomeguatrib), a chemotherapeutic agent [110], on attenuation of sepsis hyper-inflammation was also proposed following these in vitro results.

Following the in vitro results, the cecal ligation and puncture (CLP) abdominal sepsis model was tested in the littermate control (*mgmt*$^{fl/fl}$; *LysM-Cre*$^{-/-}$) and *mgmt* null mice. As such, the CLP model is used because it is a sepsis model that more resembles human conditions than the LPS injection model, as indicated by the presence of bacteremia, cytokine levels, and other parameters [67]. Interestingly, sepsis in *mgmt* null mice was less severe than in the littermate control mice, as indicated by survival, serum cytokines, and organ injury, only with antibiotics, supporting the importance of microbial control during sepsis immunotherapy [111]. However, sepsis was not worsened in *mgmt* null mice. However, macrophages in these mice might be more susceptible to injury than the control due to an abnormality in DNA repair from the loss of the MGMT enzyme.

Interestingly, sepsis attenuation in the mice without MGMT only in myeloid cells (macrophages and neutrophils), but not other cells, implying the importance of these cells in sepsis hyper-immune responses [112] and the blockage of only the MGMT enzyme in macrophages might effectively prevent severe sepsis with fewer drug complications.

However, the administration of MGMT inhibitor in CLP mice did not improve survival rate compared with the vehicle control, even with antibiotic administration, perhaps due to the drug effects on all cells in the body, not only on macrophages. Nevertheless, the MGMT inhibitor in CLP mice attenuated serum cytokines but not liver and kidney injury. Although there was no liver and kidney injury in sham mice with the MGMT inhibitor, these organs during sepsis might be more susceptible to the injury and the MGMT inhibitor effect on renal cells, and hepatocytes might be responsible for the non-improved organ injury in MGMT inhibitor-administered mice with sepsis. Perhaps, the dose adjustment, a better drug formula preparation, and/ or the selective delivery of MGMT inhibitors only on myeloid cells [113] might increase MGMT inhibitors' effectiveness and reduce complications. Despite the necessity of more studies, we reported a proof of concept to use MGMT inhibitors, the available adjunctive anti-cancer drugs, to attenuate sepsis-induced hyper-inflammation.

3.3. Clinical Aspect and Future Experiments

Overall, LPS activation during sepsis induces DNA methylation in macrophages that requires some enzymes, including MGMT, to remove the methyl groups (DNA repair) to revitalize and maintain regular macrophage functions (continuous cytokine production and inflammatory response). Then, failure of DNA repair causes a reduction in macrophage functions, especially cytokine production and hyper-inflammatory responses, which possibly turns out to be a beneficial effect on sepsis. In cancer therapy, several alkylating agents destroy cancer cells through the induction of DNA methylation, and some malignant cells resist these anti-malignant drugs partly by the increased production of DNA methylation enzymes, including MGMT, to counteract the DNA damage [90]. Hence, the administration of MGMT inhibitors and some alkylating agents enhances the anti-malignant effect of some cancers, which is used as adjuvant therapy in some types of cancer [90]. In sepsis, inflammation-induced DNA methylation in macrophages seems important for the pathophysiology of sepsis-induced organ injuries (kidney, liver, and spleen). MGMT deletion only in myeloid cells (mostly macrophages and neutrophils) attenuated the injury. Indeed, infiltration of immune cells in several organs during systemic inflammation, such as sepsis and auto-immune diseases, is one of the main pathogenesis of organ injury, and the infiltration of cells with less pro-inflammatory activities by the MGMT interference here might induce less severe injury than infiltration by the very active immune cells [14,69]. Thus, the reduction of immune cell activities, including by the MGMT blockage, might be beneficial in sepsis.

Because MGMT enzyme is an essential enzyme, not only for cancer cells but also for every cell due to the physiologic DNA methylation induced by regular cell activities, MGMT blockage might be toxic or mutagenic [110] to the cells, and the short course of treatment and/or the direct drug delivery into macrophages are interesting. Moreover, administration of MGMT blockage should be used only in a condition with a well microbial control as inflammation is necessary for organismal control, and too less inflammation might enhance secondary infection [114,115]. Therefore, although the direct detection of Q6MeG on macrophages in patients with sepsis might be the most interesting indicator for using MGMT blockage in sepsis, proper methods of O6MeG detection in patients will be needed [107,116]. For instance, high serum IL-6 and IL-1 as the hyper-inflammation markers [117,118] plus elevated Q6MeG and normal HLA-DR (a marker of sepsis immune exhaustion [114,115]) with negative blood culture (a biomarker of good control of the source of infection) might be the indicators for properly using of MGMT inhibitors in sepsis. Therefore, further experiments on MGMT inhibitors in sepsis are interesting.

4. Materials and Methods

4.1. The Proteomic Analysis

Bone marrow-derived macrophages (BMDM) were prepared from the femurs of wild-type (WT) mice using supplemented Dulbecco's Modified Eagle's Medium (DMEM) with a

20% conditioned medium of the L929 cells (ATCC CCL-1) as previously described [68,119–121]. Macrophages at 5×10^4 cells/well in supplemented DMEM (Thermo Fisher Scientific) were incubated in 5% carbon dioxide (CO_2) at 37 °C for 24 h before being treated by lipopolysaccharide (LPS; *Escherichia coli* 026:B6) (Sigma-Aldrich, St. Louis, MO, USA) at 100 ng/mL or DMEM (control) for 24 h before in-solution digestion and peptides labeling using the light reagents (CH_2O and $NaBH_3CN$) and heavy reagents (13CD2O and NaBD3CN), respectively. The pooled peptides from macrophages in the wells were fractionated using a high pH reversed-phase peptide fractionation kit (Thermo Fisher Scientific, San Jose, CA, USA) and Liquid chromatography–tandem mass spectrometry (LC-MS/MS) was performed on an EASY-nLC1000 system coupled to a Q-Exactive Orbitrap Plus mass spectrometer equipped with a nanoelectrospray ion source (Thermo Fisher Scientific, San Jose, CA, USA). The mass spectrometry (MS) raw files were searched against the Mouse Swiss-Prot Database (17,138 proteins). In parallel, the search parameters were set for fixed modifications; carbamidomethylation of cysteine (+57.02146 Da), light and heavy dimethylation of N termini and lysine (+28.031300 and +36.075670 Da), and variable modification: oxidation of methionine (15.99491 Da). The false positive discovery rate of the identified peptides based on Q-values using The Proteome Discoverer decoy database together with the Percolator algorithm was set to 1%, and the relative MS signal intensities of dimethyl labeled peptides were quantified and presented as ratios of LPS/ control. Log 2 of the ratios in triplicate was used to calculate the p-values using Student's t-test with a p-value < 0.05 as a significant difference. Then, these proteins were applied to the online DAVID Bioinformatics Resources 6.8 to investigate the enriched biological processes. The mass spectrometry proteomics data have been deposited to the ProteomeXchange Consortium via the PRIDE partner repository with the dataset identifier PXD041265. Then, the data visualization was performed using Excel and R packages. Meanwhile, KEGG pathway and Go enrichment analyses were generated by PathfindR and Shiny 0.77 (http://bioinformatics.sdstate.edu/go/, accessed on 28 May 2023), respectively.

4.2. The In Vitro Experiments

The BMDM from *mgmt* control ($mgmt^{fl/fl}$; LysM-$Cre^{-/-}$) or *mgmt* null ($mgmt^{fl/fl}$; LysM-$Cre^{cre/-}$) mice were extracted from mouse femurs and macrophages at 5×10^4 cells/well were activated by LPS (*Escherichia coli* 026:B6) (Sigma-Aldrich) at 100 ng/mL or DMEM for 24 h. Then, supernatant cytokines (TNF-α, IL-6, and IL-10) and gene expression were measured by ELISA (Invitrogen, Carlsbad, CA, USA) and quantitative real-time polymerase chain reaction (PCR), respectively, as previously described [87]. In brief, TRIzol Reagent (Invitrogen, Carlsbad, CA, USA) together with RNeasy Mini Kit (Qiagen, Hilden, Germany) was used to extract RNA from the samples, and 1 mg of total RNA was used for cDNA synthesis with iScriptreverse transcription supermix (Bio-Rad, Hercules, CA, USA) on a QuantStudio 5 real-time PCR system (Thermo Fisher Scientific) using SsoAdvanced Universal SYBR Green Supermix (Bio-Rad, Hercules, CA, USA). The gene expressed values normalized by Beta-actin (β-actin; an endogenous housekeeping gene) with the calculated fold change by the ΔΔCt method was conducted with primers listed in Table 1. In parallel, BMDM of WT mice were used. As such, WT BMDM at 5×10^4 cells/well were activated with LPS (100 ng/mL) or DMEM with or without MGMT inhibitor (Lomeguatrib) (SML0586, Sigma-Aldrich) (20 μM/well) for 24 h before the collection of supernatant and cells for supernatant cytokines and PCR as mentioned above. Notably, Lomeguatrib was dissolved in DMSO (Sigma-Aldrich) with the stock solution of 6.13 mM and was stored at −80 °C. Then, the stock solution was immediately diluted in DMEM before use to control the final DMSO concentration of less than 0.3%, according to a previous publication [122]. Additionally, several injury parameters, including cell-free DNA, malondialdehyde (MDA; a representative reactive oxygen species), and DNA break, were measured in *mgmt* null and *mgmt* control macrophages because of the influence of LPS in cell injury. As such, cell-free DNA in supernatants was measured by a PicoGreen assay kit (Invitrogen) following the manufacturer's protocol [123]. For MDA, the activated macrophages were homogenized by the Ultra-Turrax homogenizer (IKA, Staufen, Germany) and centrifuged at 12,000× g for 15 min at 4 °C to separate the supernatant. Then, malondialdehyde (MDA) in the supernatant was measured by an MDA assay kit (colorimetric) (Abcam, Cambridge, UK) according to the manufacturer's protocol for the intracellular reactive oxygen species (ROS) [124]. In parallel, immunofluorescence was used to determine DNA break. Accordingly, macrophages at 3×10^6 cells were seeded on covered glass-bottomed 6-well plates before activation with LPS (100 ng/mL) or DMEM control for 24 h. Then, the cells were fixed with 4% paraformaldehyde in Tris Buffered Saline (TBS) for 15 min, permeabilized with 0.1% triton X-100, and washed three times in 1X TBS with 0.05% Tween-20. Fixed samples were blocked with 2% bovine serum albumin in 1X TBS for 1 h at room temperature and then incubated overnight at 4 °C with phospho-histone H2A.X (Ser139) (20E3) rabbit mAb (Cell signaling). Proteins were visualized using goat anti-mouse IgG H&L tagged Alexa Flour 488 (Abcam; ab150113) (green), and actin filaments have been labeled with DY-554 phalloidin (red) and snapped by confocal laser scanning microscope (CLSM, Zeiss, Germany) at 630× magnification in 10 randomly selected fields.

Table 1. Lists of primers used in the study.

Name	Forward	Reverse
Inducible nitric oxide synthase (*iNOS*)	5′-ACCCACATCTGGCAGAATGAG-3′	5′-AGCCATGACCTTTCGCATTAG-3′
Interleukin-1β (*IL-1β*)	5′-GAAATGCCACCTTTTGACAGTG-3′	5′-TGGATGCTCTCATCAGGACAG-3′
Arginase-1 (*Arg-1*)	5′-CTTGGCTTGCTTCGGAACTC-3′	5′-GGAGAAGGCGTTTGCTTAGTT-3′
Resistin-like molecule-α1 (*Fizz-1*)	5′-GCCAGGTCCTGGAACCTTTC-3′	5′-GGAGCAGGGAGATGCAGATGA-3′
Transforming growth factor-β (*TGF-β*)	5′-CAGAGCTGCGCTTGCAGAG-3′	5′-GTCAGCAGCCGGTTACCAAG-3′
Beta-actin (*β-actin*)	5′-CGGTTCCGATGCCCTGAGGCTCTT-3′	5′-CGTCACACTTCATGATGGAATTGA-3′

4.3. Animal and Animal Model

The animal protocol (017/2562) was approved by the Institutional Animal Care and Use Committee of the Faculty of Medicine, Chulalongkorn University, Bangkok, Thailand, according to the National Institutes of Health (NIH) criteria. For the proteomic analysis, macrophages were prepared from mouse long bones (8-week-old) of wild-type (WT) male C57BL/6 mice purchased from Nomura Siam (Pathumwan, Bangkok, Thailand). For other experiments, $mgmt^{flox/flox}$ and LyM-Cre$^{Cre/Cre}$ mice were purchased from RIKEN BRC Experimental Animal Division (Ibaraki, Japan) and cross-breed to produce *mgmt* littermate control ($mgmt^{fl/fl}$; LysM-Cre$^{-/-}$) or *mgmt* null ($mgmt^{fl/fl}$; LysM-Cre$^{cre/-}$) mice in F3 of the breeding protocol. Of note, the $mgmt^{flox/flox}$ mice with the loxP sites were bred with LysM-Cre$^{Cre/Cre}$ mice and the cre recombinase mice with the control of lysozyme M targeted deleted *mgmt* only in the myeloid cells (macrophages and neutrophils). The offsprings were (i) $mgmt^{flox/flox}$ with no LysM-Cre ($mgmt^{fl/fl}$; LysM-Cre$^{-/-}$), categorized as the littermate controls or *mgmt* control, and (ii) $mgmt^{fl/fl}$; LysM-Cre$^{cre/-}$ (*mgmt* null) with positive for the Cre driver that lacks MGMT enzyme. Both groups of mice, including the conditional targeted Cre positive mice (*mgmt* null) and floxed/floxed littermate controls (*mgmt* control), were age- and gender-matched with the use of only 8–10 weeks-old male mice. To genotype these mice on the loxP sites insertion, the following primers were used; (i) LysM-cre primer; F: 5′-GAACGCACTGATTTCGACCA-3′, R: 5′-GCTAACCAGCGTTTTCGTTC-3′, (ii) *mgmt*-loxP primer F: 5′-TGGGCTTCAAATCAAGGAACAGAA-3′, R: 5′-AACTATCCTGCTCACTC-TCTGTAG-3′, and (iii) Cre recombination (for Cre activity); F: 5′-GGTGTGGATCCCAAGA-AATTGAAG-3′, R: 5′-TGTTCAAGAGTGACACACAGTCA-3′. The mice homozygous for the flox were selected and genotyped for the expression of LysM-Cre using the primers; F: 5′-CTTGGGCTGCCAGAATTCTC-3′; R: 5′-CCCAGAAATGCCAGATTACG-3′. For the test of MGMT inhibitor, only 8–10 weeks old male mice were used. Then, *mgmt* control, *mgmt* null, and WT C57BL/6 mice were used for the cecal ligation and puncture (CLP) surgery to induce sepsis or sham operation under isoflurane anesthesia following previous publications [125–127]. Briefly, the cecum was ligated at 10 cm from the cecal tip, punctured twice with a 21-gauge needle, and gently squeezed to express a small amount of fecal material before closing the abdominal wall (a midline abdominal incision) layer by layer with sutures (Nylon 4-0). After that, tramadol (25 mg/kg/dose) in 0.25 mL prewarmed normal saline solution (NSS) and imipenem/cilastatin (14 mg/kg/dose) in 0.2 mL NSS were subcutaneously administered in both frank area after surgery, and at 6 and 18 h post-CLP [6]. In the sepsis protocol without antibiotics, NSS in the same volume was used. In WT mice with CLP sepsis, an MGMT inhibitor (Lomeguatrib, SML0586, Sigma-Aldrich) at 1 mg/kg [43,128] mice in 3% dimethyl sulfoxide (DMSO) or DMSO alone (vehicle control) was subcutaneously administered at 15 min before surgery and at 6 h later (15 min before tramadol and the antibiotics). For sham-operated mice, the cecum was isolated and closed the abdomen by suturing without either ligated or punctured.

4.4. Mouse Sample Analysis

For kidney and liver injury, serum creatinine and alanine transaminase, respectively, were measured by colorimetric method using QuantiChrom™ Creatinine Assay (BioAssay System, Hayward, CA, USA) and EnzyChrom Alanine Transaminase assay (EALT-100, BioAssay), respectively [129]. Serum cell-free DNA and LPS (endotoxin) were detected by Quanti PicoGreen assay (Sigma-Aldrich) and HEK-Blue LPS Detection Kit 2 (InvivoGen™, San Diego, CA, USA), respectively. Blood bacterial abundance (bacteremia) was evaluated using the direct spread of mouse blood onto blood agar plates (Oxoid, Hampshire, UK) in serial dilutions and incubating at 37 °C for 24 h before colony enumeration. Meanwhile, ELISA (Invitrogen, Carlsbad, CA, USA) was used to detect serum cytokines (TNF-α, IL-6, and IL-10). For kidney injury determination, the injury score was semi-quantitatively evaluated on Hematoxylin and eosin (H&E) staining in 4 mm thick paraffin-embedded slides at 200× magnification by the area of injury (tubular epithelial swelling, loss of brush border, vacuolar degeneration, necrotic tubules, cast formation, and desquamation) using the following score: 0, area < 5%; 1, area 5–10%; 2, area 10–25%; 3, area 25–50%; 4, area > 50%. The liver injury score was also measured through an area of hepatic injury, defined as congestion, degenerative cellular changes, cytoplasmic vacuolization, leukocyte infiltration, or cellular necrosis, in 200x magnification of 4 mm thick H&E-stained slides using ten randomly selected fields for each animal with the following scores per examination field: 0 for an area of damage of;10%, 1 for an area of damage of 10 to 25%; 2 for damage involving 25 to 50% of the area; 3 for damage involving 50 to 75% of the area, and 4 for 75 to 100% of the area being affected [130]. In parallel, for spleen apoptosis, spleens with 10% formalin fixation were stained by anti-active caspase three antibodies (Cell Signaling Technology, Beverly, MA, USA), using immunohistochemistry, and expressed in positive cells per high-power field (200× magnification) as previously published [66,131].

4.5. Statistical Analysis

All data were analyzed with GraphPad Prism6 and demonstrated in mean ± S.E.M (standard error). The one-way analysis of variance (ANOVA) with Tukey's comparison test was used to compare among groups. Survival analysis was evaluated by the Log-rank test. A *p*-value less than 0.05 was considered significant.

5. Conclusions

There were several key findings from our data. First, LPS-induced macrophage injury, as indicated by the proteomic analysis, increased ROS (MDA), cell-free DNA, and DNA break. Second, the importance of MGMT for immune responses against LPS in macrophages using siRNA and *mgmt* null cells. Third, the MGMT influence in sepsis was demonstrated by the reduced severity in CLP sepsis of *mgmt* null mice. Fourth, the importance of effective antibiotics during immune modification therapy in sepsis as sepsis protective effect of *mgmt* null mice was lost without antibiotics. As such, the proteomic analysis of LPS-activated macrophages demonstrated possible cell injury as indicated by reduced proteins on oxidative phosphorylation with the high proteasome, perhaps correlating with mitochondrial ROS and ubiquitination of the deformed proteins, respectively. The role of MGMT enzyme to counteract LPS-induced DNA methylation was demonstrated through the attenuation of LPS-induced cell injury (DNA break and cell-free DNA) and responses (reduced cytokine production) in *mgmt* null cells compared with LPS-activated control cells. Likewise, less severe sepsis in *mgmt* null mice (MGMT loss only in myeloid cells) in CLP sepsis with antibiotics indicates the importance of myeloid cells and antibiotics in sepsis. Nevertheless, MGMT blockage, an available drug in cancer therapy, attenuated only serum cytokines but not mortality in CLP with antibiotics, implying further drug administration and delivery development.

Supplementary Materials: The following supporting information can be downloaded at: https://www.mdpi.com/article/10.3390/ijms241210175/s1.

Author Contributions: Conceptualization, A.L. and T.P.; methodology, P.P., P.V. and A.L.; formal analysis, K.S.-k., P.P. and P.V.; investigation, J.I.-A., J.M., A.B. and S.B.; resources, A.L. and T.P.; data curation P.P., P.V. and A.L.; writing—original draft preparation, K.S.-k. and A.L.; writing—review and editing T.P. and A.L.; supervision, A.L.; funding acquisition, A.L. All authors have read and agreed to the published version of the manuscript.

Funding: This research was funded by the NSRF via the Program Management Unit for Human Resources & Institutional Development, Research and Innovation (B16F640175) and (B36G660003), Rachadapisek Sompote Matching Fund (RA-MF-22/65 and RA-MF-13/66), the Thailand Science Research and Innovation Fund, Chulalongkorn University (HEA663000017 and CU_FRB65_hea (62)_125_23_55), and National Research Council of Thailand (NRCT-N41A640076 and NRCT-N34A660583) and also supported in part by the Intramural Research Program of National Institute of Allergy and Infectious Diseases (NIAID), National Institutes of Health (NIH). TP is funded by the National Research Council of Thailand (811/2563). KS was supported by the Second Century Fund (C2F) for high-efficiency PhD candidates at Chulalongkorn University.

Institutional Review Board Statement: The study was conducted in accordance with the National Institutes of Health (NIH) criteria, and the animal study protocol was approved by the Institutional Animal Care and Use Committee of the Faculty of Medicine, Chulalongkorn University, Bangkok, Thailand (CU-ACUP No. 021/2562).

Data Availability Statement: Not applicable.

Conflicts of Interest: The authors declare no conflict of interest.

References

1. Brady, J.; Horie, S.; Laffey, J.G. Role of the adaptive immune response in sepsis. *Intensive Care Med. Exp.* **2020**, *8*, 20. [CrossRef] [PubMed]
2. Sae-Khow, K.; Charoensappakit, A.; Chiewchengchol, D.; Leelahavanichkul, A. High-Dose Intravenous Ascorbate in Sepsis, a Pro-Oxidant Enhanced Microbicidal Activity and the Effect on Neutrophil Functions. *Biomedicines* **2022**, *11*, 51. [CrossRef] [PubMed]
3. Amornphimoltham, P.; Yuen, P.S.T.; Star, R.A.; Leelahavanichkul, A. Gut Leakage of Fungal-Derived Inflammatory Mediators: Part of a Gut-Liver-Kidney Axis in Bacterial Sepsis. *Dig. Dis. Sci.* **2019**, *64*, 2416–2428. [CrossRef] [PubMed]
4. Perner, A.; Rhodes, A.; Venkatesh, B.; Angus, D.C.; Martin-Loeches, I.; Preiser, J.C.; Vincent, J.L.; Marshall, J.; Reinhart, K.; Joannidis, M.; et al. Sepsis: Frontiers in supportive care, organisation and research. *Intensive Care Med.* **2017**, *43*, 496–508. [CrossRef]
5. Nemeth, K.; Leelahavanichkul, A.; Yuen, P.S.; Mayer, B.; Parmelee, A.; Doi, K.; Robey, P.G.; Leelahavanichkul, K.; Koller, B.H.; Brown, J.M.; et al. Bone marrow stromal cells attenuate sepsis via prostaglandin E(2)-dependent reprogramming of host macrophages to increase their interleukin-10 production. *Nat. Med.* **2009**, *15*, 42–49. [CrossRef]
6. Leelahavanichkul, A.; Yasuda, H.; Doi, K.; Hu, X.; Zhou, H.; Yuen, P.S.; Star, R.A. Methyl-2-acetamidoacrylate, an ethyl pyruvate analog, decreases sepsis-induced acute kidney injury in mice. *Am. J. Physiol. Physiol.* **2008**, *295*, F1825–F1835. [CrossRef]
7. Taratummarat, S.; Sangphech, N.; Vu, C.T.B.; Palaga, T.; Ondee, T.; Surawut, S.; Sereemaspun, A.; Ritprajak, P.; Leelahavanichkul, A. Gold nanoparticles attenuates bacterial sepsis in cecal ligation and puncture mouse model through the induction of M2 macrophage polarization. *BMC Microbiol.* **2018**, *18*, 85. [CrossRef]
8. Panpetch, W.; Chancharoenthana, W.; Bootdee, K.; Nilgate, S.; Finkelman, M.; Tumwasorn, S.; Leelahavanichkul, A. Lactobacillus rhamnosus L34 Attenuates Gut Translocation-Induced Bacterial Sepsis in Murine Models of Leaky Gut. *Infect. Immun.* **2018**, *86*, e00700-17. [CrossRef]

Issara-Amphorn, J.; Chancharoenthana, W.; Visitchanakun, P.; Leelahavanichkul, A. Syk Inhibitor Attenuates Polymicrobial Sepsis in FcgRIIb-Deficient Lupus Mouse Model, the Impact of Lupus Characteristics in Sepsis. *J. Innate Immun.* **2020**, *12*, 461–479. [CrossRef]
0. Dang, C.P.; Leelahavanichkul, A. Over-expression of miR-223 induces M2 macrophage through glycolysis alteration and attenuates LPS-induced sepsis mouse model, the cell-based therapy in sepsis. *PLoS ONE* **2020**, *15*, e0236038. [CrossRef]
1. Chancharoenthana, W.; Udompronpitak, K.; Manochantr, Y.; Kantagowit, P.; Kaewkanha, P.; Issara-Amphorn, J.; Leelahavanichkul, A. Repurposing of High-Dose Erythropoietin as a Potential Drug Attenuates Sepsis in Preconditioning Renal Injury. *Cells* **2021**, *10*, 3133. [CrossRef]
2. Dang, C.P.; Issara-Amphorn, J.; Charoensappakit, A.; Udompornpitak, K.; Bhunyakarnjanarat, T.; Saisorn, W.; Sae-Khow, K.; Leelahavanichkul, A. BAM15, a Mitochondrial Uncoupling Agent, Attenuates Inflammation in the LPS Injection Mouse Model: An Adjunctive Anti-Inflammation on Macrophages and Hepatocytes. *J. Innate Immun.* **2021**, *13*, 359–375. [CrossRef]
3. Mithal, L.B.; Arshad, M.; Swigart, L.R.; Khanolkar, A.; Ahmed, A.; Coates, B.M. Mechanisms and modulation of sepsis-induced immune dysfunction in children. *Pediatr. Res.* **2022**, *91*, 447–453. [CrossRef]
4. Schrijver, I.T.; Theroude, C.; Roger, T. Myeloid-Derived Suppressor Cells in Sepsis. *Front. Immunol.* **2019**, *10*, 327. [CrossRef] [PubMed]
5. Cao, C.; Ma, T.; Chai, Y.F.; Shou, S.T. The role of regulatory T cells in immune dysfunction during sepsis. *World J. Emerg. Med.* **2015**, *6*, 5–9. [CrossRef] [PubMed]
6. Vergadi, E.; Vaporidi, K.; Tsatsanis, C. Regulation of Endotoxin Tolerance and Compensatory Anti-inflammatory Response Syndrome by Non-coding RNAs. *Front. Immunol.* **2018**, *9*, 2705. [CrossRef]
7. Chancharoenthana, W.; Sutnu, N.; Visitchanakun, P.; Sawaswong, V.; Chitcharoen, S.; Payungporn, S.; Schuetz, A.; Schultz, M.J.; Leelahavanichkul, A. Critical roles of sepsis-reshaped fecal virota in attenuating sepsis severity. *Front. Immunol.* **2022**, *13*, 940935. [CrossRef] [PubMed]
8. Chancharoenthana, W.; Kamolratanakul, S.; Ariyanon, W.; Thanachartwet, V.; Phumratanaprapin, W.; Wilairatana, P.; Leelahavanichkul, A. Abnormal Blood Bacteriome, Gut Dysbiosis, and Progression to Severe Dengue Disease. *Front. Cell. Infect. Microbiol.* **2022**, *12*, 890817. [CrossRef]
9. Hiengrach, P.; Visitchanakun, P.; Tongchairawewat, P.; Tangsirisatian, P.; Jungteerapanich, T.; Ritprajak, P.; Wannigama, D.L.; Tangtanatakul, P.; Leelahavanichkul, A. Sepsis Encephalopathy Is Partly Mediated by miR370-3p-Induced Mitochondrial Injury but Attenuated by BAM15 in Cecal Ligation and Puncture Sepsis Male Mice. *Int. J. Mol. Sci.* **2022**, *23*, 5445. [CrossRef]
20. Thim-Uam, A.; Makjaroen, J.; Issara-Amphorn, J.; Saisorn, W.; Wannigama, D.L.; Chancharoenthana, W.; Leelahavanichkul, A. Enhanced Bacteremia in Dextran Sulfate-Induced Colitis in Splenectomy Mice Correlates with Gut Dysbiosis and LPS Tolerance. *Int. J. Mol. Sci.* **2022**, *23*, 1676. [CrossRef]
21. Ondee, T.; Surawut, S.; Taratummarat, S.; Hirankarn, N.; Palaga, T.; Pisitkun, P.; Pisitkun, T.; Leelahavanichkul, A. Fc Gamma Receptor IIB Deficient Mice: A Lupus Model with Increased Endotoxin Tolerance-Related Sepsis Susceptibility. *Shock* **2017**, *47*, 743–752. [CrossRef] [PubMed]
22. Seeley, J.J.; Ghosh, S. Molecular mechanisms of innate memory and tolerance to LPS. *J. Leukoc. Biol.* **2017**, *101*, 107–119. [CrossRef] [PubMed]
23. Gillen, J.; Ondee, T.; Gurusamy, D.; Issara-Amphorn, J.; Manes, N.P.; Yoon, S.H.; Leelahavanichkul, A.; Nita-Lazar, A. LPS Tolerance Inhibits Cellular Respiration and Induces Global Changes in the Macrophage Secretome. *Biomolecules* **2021**, *11*, 164. [CrossRef]
24. Lopez-Collazo, E.; del Fresno, C. Pathophysiology of endotoxin tolerance: Mechanisms and clinical consequences. *Crit. Care* **2013**, *17*, 242. [CrossRef] [PubMed]
25. Naler, L.B.; Hsieh, Y.P.; Geng, S.; Zhou, Z.; Li, L.; Lu, C. Epigenomic and transcriptomic analyses reveal differences between low-grade inflammation and severe exhaustion in LPS-challenged murine monocytes. *Commun. Biol.* **2022**, *5*, 102. [CrossRef]
26. Koos, B.; Moderegger, E.L.; Rump, K.; Nowak, H.; Willemsen, K.; Holtkamp, C.; Thon, P.; Adamzik, M.; Rahmel, T. LPS-Induced Endotoxemia Evokes Epigenetic Alterations in Mitochondrial DNA That Impacts Inflammatory Response. *Cells* **2020**, *9*, 2282. [CrossRef]
27. Lacal, I.; Ventura, R. Epigenetic Inheritance: Concepts, Mechanisms and Perspectives. *Front. Mol. Neurosci.* **2018**, *11*, 292. [CrossRef]
28. Han, M.; Jia, L.; Lv, W.; Wang, L.; Cui, W. Epigenetic Enzyme Mutations: Role in Tumorigenesis and Molecular Inhibitors. *Front. Oncol.* **2019**, *9*, 194. [CrossRef]
29. Casorelli, I.; Russo, M.T.; Bignami, M. Role of mismatch repair and MGMT in response to anticancer therapies. *Anti-Cancer Agents Med. Chem.* **2008**, *8*, 368–380. [CrossRef]
30. Sharma, S.; Salehi, F.; Scheithauer, B.W.; Rotondo, F.; Syro, L.V.; Kovacs, K. Role of MGMT in tumor development, progression, diagnosis, treatment and prognosis. *Anticancer Res.* **2009**, *29*, 3759–3768.
31. Rye, P.T.; Delaney, J.C.; Netirojjanakul, C.; Sun, D.X.; Liu, J.Z.; Essigmann, J.M. Mismatch repair proteins collaborate with methyltransferases in the repair of O(6)-methylguanine. *DNA Repair* **2008**, *7*, 170–176. [CrossRef] [PubMed]
32. Drablos, F.; Feyzi, E.; Aas, P.A.; Vaagbo, C.B.; Kavli, B.; Bratlie, M.S.; Pena-Diaz, J.; Otterlei, M.; Slupphaug, G.; Krokan, H.E. Alkylation damage in DNA and RNA--repair mechanisms and medical significance. *DNA Repair* **2004**, *3*, 1389–1407. [CrossRef] [PubMed]

33. Klapacz, J.; Pottenger, L.H.; Engelward, B.P.; Heinen, C.D.; Johnson, G.E.; Clewell, R.A.; Carmichael, P.L.; Adeleye, Y.; Andersen M.E. Contributions of DNA repair and damage response pathways to the non-linear genotoxic responses of alkylating agent *Mutat. Res. Mol. Mech. Mutagen.* **2016**, *767*, 77–91. [CrossRef] [PubMed]
34. Warren, J.J.; Forsberg, L.J.; Beese, L.S. The structural basis for the mutagenicity of O(6)-methyl-guanine lesions. *Proc. Natl. Aca Sci. USA* **2006**, *103*, 19701–19706. [CrossRef] [PubMed]
35. Kay, J.; Thadhani, E.; Samson, L.; Engelward, B. Inflammation-induced DNA damage, mutations and cancer. *DNA Repair* **201** *83*, 102673. [CrossRef]
36. Ezerskyte, M.; Paredes, J.A.; Malvezzi, S.; Burns, J.A.; Margison, G.P.; Olsson, M.; Scicchitano, D.A.; Dreij, K. O(6)-methylguanin induced transcriptional mutagenesis reduces p53 tumor-suppressor function. *Proc. Natl. Acad. Sci. USA* **2018**, *115*, 4731–473 [CrossRef]
37. Christmann, M.; Verbeek, B.; Roos, W.P.; Kaina, B. O(6)-Methylguanine-DNA methyltransferase (MGMT) in normal tissue and tumors: Enzyme activity, promoter methylation and immunohistochemistry. *Biochim. Biophys. Acta* **2011**, *1816*, 179–19 [CrossRef]
38. Kirkham, P. Oxidative stress and macrophage function: A failure to resolve the inflammatory response. *Biochem. Soc. Trans.* **200** *35*, 284–287. [CrossRef]
39. Tang, C.H.; Wei, W.; Liu, L. Regulation of DNA repair by S-nitrosylation. *Biochim. Biophys. Acta* **2012**, *1820*, 730–735. [CrossRef]
40. Colonna, M. DNA damage response impacts macrophage functions. *Blood* **2015**, *126*, 2440–2442. [CrossRef]
41. Cao, L.; Zhu, T.; Lang, X.; Jia, S.; Yang, Y.; Zhu, C.; Wang, Y.; Feng, S.; Wang, C.; Zhang, P.; et al. Inhibiting DNA Methylation Improves Survival in Severe Sepsis by Regulating NF-kappaB Pathway. *Front. Immunol.* **2020**, *11*, 1360. [CrossRef]
42. Wang, X.; Cao, Q.; Yu, L.; Shi, H.; Xue, B.; Shi, H. Epigenetic regulation of macrophage polarization and inflammation by DNA methylation in obesity. *JCI Insight* **2016**, *1*, e87748. [CrossRef] [PubMed]
43. Khan, O.A.; Ranson, M.; Michael, M.; Olver, I.; Levitt, N.C.; Mortimer, P.; Watson, A.J.; Margison, G.P.; Midgley, R.; Middleton M.R. A phase II trial of lomeguatrib and temozolomide in metastatic colorectal cancer. *Br. J. Cancer* **2008**, *98*, 1614–1618. [CrossRef] [PubMed]
44. Rumienczyk, I.; Kulecka, M.; Statkiewicz, M.; Ostrowski, J.; Mikula, M. Oncology Drug Repurposing for Sepsis Treatment *Biomedicines* **2022**, *10*, 921. [CrossRef]
45. Benjaskulluecha, S.; Boonmee, A.; Pattarakankul, T.; Wongprom, B.; Klomsing, J.; Palaga, T. Screening of compounds to identify novel epigenetic regulatory factors that affect innate immune memory in macrophages. *Sci. Rep.* **2022**, *12*, 1912. [CrossRef]
46. Pegg, A.E. Multifaceted roles of alkyltransferase and related proteins in DNA repair, DNA damage, resistance to chemotherapy and research tools. *Chem. Res. Toxicol.* **2011**, *24*, 618–639. [CrossRef] [PubMed]
47. Cheng, Y.; He, C.; Wang, M.; Ma, X.; Mo, F.; Yang, S.; Han, J.; Wei, X. Targeting epigenetic regulators for cancer therapy Mechanisms and advances in clinical trials. *Signal Transduct. Target. Ther.* **2019**, *4*, 62. [CrossRef]
48. Fernando, L.; Nguyen, V.; Hansen, T.; Golden, A.; Allen, A. Loss of proteasome subunit RPN-12 causes an increased mean lifespan at a higher temperature in *C. elegans*. *MicroPubl. Biol.* **2020**, *2020*. [CrossRef]
49. Murata, S.; Udono, H.; Tanahashi, N.; Hamada, N.; Watanabe, K.; Adachi, K.; Yamano, T.; Yui, K.; Kobayashi, N.; Kasahara, M.; et al. Immunoproteasome assembly and antigen presentation in mice lacking both PA28alpha and PA28beta. *EMBO J.* **2001**, *20*, 5898–5907. [CrossRef]
50. Flotynska, J.; Klause, D.; Kulecki, M.; Cieluch, A.; Chomicka-Pawlak, R.; Zozulinska-Ziolkiewicz, D.; Uruska, A. Higher NADH Dehydrogenase [Ubiquinone] Iron-Sulfur Protein 8 (NDUFS8) Serum Levels Correlate with Better Insulin Sensitivity in Type 1 Diabetes. *Curr. Issues Mol. Biol.* **2022**, *44*, 3872–3883. [CrossRef]
51. Yu, J.; Hederstedt, L.; Piggot, P.J. The cytochrome bc complex (menaquinone:cytochrome c reductase) in Bacillus subtilis has a nontraditional subunit organization. *J. Bacteriol.* **1995**, *177*, 6751–6760. [CrossRef]
52. Zhao, Y.; Sun, X.; Nie, X.; Sun, L.; Tang, T.S.; Chen, D.; Sun, Q. COX5B regulates MAVS-mediated antiviral signaling through interaction with ATG5 and repressing ROS production. *PLoS Pathog.* **2012**, *8*, e1003086. [CrossRef] [PubMed]
53. Antoniel, M.; Jones, K.; Antonucci, S.; Spolaore, B.; Fogolari, F.; Petronilli, V.; Giorgio, V.; Carraro, M.; Di Lisa, F.; Forte, M.; et al. The unique histidine in OSCP subunit of F-ATP synthase mediates inhibition of the permeability transition pore by acidic pH. *EMBO Rep.* **2018**, *19*, 257–268. [CrossRef] [PubMed]
54. Sweder, K.; Madura, K. Regulation of repair by the 26S proteasome. *J. Biomed. Biotechnol.* **2002**, *2*, 94–105. [CrossRef] [PubMed]
55. Sakai, W.; Yuasa-Sunagawa, M.; Kusakabe, M.; Kishimoto, A.; Matsui, T.; Kaneko, Y.; Akagi, J.I.; Huyghe, N.; Ikura, M.; Ikura, T.; et al. Functional impacts of the ubiquitin-proteasome system on DNA damage recognition in global genome nucleotide excision repair. *Sci. Rep.* **2020**, *10*, 19704. [CrossRef]
56. Lopes, A.F.C. Mitochondrial metabolism and DNA methylation: A review of the interaction between two genomes. *Clin. Epigenet.* **2020**, *12*, 182. [CrossRef] [PubMed]
57. Chen, J.; Wu, Y.; Sun, Y.; Dong, X.; Wang, Z.; Zhang, Z.; Xiao, Y.; Dong, G. Bacterial Lipopolysaccharide Induced Alterations of Genome-Wide DNA Methylation and Promoter Methylation of Lactation-Related Genes in Bovine Mammary Epithelial Cells. *Toxins* **2019**, *11*, 298. [CrossRef]
58. Angrisano, T.; Pero, R.; Brancaccio, M.; Coretti, L.; Florio, E.; Pezone, A.; Calabro, V.; Falco, G.; Keller, S.; Lembo, F.; et al. Cyclical DNA Methylation and Histone Changes Are Induced by LPS to Activate COX-2 in Human Intestinal Epithelial Cells. *PLoS ONE* **2016**, *11*, e0156671. [CrossRef] [PubMed]

59. Podhorecka, M.; Skladanowski, A.; Bozko, P. H2AX Phosphorylation: Its Role in DNA Damage Response and Cancer Therapy. *J. Nucleic Acids* **2010**, *2010*, 920161. [CrossRef]
60. Seminotti, B.; Amaral, A.U.; Grings, M.; Ribeiro, C.A.J.; Leipnitz, G.; Wajner, M. Lipopolysaccharide-Elicited Systemic Inflammation Induces Selective Vulnerability of Cerebral Cortex and Striatum of Developing Glutaryl-CoA Dehydrogenase Deficient (Gcdh(-/-)) Mice to Oxidative Stress. *Neurotox. Res.* **2020**, *38*, 1024–1036. [CrossRef]
61. Joo, Y.M.; Chae, M.K.; Hwang, S.Y.; Jin, S.C.; Lee, T.R.; Cha, W.C.; Jo, I.J.; Sim, M.S.; Song, K.J.; Jeong, Y.K.; et al. Impact of timely antibiotic administration on outcomes in patients with severe sepsis and septic shock in the emergency department. *Clin. Exp. Emerg. Med.* **2014**, *1*, 35–40. [CrossRef]
62. Leentjens, J.; Kox, M.; van der Hoeven, J.G.; Netea, M.G.; Pickkers, P. Immunotherapy for the adjunctive treatment of sepsis: From immunosuppression to immunostimulation. Time for a paradigm change? *Am. J. Respir. Crit. Care Med.* **2013**, *187*, 1287–1293. [CrossRef] [PubMed]
63. Tungsanga, S.; Panpetch, W.; Bhunyakarnjanarat, T.; Udompornpitak, K.; Katavetin, P.; Chancharoenthana, W.; Chatthanathon, P.; Somboonna, N.; Tungsanga, K.; Tumwasorn, S.; et al. Uremia-Induced Gut Barrier Defect in 5/6 Nephrectomized Mice Is Worsened by Candida Administration through a Synergy of Uremic Toxin, Lipopolysaccharide, and (1→3)-β-D-Glucan, but Is Attenuated by Lacticaseibacillus rhamnosus L34. *Int. J. Mol. Sci.* **2022**, *23*, 2511. [CrossRef] [PubMed]
64. Visitchanakun, P.; Panpetch, W.; Saisorn, W.; Chatthanathon, P.; Wannigama, D.L.; Thim-Uam, A.; Svasti, S.; Fucharoen, S.; Somboonna, N.; Leelahavanichkul, A. Increased susceptibility to dextran sulfate-induced mucositis of iron-overload beta-thalassemia mice, another endogenous cause of septicemia in thalassemia. *Clin. Sci.* **2021**, *135*, 1467–1486. [CrossRef] [PubMed]
65. Boonhai, S.; Bootdee, K.; Saisorn, W.; Takkavatakarn, K.; Sitticharoenchai, P.; Tungsanga, S.; Tiranathanagul, K.; Leelahavanichkul, A. TMAO reductase, a biomarker for gut permeability defect induced inflammation, in mouse model of chronic kidney disease and dextran sulfate solution-induced mucositis. *Asian Pac. J. Allergy Immunol.* **2021**. Epub ahead of print. [CrossRef]
66. Hiengrach, P.; Panpetch, W.; Chindamporn, A.; Leelahavanichkul, A. Macrophage depletion alters bacterial gut microbiota partly through fungal overgrowth in feces that worsens cecal ligation and puncture sepsis mice. *Sci. Rep.* **2022**, *12*, 9345. [CrossRef] [PubMed]
67. Doi, K.; Leelahavanichkul, A.; Yuen, P.S.; Star, R.A. Animal models of sepsis and sepsis-induced kidney injury. *J. Clin. Investig.* **2009**, *119*, 2868–2878. [CrossRef]
68. Binmama, S.; Dang, C.P.; Visitchanakun, P.; Hiengrach, P.; Somboonna, N.; Cheibchalard, T.; Pisitkun, P.; Chindamporn, A.; Leelahavanichkul, A. Beta-Glucan from S. cerevisiae Protected AOM-Induced Colon Cancer in cGAS-Deficient Mice Partly through Dectin-1-Manipulated Macrophage Cell Energy. *Int. J. Mol. Sci.* **2022**, *23*, 10951. [CrossRef]
69. Charoensappakit, A.; Sae-Khow, K.; Leelahavanichkul, A. Gut Barrier Damage and Gut Translocation of Pathogen Molecules in Lupus, an Impact of Innate Immunity (Macrophages and Neutrophils) in Autoimmune Disease. *Int. J. Mol. Sci.* **2022**, *23*, 8223. [CrossRef]
70. Marshall, R.S.; Vierstra, R.D. Dynamic Regulation of the 26S Proteasome: From Synthesis to Degradation. *Front. Mol. Biosci.* **2019**, *6*, 40. [CrossRef]
71. Rousseau, A.; Bertolotti, A. Regulation of proteasome assembly and activity in health and disease. *Nat. Rev. Mol. Cell Biol.* **2018**, *19*, 697–712. [CrossRef] [PubMed]
72. Silswal, N.; Reis, J.; Qureshi, A.A.; Papasian, C.; Qureshi, N. of Mice and Men: Proteasome's Role in LPS-Induced Inflammation and Tolerance. *Shock* **2017**, *47*, 445–454. [CrossRef] [PubMed]
73. Reis, J.; Guan, X.Q.; Kisselev, A.F.; Papasian, C.J.; Qureshi, A.A.; Morrison, D.C.; Van Way, C.W., 3rd; Vogel, S.N.; Qureshi, N. LPS-induced formation of immunoproteasomes: TNF-alpha and nitric oxide production are regulated by altered composition of proteasome-active sites. *Cell Biochem. Biophys.* **2011**, *60*, 77–88. [CrossRef]
74. Hansen, M.E.; Simmons, K.J.; Tippetts, T.S.; Thatcher, M.O.; Saito, R.R.; Hubbard, S.T.; Trumbull, A.M.; Parker, B.A.; Taylor, O.J.; Bikman, B.T. Lipopolysaccharide Disrupts Mitochondrial Physiology in Skeletal Muscle via Disparate Effects on Sphingolipid Metabolism. *Shock* **2015**, *44*, 585–592. [CrossRef] [PubMed]
75. Strichman-Almashanu, L.Z.; Lee, R.S.; Onyango, P.O.; Perlman, E.; Flam, F.; Frieman, M.B.; Feinberg, A.P. A genome-wide screen for normally methylated human CpG islands that can identify novel imprinted genes. *Genome Res.* **2002**, *12*, 543–554. [CrossRef] [PubMed]
76. Saxonov, S.; Berg, P.; Brutlag, D.L. A genome-wide analysis of CpG dinucleotides in the human genome distinguishes two distinct classes of promoters. *Proc. Natl. Acad. Sci. USA* **2006**, *103*, 1412–1417. [CrossRef]
77. Ezponda, T.; Licht, J.D. Molecular pathways: Deregulation of histone h3 lysine 27 methylation in cancer-different paths, same destination. *Clin. Cancer Res.* **2014**, *20*, 5001–5008. [CrossRef]
78. Nichol, J.N.; Dupere-Richer, D.; Ezponda, T.; Licht, J.D.; Miller, W.H., Jr. H3K27 Methylation: A Focal Point of Epigenetic Deregulation in Cancer. *Adv. Cancer Res.* **2016**, *131*, 59–95. [CrossRef]
79. Yue, D.; Wang, Z.; Yang, Y.; Hu, Z.; Luo, G.; Wang, F. EZH2 inhibitor GSK343 inhibits sepsis-induced intestinal disorders. *Exp. Ther. Med.* **2021**, *21*, 437. [CrossRef]
80. Jin, B.; Li, Y.; Robertson, K.D. DNA methylation: Superior or subordinate in the epigenetic hierarchy? *Genes Cancer* **2011**, *2*, 607–617. [CrossRef]
81. Du, Q.; Luu, P.L.; Stirzaker, C.; Clark, S.J. Methyl-CpG-binding domain proteins: Readers of the epigenome. *Epigenomics* **2015**, *7*, 1051–1073. [CrossRef] [PubMed]

82. Moore, L.D.; Le, T.; Fan, G. DNA methylation and its basic function. *Neuropsychopharmacology* **2013**, *38*, 23–38. [CrossRef] [PubMed]
83. Thomas, B.; Matson, S.; Chopra, V.; Sun, L.; Sharma, S.; Hersch, S.; Rosas, H.D.; Scherzer, C.; Ferrante, R.; Matson, W. A novel method for detecting 7-methyl guanine reveals aberrant methylation levels in Huntington disease. *Anal. Biochem.* **2013**, *436*, 112–120. [CrossRef] [PubMed]
84. Fan, C.H.; Liu, W.L.; Cao, H.; Wen, C.; Chen, L.; Jiang, G. O6-methylguanine DNA methyltransferase as a promising target for the treatment of temozolomide-resistant gliomas. *Cell Death Dis.* **2013**, *4*, e876. [CrossRef] [PubMed]
85. Yu, W.; Zhang, L.; Wei, Q.; Shao, A. O(6)-Methylguanine-DNA Methyltransferase (MGMT): Challenges and New Opportunities in Glioma Chemotherapy. *Front. Oncol.* **2019**, *9*, 1547. [CrossRef]
86. Gopisetty, G.; Ramachandran, K.; Singal, R. DNA methylation and apoptosis. *Mol. Immunol.* **2006**, *43*, 1729–1740. [CrossRef]
87. Kunanopparat, A.; Leelahavanichkul, A.; Visitchanakun, P.; Kueanjinda, P.; Phuengmaung, P.; Sae-Khow, K.; Boonmee, A.; Benjaskulluecha, S.; Palaga, T.; Hirankarn, N. The Regulatory Roles of Ezh2 in Response to Lipopolysaccharide (LPS) in Macrophages and Mice with Conditional Ezh2 Deletion with LysM-Cre System. *Int. J. Mol. Sci.* **2023**, *24*, 5363. [CrossRef]
88. Zhang, Q.; Sun, H.; Zhuang, S.; Liu, N.; Bao, X.; Liu, X.; Ren, H.; Lv, D.; Li, Z.; Bai, J.; et al. Novel pharmacological inhibition of EZH2 attenuates septic shock by altering innate inflammatory responses to sepsis. *Int. Immunopharmacol.* **2019**, *76*, 105899. [CrossRef]
89. Zhao, D.; Li, Z.; Liu, X.; Liu, N.; Bao, X.; Sun, H.; Meng, Q.; Ren, H.; Bai, J.; Zhou, X.; et al. Lymphocyte expression of EZH2 is associated with mortality and secondary infectious complications in sepsis. *Int. Immunopharmacol.* **2020**, *89*, 107042. [CrossRef]
90. Das, A.; Henderson, F.C., Jr.; Alshareef, M.; Porto, G.B.F.; Kanginakudru, I.; Infinger, L.K.; Vandergrift, W.A., 3rd; Lindhorst, S.M.; Varma, A.K.; Patel, S.J.; et al. MGMT-inhibitor in combination with TGF-betaRI inhibitor or CDK 4/6 inhibitor increases temozolomide sensitivity in temozolomide-resistant glioblastoma cells. *Clin. Transl. Oncol.* **2021**, *23*, 612–619. [CrossRef]
91. Davies, R.; O'Dea, K.; Gordon, A. Immune therapy in sepsis: Are we ready to try again? *J. Intensive Care Soc.* **2018**, *19*, 326–344. [CrossRef]
92. Ruenjaiman, V.; Butta, P.; Leu, Y.W.; Pongpanich, M.; Leelahavanichkul, A.; Kueanjinda, P.; Palaga, T. Profile of Histone H3 Lysine 4 Trimethylation and the Effect of Lipopolysaccharide/Immune Complex-Activated Macrophages on Endotoxemia. *Front. Immunol.* **2019**, *10*, 2956. [CrossRef] [PubMed]
93. Leelahavanichkul, A.; Somparn, P.; Bootprapan, T.; Tu, H.; Tangtanatakul, P.; Nuengjumnong, R.; Worasilchai, N.; Tiranathanagul, K.; Eiam-ong, S.; Levine, M.; et al. High-dose ascorbate with low-dose amphotericin B attenuates severity of disease in a model of the reappearance of candidemia during sepsis in the mouse. *Am. J. Physiol. Regul. Integr. Comp. Physiol.* **2015**, *309*, R223–R234. [CrossRef]
94. Vu, C.T.B.; Thammahong, A.; Leelahavanichkul, A.; Ritprajak, P. Alteration of macrophage immune phenotype in a murine sepsis model is associated with susceptibility to secondary fungal infection. *Asian Pac. J. Allergy Immunol.* **2022**, *40*, 162–171. [CrossRef]
95. Vu, C.T.B.; Thammahong, A.; Yagita, H.; Azuma, M.; Hirankarn, N.; Ritprajak, P.; Leelahavanichkul, A. Blockade of PD-1 Attenuated Postsepsis Aspergillosis via the Activation of IFN-gamma and the Dampening of IL-10. *Shock* **2020**, *53*, 514–524. [CrossRef] [PubMed]
96. Aleksandrov, R.; Hristova, R.; Stoynov, S.; Gospodinov, A. The Chromatin Response to Double-Strand DNA Breaks and Their Repair. *Cells* **2020**, *9*, 1853. [CrossRef]
97. White, R.R.; Vijg, J. Do DNA Double-Strand Breaks Drive Aging? *Mol. Cell* **2016**, *63*, 729–738. [CrossRef] [PubMed]
98. Cannan, W.J.; Pederson, D.S. Mechanisms and Consequences of Double-Strand DNA Break Formation in Chromatin. *J. Cell. Physiol.* **2016**, *231*, 3–14. [CrossRef]
99. Glukhov, I.L.; Sirota, N.P.; Kuznetsova, E.A. DNA damage in human mononuclear cells induced by bacterial endotoxin. *Bull. Exp. Biol. Med.* **2008**, *146*, 301–303. [CrossRef]
100. Kietzmann, T.; Petry, A.; Shvetsova, A.; Gerhold, J.M.; Gorlach, A. The epigenetic landscape related to reactive oxygen species formation in the cardiovascular system. *Br. J. Pharmacol.* **2017**, *174*, 1533–1554. [CrossRef]
101. Alhmoud, J.F.; Woolley, J.F.; Al Moustafa, A.E.; Malki, M.I. DNA Damage/Repair Management in Cancers. *Cancers* **2020**, *12*, 10502. [CrossRef] [PubMed]
102. Wyatt, M.D.; Pittman, D.L. Methylating agents and DNA repair responses: Methylated bases and sources of strand breaks. *Chem. Res. Toxicol.* **2006**, *19*, 1580–1594. [CrossRef] [PubMed]
103. Griffin, S.; Branch, P.; Xu, Y.Z.; Karran, P. DNA mismatch binding and incision at modified guanine bases by extracts of mammalian cells: Implications for tolerance to DNA methylation damage. *Biochemistry* **1994**, *33*, 4787–4793. [CrossRef] [PubMed]
104. Xaus, J.; Comalada, M.; Valledor, A.F.; Lloberas, J.; Lopez-Soriano, F.; Argiles, J.M.; Bogdan, C.; Celada, A. LPS induces apoptosis in macrophages mostly through the autocrine production of TNF-alpha. *Blood* **2000**, *95*, 3823–3831. [CrossRef]
105. Canton, M.; Sanchez-Rodriguez, R.; Spera, I.; Venegas, F.C.; Favia, M.; Viola, A.; Castegna, A. Reactive Oxygen Species in Macrophages: Sources and Targets. *Front. Immunol.* **2021**, *12*, 734229. [CrossRef]
106. Luo, M.; Bao, Z.; Xu, F.; Wang, X.; Li, F.; Li, W.; Chen, Z.; Ying, S.; Shen, H. Unrepaired DNA damage in macrophages causes elevation of particulate matter- induced airway inflammatory response. *Aging* **2018**, *10*, 549–560. [CrossRef]
107. Kraus, A.; McKeague, M.; Seiwert, N.; Nagel, G.; Geisen, S.M.; Ziegler, N.; Trantakis, I.A.; Kaina, B.; Thomas, A.D.; Sturla, S.J.; et al. Immunological and mass spectrometry-based approaches to determine thresholds of the mutagenic DNA adduct O(6)-methylguanine in vivo. *Arch. Toxicol.* **2019**, *93*, 559–572. [CrossRef]

08. Aguilera, A.; Gomez-Gonzalez, B. Genome instability: A mechanistic view of its causes and consequences. *Nat. Rev. Genet.* **2008**, *9*, 204–217. [CrossRef]
09. Snezhkina, A.V.; Kudryavtseva, A.V.; Kardymon, O.L.; Savvateeva, M.V.; Melnikova, N.V.; Krasnov, G.S.; Dmitriev, A.A. ROS Generation and Antioxidant Defense Systems in Normal and Malignant Cells. *Oxidative Med. Cell. Longev.* **2019**, *2019*, 6175804. [CrossRef]
10. Ranson, M.; Hersey, P.; Thompson, D.; Beith, J.; McArthur, G.A.; Haydon, A.; Davis, I.D.; Kefford, R.F.; Mortimer, P.; Harris, P.A.; et al. Randomized trial of the combination of lomeguatrib and temozolomide compared with temozolomide alone in chemotherapy naive patients with metastatic cutaneous melanoma. *J. Clin. Oncol.* **2007**, *25*, 2540–2545. [CrossRef]
11. Chancharoenthana, W.; Kamolratanakul, S.; Schultz, M.J.; Leelahavanichkul, A. The leaky gut and the gut microbiome in sepsis—Targets in research and treatment. *Clin. Sci.* **2023**, *137*, 645–662. [CrossRef]
12. Arango Duque, G.; Descoteaux, A. Macrophage cytokines: Involvement in immunity and infectious diseases. *Front. Immunol.* **2014**, *5*, 491. [CrossRef] [PubMed]
13. Yu, H.; Yang, Z.; Li, F.; Xu, L.; Sun, Y. Cell-mediated targeting drugs delivery systems. *Drug Deliv.* **2020**, *27*, 1425–1437. [CrossRef] [PubMed]
14. Greco, M.; Mazzei, A.; Suppressa, S.; Palumbo, C.; Verri, T.; Lobreglio, G. Human Leukocyte Antigen-DR Isotype Expression in Monocytes and T Cells Interferon-Gamma Release Assay in Septic Patients and Correlation with Clinical Outcome. *J. Clin. Med. Res.* **2021**, *13*, 293–303. [CrossRef] [PubMed]
15. Mansfield, S.; Griessl, M.; Gutknecht, M.; Cook, C.H. Sepsis and cytomegalovirus: Foes or conspirators? *Med. Microbiol. Immunol.* **2015**, *204*, 431–437. [CrossRef]
16. Aloisi, C.M.N.; Sturla, S.J.; Gahlon, H.L. A gene-targeted polymerase-mediated strategy to identify O(6)-methylguanine damage. *Chem. Commun.* **2019**, *55*, 3895–3898. [CrossRef]
17. Song, J.; Park, D.W.; Moon, S.; Cho, H.J.; Park, J.H.; Seok, H.; Choi, W.S. Diagnostic and prognostic value of interleukin-6, pentraxin 3, and procalcitonin levels among sepsis and septic shock patients: A prospective controlled study according to the Sepsis-3 definitions. *BMC Infect. Dis.* **2019**, *19*, 968. [CrossRef]
18. Wang, L.; Zhao, H.; Wang, D. Inflammatory cytokine expression in patients with sepsis at an intensive care unit. *Exp. Ther. Med.* **2018**, *16*, 2126–2131. [CrossRef]
19. Hiengrach, P.; Visitchanakun, P.; Finkelman, M.A.; Chancharoenthana, W.; Leelahavanichkul, A. More Prominent Inflammatory Response to Pachyman than to Whole-Glucan Particle and Oat-beta-Glucans in Dextran Sulfate-Induced Mucositis Mice and Mouse Injection through Proinflammatory Macrophages. *Int. J. Mol. Sci.* **2022**, *23*, 4026. [CrossRef]
20. Udompornpitak, K.; Charoensappakit, A.; Sae-Khow, K.; Bhunyakarnjanarat, T.; Dang, C.P.; Saisorn, W.; Visitchanakun, P.; Phuengmaung, P.; Palaga, T.; Ritprajak, P.; et al. Obesity Exacerbates Lupus Activity in Fc Gamma Receptor IIb Deficient Lupus Mice Partly through Saturated Fatty Acid-Induced Gut Barrier Defect and Systemic Inflammation. *J. Innate Immun.* **2022**, *15*, 240–261. [CrossRef]
121. Singkham-In, U.; Phuengmaung, P.; Makjaroen, J.; Saisorn, W.; Bhunyakarnjanarat, T.; Chatsuwan, T.; Chirathaworn, C.; Chancharoenthana, W.; Leelahavanichkul, A. Chlorhexidine Promotes Psl Expression in Pseudomonas aeruginosa That Enhances Cell Aggregation with Preserved Pathogenicity Demonstrates an Adaptation against Antiseptic. *Int. J. Mol. Sci.* **2022**, *23*, 8308. [CrossRef] [PubMed]
122. Kirstein, A.; Schilling, D.; Combs, S.E.; Schmid, T.E. Lomeguatrib Increases the Radiosensitivity of MGMT Unmethylated Human Glioblastoma Multiforme Cell Lines. *Int. J. Mol. Sci.* **2021**, *22*, 6781. [CrossRef] [PubMed]
123. Saisorn, W.; Saithong, S.; Phuengmaung, P.; Udompornpitak, K.; Bhunyakarnjanarat, T.; Visitchanakun, P.; Chareonsappakit, A.; Pisitkun, P.; Chiewchengchol, D.; Leelahavanichkul, A. Acute Kidney Injury Induced Lupus Exacerbation through the Enhanced Neutrophil Extracellular Traps (and Apoptosis) in Fcgr2b Deficient Lupus Mice with Renal Ischemia Reperfusion Injury. *Front. Immunol.* **2021**, *12*, 669162. [CrossRef] [PubMed]
124. Tzafrir, T.; Wainstock, T.; Sheiner, E.; Miodownik, S.; Pariente, G. Fetal Growth Restriction and Long-Term Cardiovascular Morbidity of Offspring in Dichorionic-Diamniotic Twin Pregnancies. *J. Clin. Med.* **2023**, *12*, 1628. [CrossRef]
125. Visitchanakun, P.; Tangtanatakul, P.; Trithiphen, O.; Soonthornchai, W.; Wongphoom, J.; Tachaboon, S.; Srisawat, N.; Leelahavanichkul, A. Plasma miR-370-3P as a Biomarker of Sepsis-Associated Encephalopathy, the Transcriptomic Profiling Analysis of Microrna-Arrays from Mouse Brains. *Shock* **2020**, *54*, 347–357. [CrossRef] [PubMed]
126. Issara-Amphorn, J.; Surawut, S.; Worasilchai, N.; Thim-Uam, A.; Finkelman, M.; Chindamporn, A.; Palaga, T.; Hirankarn, N.; Pisitkun, P.; Leelahavanichkul, A. The Synergy of Endotoxin and (1-->3)-beta-D-Glucan, from Gut Translocation, Worsens Sepsis Severity in a Lupus Model of Fc Gamma Receptor IIb-Deficient Mice. *J. Innate Immun.* **2018**, *10*, 189–201. [CrossRef]
127. Panpetch, W.; Somboonna, N.; Bulan, D.E.; Issara-Amphorn, J.; Finkelman, M.; Worasilchai, N.; Chindamporn, A.; Palaga, T.; Tumwasorn, S.; Leelahavanichkul, A. Oral administration of live- or heat-killed Candida albicans worsened cecal ligation and puncture sepsis in a murine model possibly due to an increased serum (1-->3)-beta-D-glucan. *PLoS ONE* **2017**, *12*, e0181439. [CrossRef]
128. Ranson, M.; Middleton, M.R.; Bridgewater, J.; Lee, S.M.; Dawson, M.; Jowle, D.; Halbert, G.; Waller, S.; McGrath, H.; Gumbrell, L.; et al. Lomeguatrib, a potent inhibitor of O6-alkylguanine-DNA-alkyltransferase: Phase I safety, pharmacodynamic, and pharmacokinetic trial and evaluation in combination with temozolomide in patients with advanced solid tumors. *Clin. Cancer Res.* **2006**, *12*, 1577–1584. [CrossRef]

129. Panpetch, W.; Phuengmaung, P.; Hiengrach, P.; Issara-Amphorn, J.; Cheibchalard, T.; Somboonna, N.; Tumwasorn, S.; Leela havanichkul, A. Candida Worsens Klebsiella pneumoniae Induced-Sepsis in a Mouse Model with Low Dose Dextran Sulfate Solution through Gut Dysbiosis and Enhanced Inflammation. *Int. J. Mol. Sci.* **2022**, *23*, 7050. [CrossRef]
130. Panpetch, W.; Kullapanich, C.; Dang, C.P.; Visitchanakun, P.; Saisorn, W.; Wongphoom, J.; Wannigama, D.L.; Thim-Uam, A Patarakul, K.; Somboonna, N.; et al. Candida Administration Worsens Uremia-Induced Gut Leakage in Bilateral Nephrectomy Mice, an Impact of Gut Fungi and Organismal Molecules in Uremia. *mSystems* **2021**, *6*, e01187-20. [CrossRef]
131. Makjaroen, J.; Thim-Uam, A.; Dang, C.P.; Pisitkun, T.; Somparn, P.; Leelahavanichkul, A. A Comparison Between 1 Day versu 7 Days of Sepsis in Mice with the Experiments on LPS-Activated Macrophages Support the Use of Intravenous Immunoglobulin for Sepsis Attenuation. *J. Inflamm. Res.* **2021**, *14*, 7243–7263. [CrossRef] [PubMed]

International Journal of
Molecular Sciences

MDPI

Article

Less Severe Sepsis in Cecal Ligation and Puncture Models with and without Lipopolysaccharide in Mice with Conditional *Ezh2*-Deleted Macrophages (LysM-Cre System)

Pornpimol Phuengmaung [1,2], Phuriwat Khiewkamrop [3,4], Jiradej Makjaroen [5], Jiraphorn Issara-Amphorn [6], Atsadang Boonmee [7], Salisa Benjaskulluecha [7], Patcharee Ritprajak [8], Aleksandra Nita-Lazar [6], Tanapat Palaga [3,7], Nattiya Hirankarn [2,3,*] and Asada Leelahavanichkul [1,2,9,*]

1 Center of Excellence in Translational Research in Inflammation and Immunology (CETRII), Faculty of Medicine, Chulalongkorn University, Bangkok 10330, Thailand; pphuengmaung@gmail.com
2 Department of Microbiology, Faculty of Medicine, Chulalongkorn University, Bangkok 10330, Thailand
3 Center of Excellence in Immunology and Immune-Mediated Diseases, Chulalongkorn University, Bangkok 10330, Thailand; phuriwat.khi@gmail.com (P.K.); tanapat.palaga@gmail.com (T.P.)
4 Medical Microbiology, Interdisciplinary and International Program, Graduate School, Chulalongkorn University, Bangkok 10330, Thailand
5 Center of Excellence in Systems Biology, Research Affairs, Faculty of Medicine, Chulalongkorn University, Bangkok 10330, Thailand; jiradejmak@gmail.com
6 Functional Cellular Networks Section, Laboratory of Immune System Biology, National Institute of Allergy and Infectious Diseases, National Institutes of Health, Bethesda, MD 20892, USA; jiraphorn.issara-amphorn@nih.gov (J.I.-A.); nitalazarau@niaid.nih.gov (A.N.-L.)
7 Department of Microbiology, Faculty of Science, Chulalongkorn University, Bangkok 10330, Thailand; atsadang88@gmail.com (A.B.); salisafaii@gmail.com (S.B.)
8 Research Unit in Integrative Immuno-Microbial Biochemistry and Bioresponsive Nanomaterials, Department of Microbiology, Faculty of Dentistry, Chulalongkorn University, Bangkok 10330, Thailand; p.ritprajak@gmail.com
9 Division of Nephrology, Department of Medicine, Faculty of Medicine, Chulalongkorn University, Bangkok 10330, Thailand
* Correspondence: nattiya.h@chula.ac.th (N.H.); aleelahavanit@gmail.com (A.L.); Tel.: +66-22-564-132 (N.H. & A.L.)

Citation: Phuengmaung, P.; Khiewkamrop, P.; Makjaroen, J.; Issara-Amphorn, J.; Boonmee, A.; Benjaskulluecha, S.; Ritprajak, P.; Nita-Lazar, A.; Palaga, T.; Hirankarn, N.; et al. Less Severe Sepsis in Cecal Ligation and Puncture Models with and without Lipopolysaccharide in Mice with Conditional *Ezh2*-Deleted Macrophages (LysM-Cre System). *Int. J. Mol. Sci.* **2023**, *24*, 8517. https://doi.org/10.3390/ijms24108517

Academic Editor: Paolo Iadarola

Received: 8 April 2023
Revised: 2 May 2023
Accepted: 7 May 2023
Published: 10 May 2023

Abstract: Despite a previous report on less inflammatory responses in mice with an absence of the enhancer of zeste homologue 2 (Ezh2), a histone lysine methyltransferase of epigenetic regulation, using a lipopolysaccharide (LPS) injection model, proteomic analysis and cecal ligation and puncture (CLP), a sepsis model that more resembles human conditions was devised. As such, analysis of cellular and secreted protein (proteome and secretome) after a single LPS activation and LPS tolerance in macrophages from Ezh2 null ($Ezh2^{flox/flox}$; LysM-$Cre^{cre/-}$) mice (Ezh2 null) and the littermate control mice ($Ezh2^{fl/fl}$; LysM-$Cre^{-/-}$) (Ezh2 control) compared with the unstimulated cells from each group indicated fewer activities in Ezh2 null macrophages, especially by the volcano plot analysis. Indeed, supernatant IL-1β and expression of genes in pro-inflammatory M1 macrophage polarization (*IL-1β* and *iNOS*), *TNF-α*, and *NF-κB* (a transcription factor) were lower in Ezh2 null macrophages compared with the control. In LPS tolerance, downregulated *NF-κB* compared with the control was also demonstrated in Ezh2 null cells. In CLP sepsis mice, those with CLP alone and CLP at 2 days after twice receiving LPS injection, representing sepsis and sepsis after endotoxemia, respectively, symptoms were less severe in Ezh2 null mice, as indicated by survival analysis and other biomarkers. However, the Ezh2 inhibitor improved survival only in CLP, but not LPS with CLP. In conclusion, an absence of Ezh2 in macrophages resulted in less severe sepsis, and the use of an Ezh2 inhibitor might be beneficial in sepsis.

Keywords: sepsis; lipopolysaccharide; macrophages; epigenetics; Ezh2

1. Introduction

Sepsis is a potentially life-threatening condition in response to severe infection regardless of the organismal causes of the infection [1–3], which is roughly divided into the hyperinflammation stage and immune exhaustion (immune paralysis) phase [4,5]. Sepsis-induced hyperinflammation is a well-known cause of sepsis mortality, partly through hypercytokinemia-mediated septic shock, especially at an early phase of sepsis [6]. Meanwhile, sepsis-induced immune exhaustion is developed at the same time or shortly after the hyperinflammation, at least in part, due to immune cell death from overwhelming responses against several stimulators from the pathogens and hosts, referred to as pathogen-associated molecular patterns (PAMPs) and damage-associated molecular patterns (DAMPs), respectively [6]. Subsequently, an inadequate response to control the organism during sepsis-induced immune exhaustion results in another episode of secondary infection and another episode of septic shock from the different pathogens [7]. Due to the opposite direction of immune responses in different phases of sepsis, different strategic treatments are necessary. As such, an anti-inflammatory treatment might be beneficial for sepsis-hyperinflammation to attenuate the unnecessary overwhelming immune responses that might be harmful to the host. Meanwhile, an escalation of immune responsiveness during immune exhaustion may be helpful to enhance the microbial control ability of the host to prevent secondary infections [8–15]. Despite the successful decrease in short-term sepsis mortality due to improved supportive care, immune exhaustion-induced secondary infection seems to become more common [16]. Indeed, immune cell apoptosis, myeloid-derived suppressor cells, regulatory T cells, and lipopolysaccharide (LPS) tolerance are mentioned as underlying mechanisms of sepsis-induced immune exhaustion [17–20]. Among these topics, data on LPS tolerance in sepsis are relatively fewer compared with those on other mechanisms. The presence of LPS, a major molecule of Gram-negative bacteria, in blood circulation during sepsis (endotoxemia) is common due to Gram-negative bacterial infection and/or the translocation of LPS from the intestine into the blood circulation, referred to as "leaky gut", and is a common cause of endotoxemia [21–23]. Because of the highest abundance of Gram-negative bacteria in the gut compared with other organisms, endotoxemia from a leaky gut is mentioned in several conditions with gut barrier defects, including sepsis [1–3]. Subsequently, an adaptation to the prolonged LPS stimulations in sepsis may initiate LPS tolerance [24,25]. Among several models of sepsis, cecal ligation and puncture (CLP) is a standard model used for sepsis hyper-inflammation [14,23,26–29] that more resembles the human condition than a single LPS injection [30]. Meanwhile, sepsis immune exhaustion consists of several models, based on increased susceptibility of the secondary infection [31,32], and the more severe sepsis in CLP surgery after LPS tolerance, using twice-administered LPS injection, compared with CLP without LPS pre-conditioning is previously mentioned [25]. Indeed, the more severe sepsis, especially bacteremia, in CLP after LPS tolerance compared with CLP alone is matched with the common characteristic of the worsened infection in sepsis-induced immune exhaustion compared with the normal immune regulation [33]. Then, CLP and CLP after LPS tolerance were used as models of hyperinflammation and LPS tolerance-associated immune exhaustion with sepsis, respectively.

Interestingly, the tolerance against LPS, especially in monocytes or macrophages, is possibly due to the epigenetic modifications, chromatin remodeling, and interferences in cell energy status [34–36], among which epigenetic alteration is the most extensively studied [37,38]. Epigenetics is the phenotypic alterations without the changes in the DNA sequence for the switch "on" and "off" of DNA transcription through (i) the modifications of DNA and/or histone through several enzymes [3,39], and (ii) noncoding RNA (microRNA) [2]. Among all, the methylation at histone 3 lysine 27 (H3K27) is one of the most common epigenetic processes through histone changes for several cell activities, including after LPS activation [40]. In LPS-activated macrophages, the insertion of methyl groups at lysine 27 on histone 3 (H3K27) by histone demethylase is controlled by the polycomb repressor complex group 2 (PCR2), a repressor molecule consisting of several subunits, including Ezh2 (histone-lysine N-methyltransferase-2 or enhancer of zeste homolog) [41,42], to switch off the DNA transcription (reduce cytokine production) through this histone modification [43,44]. Due to the important Ezh2 catalytic activity on PCR2, Ezh2 overexpression enhances PCR2 inhibitory function with an anti-inflammatory effect [45,46] and Ezh2 deletion should enhance pro-inflammatory responses [47]. Indeed, the enhanced pro-inflammatory effect of Ezh2 blockage is mentioned via (i) increased tumoricidal impact of tazemetostat (an Ezh2 inhibitor) [48] and (ii) worsened colitis after *Ezh2* downregulation [49]. In contrast, the anti-inflammatory property of Ezh2 blockage is also mentioned in the atherosclerosis model [50]. Hence, Ezh2 not only downregulates pro-inflammatory cytokines but also can decrease the anti-inflammatory process, partly through the downregulation of the suppressor of cytokine signaling 3 (*Socs3*) [51]. Ezh2 causes histone methylation that block transcription of both pro-inflammatory genes (cytokines) and anti-inflammatory genes (Sosc3) and the impact of Ezh2 might be different among various genes and cell types. Indeed, in mice with conditional Ezh2 deletion by the LysM-Cre system, Ezh2 deletion only in the myeloid cells (monocytes, macrophages, and neutrophils) demonstrates less severe responses against a single LPS injection, but not LPS tolerance (two LPS injections), and bone marrow-derived macrophages from these mice indicated less potent LPS responses (lower supernatant cytokines) and less severe LPS tolerance (higher supernatant cytokines) compared with the control cells [39]. Although Ezh2 impacts on sepsis are still inconclusive, Ezh2 is one of the upregulated genes in LPS-tolerant macrophages [52] which is improved by the Ezh2 inhibitor (enhanced TNF-α expression) [53] as an interesting control of macrophage through epigenetic manipulation [13,54].

Here, the influence of Ezh2 on LPS was further explored through proteomic analysis and tested in a model with sepsis hyper-inflammatory responses (CLP) and a sepsis model after LPS tolerance (twice-administered LPS injection before CLP surgery) using the conditional Ezh2 deletion mice and an Ezh2 inhibitor.

2. Results

2.1. *Proteomic Analysis of Lipopolysaccharide (LPS)-Induced Macrophages from Control and Ezh2 Null Mice*

The difference in bone marrow-derived macrophage (BMDM) after activation with three protocols, namely control, a single LPS stimulation, and LPS tolerance (Figure 1A), of macrophages from Ezh2 control (Ezh$^{fl/fl}$; LysM-Cre$^{-/-}$) or Ezh2 null (Ezh$^{fl/fl}$; LysM-Cre$^{cre/-}$) mice were analyzed by proteome and secretome, using the cells and cell supernatant, respectively. With a single LPS stimulation, there were prominent alterations in the peptides of Ezh2 control macrophages compared with the neutral state of Ezh2 control cells as indicated by the up- and downregulation at 188 and 119 proteins, respectively (Figure 1B, left upper). Meanwhile, the values of LPS-activated Ezh2 null cells were 32 and 9 proteins, respectively (Figure 1B, lower left), suggesting possible less activity of Ezh2 null macrophages compared with that in the cells from littermate control mice. Likewise, the up- and downregulated molecules in LPS tolerance of Ezh2 control macrophages compared with the neutral state were 392 and 296 proteins, respectively (Figure 1B, upper right), while for the Ezh2 null cells they were 107 and 68 proteins, respectively (Figure 1B, lower right). With the fold enrichment pathway analysis (ShinyGo 0.77), most of the proteins were correlated to immune response pathways and cell energy status in macrophages from both mouse strains with either single or twice LPS stimulation (Figure 2). The proteins with aerobic respiration and interferon-gamma responses were the groups with the highest fold enrichment in Ezh2 control cells and Ezh2 null macrophages, respectively, after a single LPS activation (Figure 2, left). In LPS tolerance, the proteins with interferon-gamma responses and the negative regulation of innate immune responses were the groups with the highest fold enrichment in Ezh2 control cells and Ezh2 null macrophages, respectively (Figure 2, right). These data implied similar downstream LPS responses in macrophages of both mouse strains. Although there was no direct comparison between Ezh2 control macrophages versus Ezh2 null cells, the Venn diagram analysis, using the Venny 2.1 program (https://bioinfogp.cnb.csic.es/tools/venny) accessed on 15 March 2023, from the list of proteins roughly indicated differences in the deviation of the macrophage proteome away from the resting control condition (activated cells versus control cells) (Figure 3). With a single LPS, 307 and 42 proteins were up- or downregulated in macrophages of littermate control mice (LPS Ezh2 control vs. Ezh2 control) and Ezh2 null mice (LPS Ezh2 null vs. Ezh2 null), respectively, with 32 unique proteins presenting only in the latter group (Figure 3, upper). Because the unique proteins in Ezh2 null macrophages might be responsible for the phenotypic differences between Ezh2 null versus Ezh2 control cells after LPS activation, these proteins were further evaluated by the ShinyGO 0.77 program. Interestingly, these 32 proteins were the member of only 2 pathways, including nuclear factor-κB (NF-κB) and Toll-like receptor (TLR) pathways (Figure 3, upper). Similarly, after LPS tolerance, 1248 and 1269 proteins were altered from the neutral state (up- or downregulation) in the proteome of macrophages from littermate control mice (Ezh2 control) and Ezh2 null mice, respectively, with 104 unique peptides presenting only in Ezh2 null macrophages (Figure 3, lower) which also were mostly associated with cell energy status and immune responses (Figure 3, lower). Details of the proteins from macrophages uniquely elevated in Ezh2 null cells but not in Ezh2 control macrophages are indicated in Supplementary Tables S1 and S2.

As expected, analysis of the secreted proteins from supernatant (secretome analysis) that deviated from the neutral status demonstrated the lessor proteins (Figure 4) compared with the analysis from the cell lysate (Figure 2). There were 6 and 23 up- and downregulated proteins, respectively, in the secretome of LPS Ezh2 control vs. Ezh2 control and 2 and 22 up- and downregulated proteins, respectively, in the secretome of LPS Ezh2 null vs Ezh2 null (Figure 4, left). Meanwhile, there were 26 and 66 up- and downregulated proteins, respectively, in the secretome of LPS/LPS Ezh2 control vs. Ezh2 control and 8 and 1 up- and downregulated proteins, respectively, in the secretome of LPS/LPS Ezh2 null vs. Ezh2 null (Figure 4, right). Additionally, The Venn diagram demonstrated 106 and 11 overlapped proteins after the activation compared with the neutral state in the cells from each mouse strain after a single LPS and LPS tolerance, respectively (Figure 5). The fold enrichment pathway analysis of the unique proteins in Ezh2 null macrophages indicated an involvement in cell energy status and responses against infection in a single LPS stimulation (Figure 5, upper), while mostly involved in responses to infection in the LPS tolerance group (Figure 5, lower). Details of the proteins in secretome analysis that were uniquely elevated in Ezh2 null cells but not in Ezh2 control macrophages are indicated in Supplementary Tables S3 and S4. These data implied fewer activities of Ezh2 null macrophages compare with the control cells after activation by either LPS or LPS tolerance that might be responsible for the phenotypic responses against LPS.

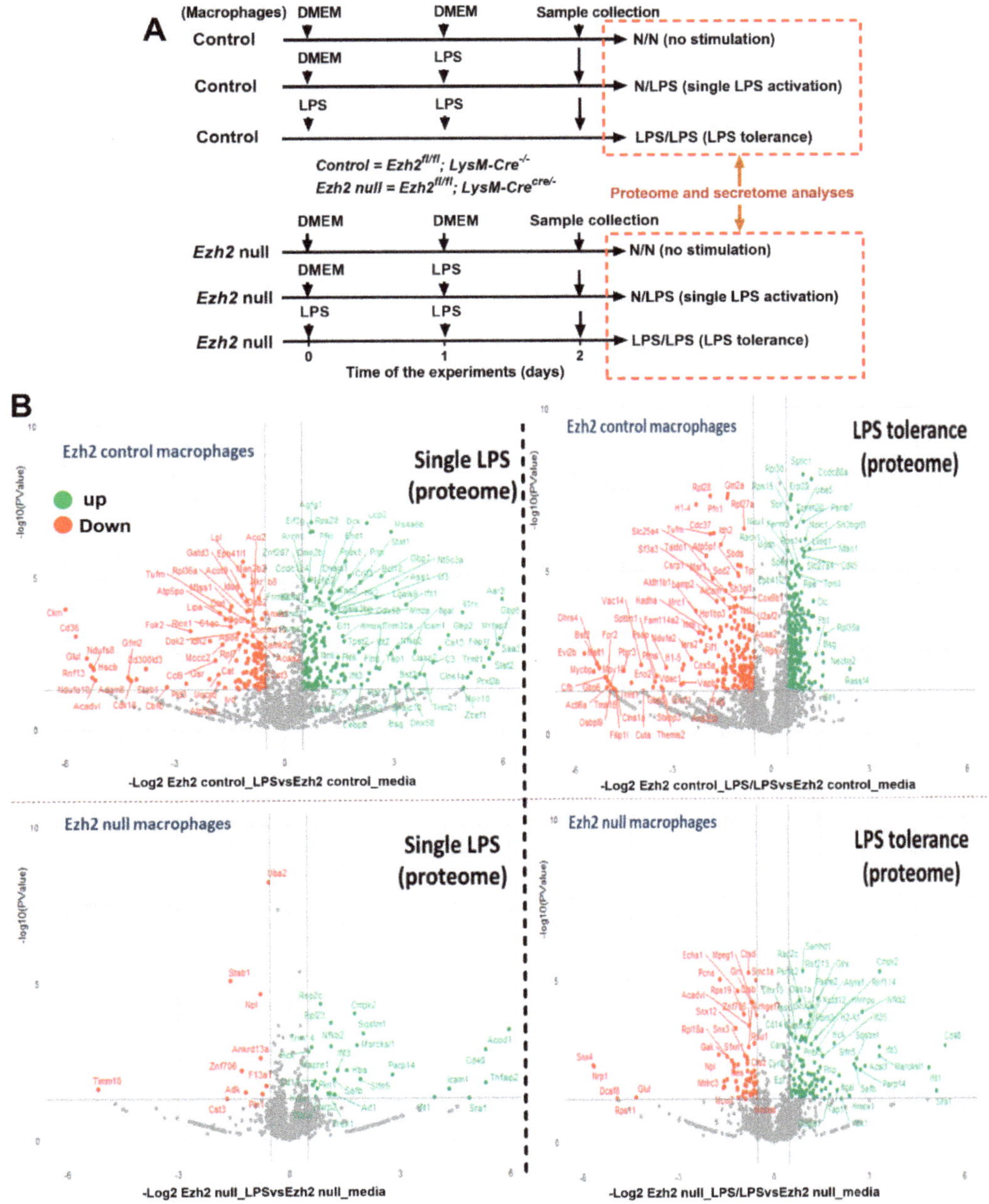

Figure 1. Schema of the experiments using bone marrow-derived macrophages (macrophages) from control mice (Ezh$^{fl/fl}$; LysM-Cre$^{-/-}$) or Ezh2 null mice (Ezh$^{fl/fl}$; LysM-Cre$^{cre/-}$) after activation by lipopolysaccharide (LPS) in a single protocol (N/LPS) that started with the culture media (DMEM) followed by LPS 24 h later or LPS tolerance (LPS/LPS) by the repeated LPS stimulations or no stimulation control (N/N) with DMEM incubation (**A**) is demonstrated. The volcano plot indicating the up- and downregulated proteins, in green and red color, respectively, as compared between the activated cells versus the non-stimulated cells (**B**) is demonstrated. Macrophages were isolated from 3 different mice for the triplicate analysis.

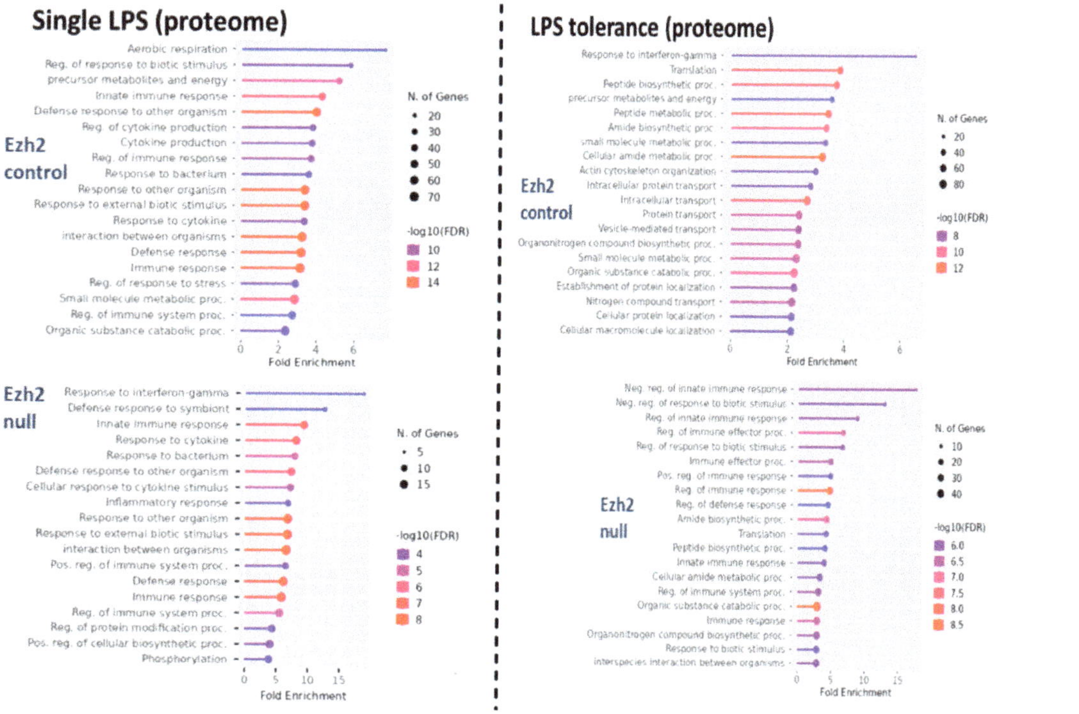

Figure 2. Fold enrichment analysis of cellular proteins in macrophages (proteome) from Ezh2 control ($Ezh^{fl/fl}$; LysM-$Cre^{-/-}$) (Ezh2 control) or Ezh2 null mice ($Ezh^{fl/fl}$; LysM-$Cre^{cre/-}$) (Ezh2 null) after a single lipopolysaccharide (LPS) activation (**left side**) or LPS tolerance (**right side**) between Ezh2 control cells or Ezh2 null macrophages with versus without activation is demonstrated.

2.2. Less Prominent M1 Macrophage Polarization in LPS-Activated Ezh2 Null Cells with the Downregulation of NF-κB after A Single and Twice LPS Stimulation

Then, macrophages from Ezh2 control ($Ezh^{fl/fl}$; LysM-$Cre^{-/}$) and Ezh2 null ($Ezh^{fl/fl}$; LysM-$Cre^{cre/-}$) mice were tested. Here, the supernatant was removed and the cells were washed 1 day post-incubation before the difference between single and twice LPS stimulations was determined at 2 days to control the duration of culture in both groups (Figure 6A). As such, both a single (N/LPS) and LPS tolerance (LPS/LPS) upregulated M1 macrophage polarization compared with control, as determined by supernatant interleukin (IL)-1β with upregulated *IL-1β* and inducible nitric oxide synthase (*iNOS*), without an alteration in genes of M2 polarization, including resistin-like-α (Retnla or *Fizz-1*), arginase-1 (*Arg-1*), and transforming growth factor beta (*TGF-β*) (Figure 6B–F). In addition, the characteristic of LPS tolerance, a less potent response to the following LPS stimulations compared with the first response to LPS [55–57] was demonstrated in both control and *Ezh2* null cells, as the gene expression of *TNF-α* and *IL-6* (but not *IL-10*) in N/LPS were more prominent than LPS tolerance (Figure 6G–I). However, the expression of genes for pro-inflammatory molecule nuclear factor kappa B (*NF-kB*) was similarly higher than in the control group (N/N) (Figure 6J). On the other hand, there was a less prominent M1 polarization (pro-inflammatory macrophages), as indicated by *IL-1β* and *iNOS*, together with less pro-inflammatory responses (*TNF-α*, *IL-6*, and *NF-κB*) in Ezh2 null macrophages compared with control cells after a single LPS stimulation (N/LPS) (Figure 6C–J) supporting a pro-inflammatory effect of Ezh2 on macrophages [39]. In LPS tolerance (LPS/LPS), there was a non-difference in supernatant IL-1β, macrophage polarization, and cytokine genes between Ezh2 null macrophages and control cells, despite lower *NF-κB* expression in Ezh2 null macrophages (Figure 6B–J). Between Ezh2 null macrophages with a single LPS and

LPS tolerance, all of these parameters were similar (Figure 6B–J). Despite lower *NF-kB* (a pro-inflammation molecule) in Ezh2 null macrophages with LPS tolerance compared with control cells (Figure 6J), supernatant IL-1β (Figure 6B) and expression of cytokine genes (Figure 6G–I) in Ezh2 null macrophages were similar to that in control cells, implying a limited Ezh2 impact on the control of macrophage responses during LPS tolerance. Although there was a limited impact on LPS tolerance, these data supported an anti-inflammatory response of Ezh2 null macrophages after a single LPS stimulation that might be useful as an anti-inflammation in sepsis.

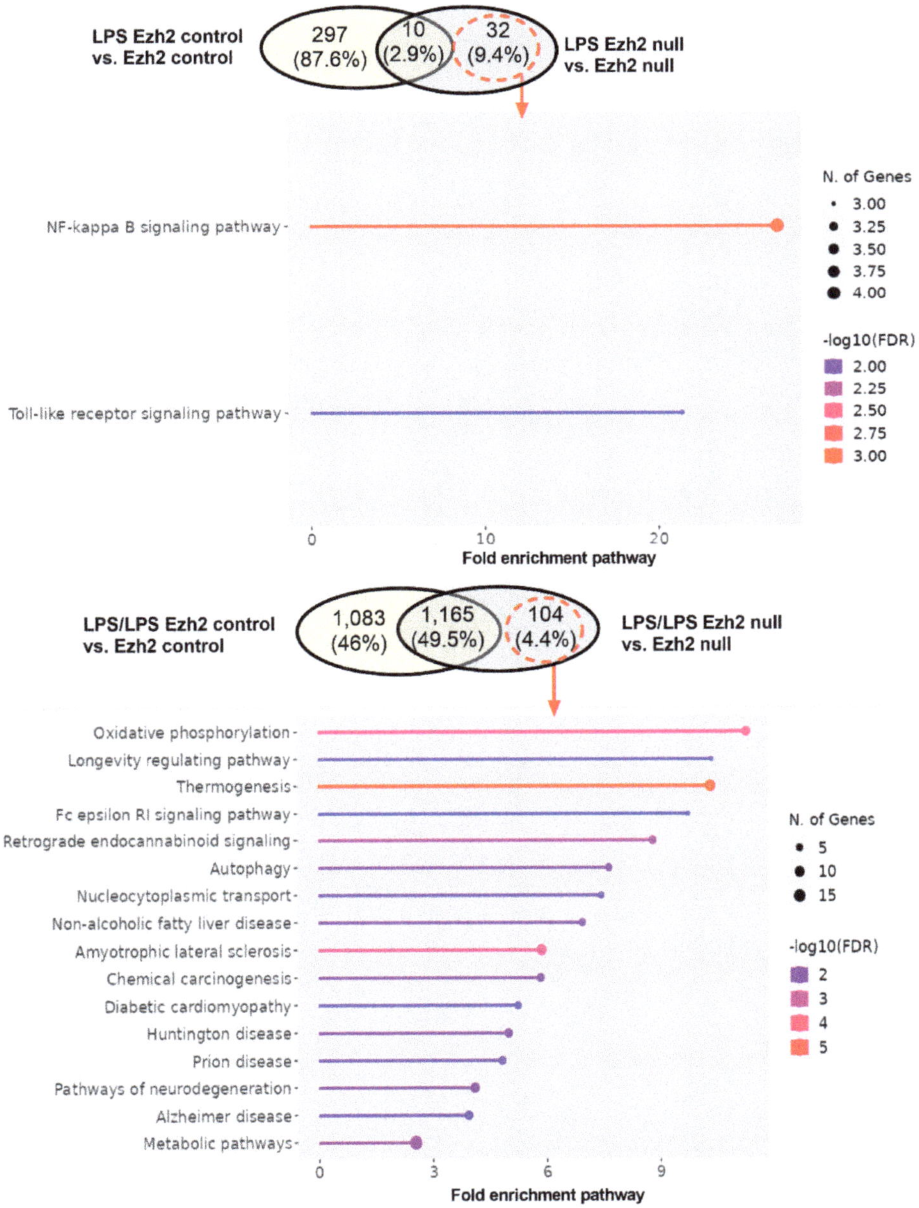

Figure 3. The Venn diagrams of proteome analysis from macrophages of Ezh2 control (Ezh$^{fl/fl}$; LysM-Cre$^{-/-}$) (Ezh2 control) or Ezh2 null mice (Ezh$^{fl/fl}$; LysM-Cre$^{cre/-}$) (Ezh2 null) after a single

lipopolysaccharide (LPS) activation (**upper**) or LPS tolerance (**lower**) between Ezh2 control cells o Ezh2 null macrophages versus the non-activated control of each group are demonstrated. The fol enrichment pathway of the unique proteins that presented only in LPS-stimulated macrophages bu not in the Ezh2 control cells (dashed circles with arrows) is also demonstrated. Notably, percentage in the Venn diagram are the number of proteins in each part divided by the total number of protein from both groups.

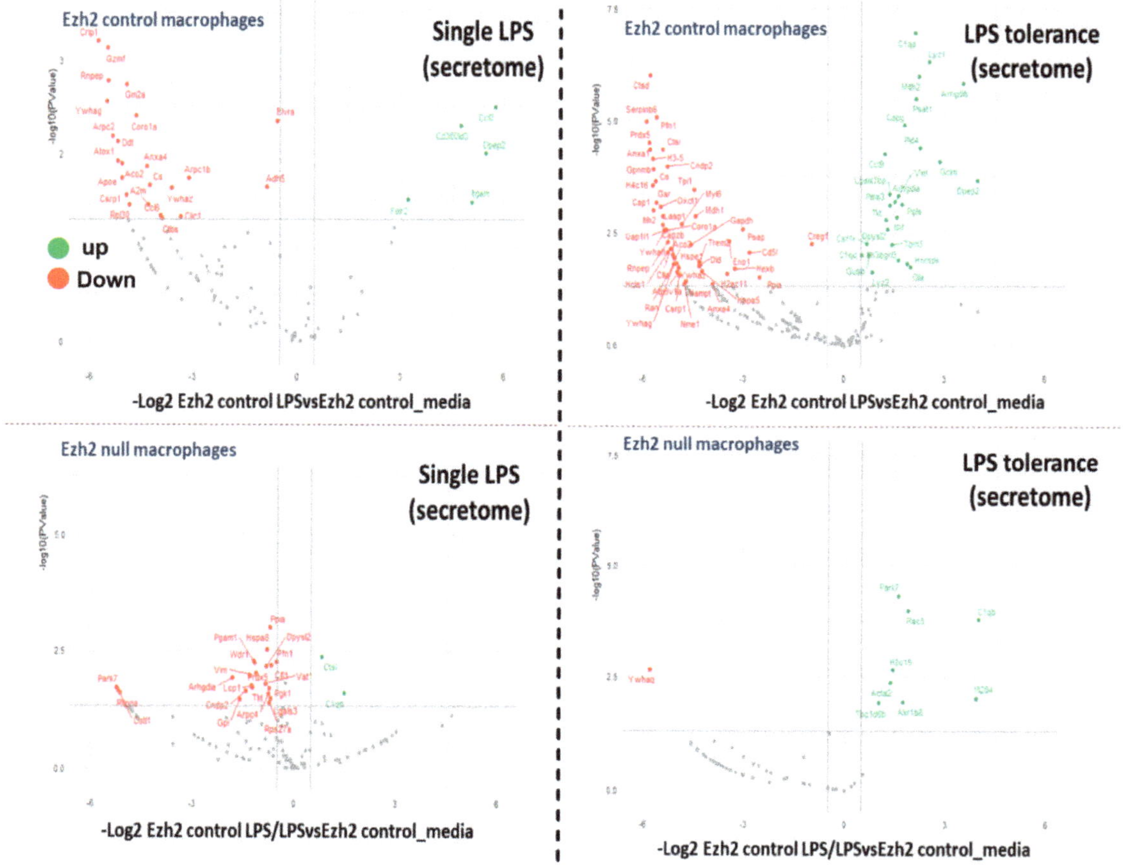

Figure 4. The volcano plots of secretome analysis from macrophages of Ezh2 control (Ezh$^{fl/fl}$; LysM-Cre$^{-/-}$) (Ezh2 control) or Ezh2 null mice (Ezh$^{fl/fl}$; LysM-Cre$^{cre/-}$) (Ezh2 null) after a single lipopolysaccharide (LPS) activation (**left side**) or LPS tolerance (**right side**) between Ezh2 control cells (**upper**) or Ezh2 null macrophages (**lower**) with versus without the activations are demonstrated.

Figure 5. The Venn diagrams of secretome analysis from macrophages of Ezh2 control ($Ezh^{fl/fl}$; $LysM\text{-}Cre^{-/-}$) (Ezh2 control) or Ezh2 null mice ($Ezh^{fl/fl}$; $LysM\text{-}Cre^{cre/-}$) (Ezh2 null) after a single lipopolysaccharide (LPS) activation (**upper**) or LPS tolerance (**lower**) between Ezh2 control cells or Ezh2 null macrophages versus the non-activated control of each group are demonstrated. The fold enrichment pathway of the unique proteins that presented only in LPS-stimulated macrophages but not in the Ezh2 control cells (dashed circles with arrows) is also demonstrated. Notably, percentages in the Venn diagram are the number of proteins in each part divided by the total number of proteins from both groups.

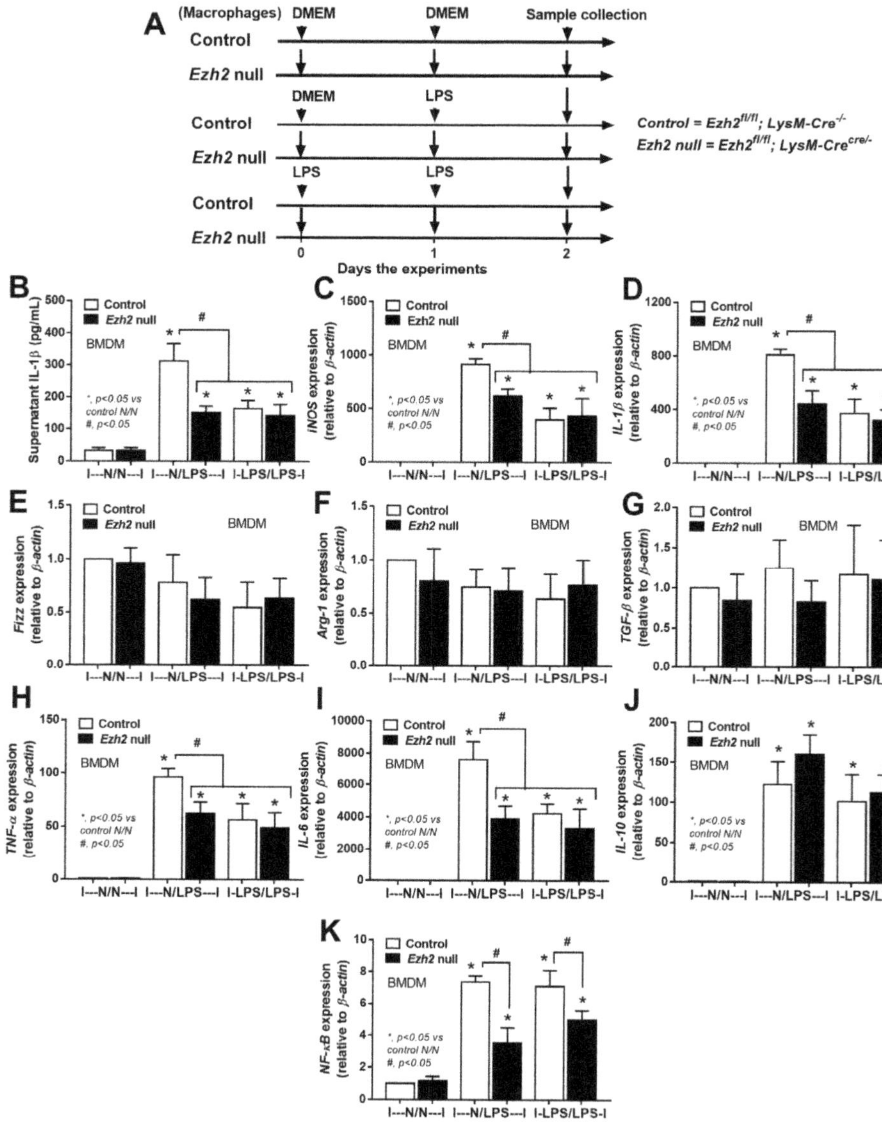

Figure 6. The schema of the experiments in bone marrow-derived macrophages (BMDM) from Ezh2 control (Ezh$^{fl/fl}$; LysM-Cre$^{-/-}$) or Ezh2 null (Ezh$^{fl/fl}$; LysM-Cre$^{cre/-}$) mice after activation by lipopolysaccharide (LPS) in a single protocol (N/LPS) that started with the culture media followed by LPS at 24 h later or LPS tolerance (LPS/LPS) by the twice LPS stimulations or control (N/N) using the twice culture media incubation (**A**) is demonstrated. The characteristics of these macrophages with different protocols, as indicated by supernatant IL-1β (**B**) with the expression of genes for M1 macrophage polarization (*IL-1β* and *iNOS*) and M2 polarization (*Fizz-1*, *Arg-1*, and *TGF-β*) (**C–G**), inflammatory genes (*TNF-α*, *IL-6*, and *IL-10*) (**H–J**), and inflammatory mediators (*NF-κB*) (**K**) are demonstrated. Triplicated independent experiments were performed. Mean ± SEM is presented with the one-way ANOVA followed by Tukey's analysis (*, $p < 0.05$ vs. WT N/N and #, $p < 0.05$).

2.3. Characteristics of Ezh2 Control (Ezh$^{fl/fl}$; LysM-Cre$^{-/-}$) or Ezh2 Null (Ezh$^{fl/fl}$; LysM-Cr $^{cre/-}$) Mice after Cecal Ligation and Puncture (CLP) and LPS Tolerance before CLP Surgery

To investigate the impact of Ezh2 in macrophages on hyper-inflammatory sepsis and sepsis after LPS tolerance, CLP after PBS injection (CLP) and CLP after twice LPS administration (LPS-CLP), respectively, in Ezh2 control (Ezh$^{fl/fl}$; LysM-Cre$^{-/-}$) and Ezh2 null (Ezh$^{fl/fl}$; LysM-Cr $^{cre/-}$) mice was performed (Figure 7A). In the survival analysis, CLP after LPS tolerance (LPS-CLP) in control mice showed the most severe sepsis as all mice died within 72 h post-surgery (Figure 7A). The more severe sepsis in LPS-CLP mice compared with CLP, especially in the control mice (Figure 7A), implied a possible impact of LPS tolerance on a defect of microbial control. Interestingly, the best survival rate of Ezh2 null mice with CLP and the better survival rate after LPS-CLP of Ezh2 null compared with control mice (Figure 7A) indicated a possible beneficial impact of Ezh2 blockage in macrophages during sepsis. Among the control group, LPS-CLP demonstrated more severe sepsis than CLP alone as indicated by cell-free DNA, bacteremia, and pro-inflammatory cytokines (TNF-α and IL-6), but not other parameters (serum creatinine, alanine transaminase, renal histology score, spleen apoptosis, endotoxemia, and IL-10) (Figures 7C–L and 8). These data indicated more severe sepsis after LPS tolerance compared with no LPS tolerance in the Ezh2 control mice, possibly due to immune exhaustion. However, there was no difference in most of the sepsis severity biomarkers between Ezh2 null mice with LPS-CLP and those with CLP alone, except the higher histology score in LPS-CLP Ezh2 null mice (Figure 7C–L). The data implied no or less immune exhaustion after LPS tolerance in Ezh2 null mice, as mentioned in previous publications [39,58]. Between Ezh2 null versus Ezh2 control mice with CLP-induced sepsis hyperinflammation (CLP) and LPS tolerance with subsequent sepsis (LPS-CLP), sepsis severity was more severe in Ezh2 control mice as indicated by survival analysis, organ injury (kidney and liver), cell-free DNA, endotoxemia, bacteremia, and serum cytokines (TNF-α and IL-6, but not IL-10) (Figure 7C–L), supporting a beneficial impact of Ezh2 deletion in macrophages during sepsis in both conditions.

2.4. Ezh2 Inhibitor Attenuated Cecal Ligation and Puncture (CLP) Sepsis in Wild-Type (WT) Mice with Less Impact on CLP after LPS Tolerance

Due to the reduced sepsis severity of Ezh2 null over Ezh2 control mice from our data and the control of inhibitory Socs3 by Ezh2 gene from previous publications [59,60], an Ezh2 inhibitor was further tested, similar to the experiments on Ezh2 null mice mentioned above (Figure 9A). In WT mice, CLP after LPS tolerance (LPS-CLP) also demonstrated the highest mortality rate as all mice died within 96 h of the observation, while approximately 25% of the mice survived at 96 h post-surgery with CLP alone (Figure 9B), supporting more severe sepsis after LPS tolerance similar to Ezh2 control mice (Figure 7A–L). However, Ezh2 inhibitor attenuated disease severity only in CLP alone, but not LPS-CLP (Figure 9B), perhaps because of the more severe sepsis in LPS-CLP compared with CLP alone, as indicated by higher cell-free DNA and serum cytokines (TNF-α and IL-6) in vehicle-administered LPS-CLP mice compared with the CLP alone group (Figure 9E,H,I). In CLP alone, the Ezh2 inhibitor attenuated kidney damage (serum creatinine) and serum cytokines (TNF-α, IL-6, and IL-10) but not cell-free DNA, endotoxemia, and bacteremia (Figure 9B–J). On the other hand, the Ezh2 inhibitor did not attenuate LPS-CLP mice, as indicated by the non-difference in survival analysis, organ damage (kidney and liver), endotoxemia, and bacteremia (Figure 9B–D,F,G), despite the reduction of cell-free DNA and serum cytokines (TNF-α, IL-6, and IL-10) in Ezh2 inhibitor-administered mice (Figure 9E,H–J).

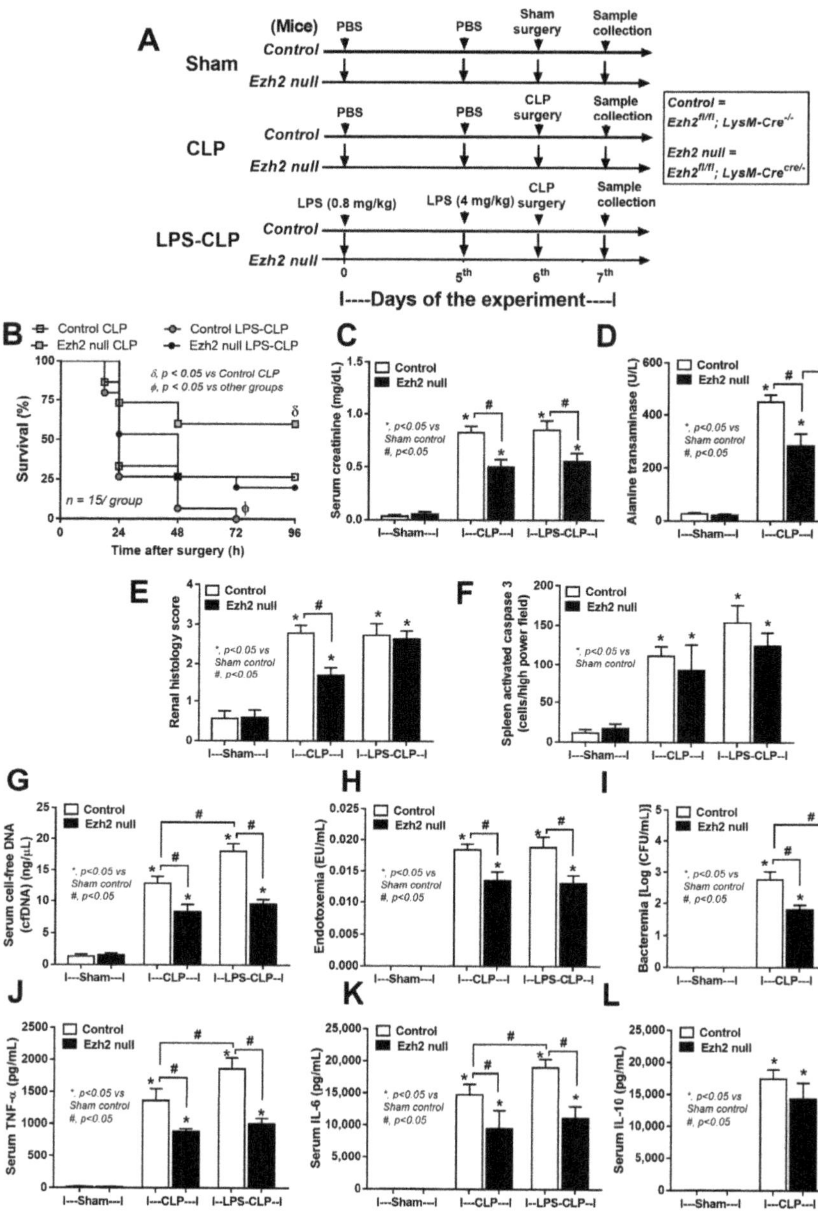

Figure 7. The schema of the experiments in Ezh2 control (Ezh$^{fl/fl}$; LysM-Cre$^{-/-}$) (Control) or Ezh2 null (Ezh$^{fl/fl}$; LysM-Cre$^{cre/-}$) mice (Ezh2 null) in sham, cecal ligation and puncture surgery (CLP), and lipopolysaccharide (LPS) tolerance before CLP surgery (LPS-CLP). The protocol start by injection of phosphate buffer solution (PBS) or LPS intraperitoneal (ip) injection (0.8 mg/kg) followed by PBS or LPS (ip 4 mg/kg) at 5th day before sham or CLP surgery at 6th day, and sacrifice with sample collection at 7th day of the experiment (**A**). Characteristics of these mice as indicated by survival analysis (**B**), kidney injury (serum creatinine) (**C**), liver damage (alanine transaminase) (**D**), renal injury score (**E**), spleen apoptosis (**F**), cell-free DNA (**G**), endotoxemia (**H**), bacteremia (**I**), and serum cytokines (TNF-α, IL-6, and IL-10) (**J–L**) are demonstrated (n = 15/group for B and n = 5–7/group for (**C–J**)). Mean ± SEM is presented with the one-way ANOVA followed by Tukey's analysis (*, $p < 0.05$ vs. Sham control and #, $p < 0.05$).

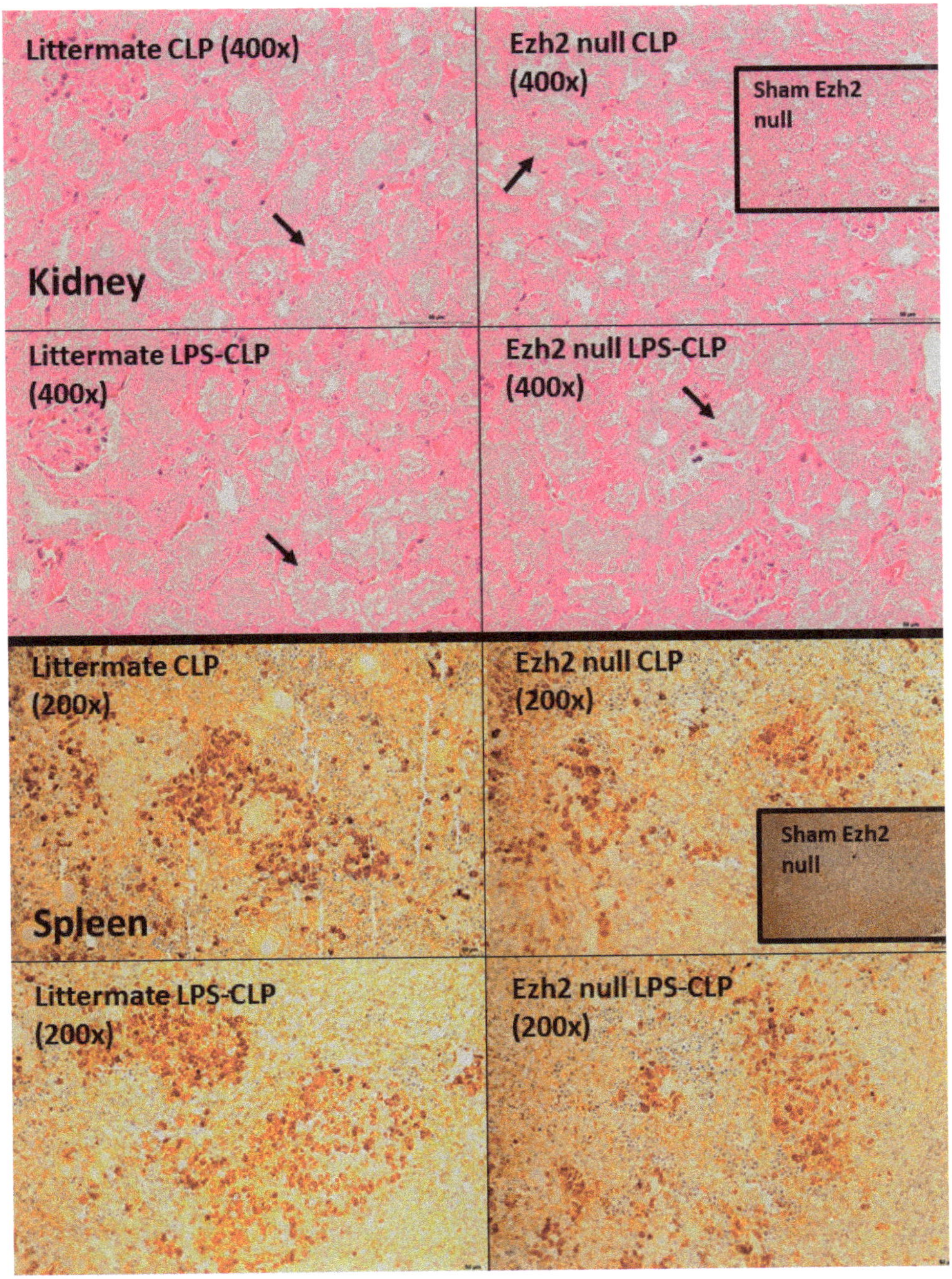

Figure 8. Representative pictures of renal histology with hematoxylin and eosin (H&E) stain (**upper**) and activated caspase 3 spleen apoptosis (**lower**) of Ezh2 control ($Ezh^{fl/fl}$; LysM-$Cre^{-/-}$) (littermate) or Ezh2 null ($Ezh^{fl/fl}$; LysM-$Cre^{cre/-}$) mice (Ezh2 null) in cecal ligation and puncture surgery (CLP), and lipopolysaccharide (LPS) tolerance before CLP surgery (LPS-CLP) are shown. Only the sham of Ezh2 null mice, but not the sham of littermate control, in renal histology and spleen apoptosis are demonstrated in the inset pictures due to the non-difference between both shams. Arrows indicate an example of renal tubular cell injury.

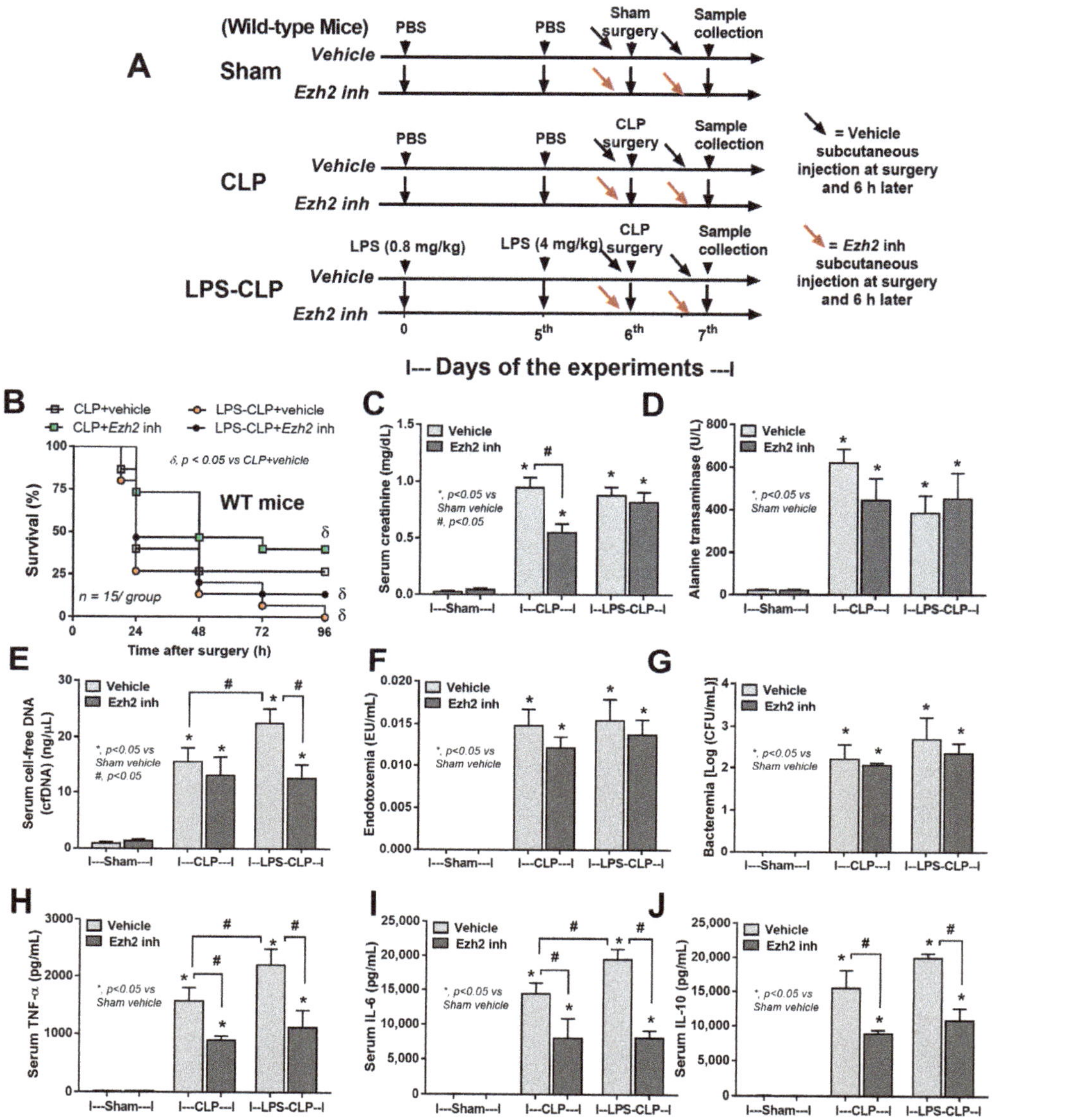

Figure 9. The schema of the experiments in wild-type mice in sham, cecal ligation and puncture surgery (CLP), and lipopolysaccharide (LPS) tolerance before CLP surgery (LPS-CLP). The protocol start by injection of phosphate buffer solution (PBS) or LPS intraperitoneal (ip) injection (0.8 mg/kg) followed by PBS or LPS (ip 4 mg/kg) at 5th day before sham or CLP surgery at 6th day, and sacrifice with sample collection at 7th day of the experiment. Mice were subcutaneously administered with vehicle or an Ezh2 inhibitor at 0 and 6 h post-surgery on the 6th day of protocol (**A**). Characteristics of these mice as indicated by survival analysis (**B**), kidney injury (serum creatinine) (**C**), liver damage (alanine transaminase) (**D**), cell-free DNA (**E**), endotoxemia (**F**), bacteremia (**G**), and serum cytokines (TNF-α, IL-6, and IL-10) (**H–J**) are demonstrated (n = 15/group for B and n = 5–7/group for (**C–J**)). Mean ± SEM is presented with the one-way ANOVA followed by Tukey's analysis (*, $p < 0.05$ vs. Sham vehicle control and #, $p < 0.05$).

3. Discussion

Lipopolysaccharide (LPS) is one of the potent activators of macrophages, as indicated by the up- and downregulation of several proteins compared with the neutral state of the cells. The macrophages from Ezh2 littermate control or Ezh2 null mice without activation were used as the controls to see the deviation away from the neutral state due to the possible differences between littermate control and Ezh2 null mice. Less profound activities of Ezh2 null macrophages were indicated by a lower number of proteins with similar functions (immune responses, cytokines, and cell energy) between control and Ezh2 null cells, implying similar downstream signals. Because of (i) the possibly reduced adverse effects of selective inhibition only on macrophages but not all cells in the body [61], (ii) endotoxemia in several conditions [62–64], partly through gut barrier damage [1,22,24,26] with profound LPS recognition by macrophages [65,66] and (iii) the epigenetic regulation in LPS responses [67], LPS was used to test Ezh2 null macrophages. Notably, Ezh2 reduced chromatin accessibility by adding methyl marks at the tail of histone H3, and the presence of trimethylation of H3K27 (H3K27me3) at promoter regions [68] is mentioned in sepsis [69]. Indeed, the increased mortality in patients with high *Ezh2* and H3K27 is mentioned [70,71] and inhibition of Ezh2 [72] might be beneficial. M1 macrophage polarization (*IL-1β* and *iNOS*) along with pro-inflammatory molecules (*TNF-α*, *IL-6*, and *NF-κB*) were less prominent in LPS-activated Ezh2 null macrophages than the control (Figure 6), possibly correlated with more profound suppressor of cytokine signaling 3 (*Socs3*; an anti-inflammatory molecule) as mentioned in our previous publication [39]. Perhaps *Ezh2* more potently inhibits some molecules than others. In a single LPS stimulation, Ezh2 more profoundly suppressed *Socs3* than *NF-κB* (a transcriptional factor for several cytokines), resulting in less inhibition of *NF-κB* with high cytokine production. Without Ezh2, elevated *Socs3* downregulated *NF-κB* and led to lower *TNF-α* and *IL-6* expression after a single LPS stimulation. Indeed, cytosolic *Socs3* inhibits the NF-κB-dependent inflammatory genes through enhanced ubiquitination and proteasomal degradation [44,59]. Additionally, Ezh2 blockage enhances anti-inflammatory *Socs3* that inhibits hyperinflammation in sepsis (here and others), multiple sclerosis, and glucose-activated peritoneal fibrosis [60,69,73]. In contrast, suppression of Ezh2 might enhance pro-inflammatory genes, including *NF-κB*, that possibly worsens inflammatory bowel diseases and muscle cell apoptosis in sepsis [49,74–76]. Although Socs3 can inhibit both anti- and pro-inflammatory cytokines [77], possibly driven by different molecules [78], Socs3 seems to have less impact on anti-inflammatory IL-10 as serum TNF-α and IL-6, but not IL-10, were lower in LPS-injected Ezh2 null mice [39]. Accordingly, Socs3 might be correlated with IL-10 in macrophages responses because (i) IL-10 directly upregulates *Socs3* [79], (ii) Socs3 and IL-10 are simultaneously used to inhibit inflammation [80,81], and (iii) Ezh2 inhibitor (EPZ-6438) upregulated *IL-10* [82].

In mice, the deletion of Ezh2 only in myeloid cells in Ezh2 null mice was enough to induce less severe sepsis, similar to the LPS injection model [39] and pneumococcal sepsis model [70,83], highlighting the major impact of macrophages in the cytokine production [84]. Then, blockage of cytokine production only on the myeloid cells, especially macrophages, might be beneficial. Despite the improved sepsis mortality with the previously known hepatic protection of Ezh2 inhibitor [85], liver enzyme and bacteremia were not different from sepsis in Ezh2 control mice (Figure 7) implying a possible hepatoxicity and incomplete Ezh2 blockage of the selected inhibitor (GSK126). Notably, injection of Ezh2 inhibitor [86] in mice inhibits Ezh2 in all cells which might be harmful in some cell types and the selective Ezh2 blockage only in macrophages might reduce the adverse effect. However, the balance between the pro- and anti-inflammatory directions (a yin–yang effect) after Ezh2 blockage might be different in individual patients.

Because leaky gut causes endotoxemia [1,66] and relatively low inflammation (repeat or chronic exposure of LPS), or LPS tolerance [24,87] possibly causes more severe sepsis [25] and/or secondary infection [31,88,89], CLP surgery after LPS tolerance (LPS-CLP) might be different from CLP without LPS priming (CLP). In control mice, LPS-CLP was more severe than in those with CLP alone, possibly due to inadequate inflammation to control organisms at an early phase of sepsis [25] leading to a higher blood bacterial burden and higher cytokines than in the CLP alone group. Because chronic endotoxemia might induce LPS tolerance [25,57], sepsis in these individuals, such as obesity and uremia, might be more severe than the non-endotoxemia cases [90–95], partly, due to LPS tolerance. Despite the higher disease severity of LPS-CLP over CLP alone, Ezh2 null mice still demonstrated less severe sepsis compared with LPS-CLP in control mice, supporting the anti-inflammatory effect of Ezh2 deletion in macrophages. Although the Ezh2 inhibitor did not reduce LPS-CLP mortality, there was less severe systemic inflammation, implying a possible benefit of the dose adjustment. More studies are interesting. Despite the broad spectrum antiviral of Ezh1/2 inhibitors [96], the influence of Ezh1/2 blockage on bacterial infection needs further testing. In cancer, Ezh2 blockage might induce anti-inflammatory macrophages, resulting in less severe sepsis or more infection susceptibility due to the inadequate inflammation to control organisms. Importantly, an effective antibiotic with good microbial control is a main strategy for the treatment of sepsis-induced hyper-inflammation [97]. Hence, our results are a proof of concept to use clinically available Ezh2 inhibitors in sepsis, especially the blockage of Ezh2 specifically only in macrophages.

Several limitations should be mentioned. First, only male mice were used and the impact of gender differences in sepsis might affect the translation of the results. Second, details of mechanistic pathways, including the Western blot analysis, were not performed. Despite a proof of concept for the translational use of Ezh2 inhibitors in clinical sepsis, more experiments would be interesting. Third, the results need to be validated in human situations before a solid conclusion. Nevertheless, we concluded that Ezh2 inhibitors, an available anticancer treatment, might be beneficial in some situations of sepsis. More studies are warranted.

4. Materials and Methods

4.1. Animal

The Institutional Animal Care and Use Committee of the Faculty of Medicine, Chulalongkorn University, Bangkok, Thailand approved the protocol (No. 017/2562) according to the National Institutes of Health (NIH) criteria. Wild-type (WT) 8-week-old C57BL/6 male mice were purchased from Nomura Siam, Pathumwan, Bangkok, Thailand. Meanwhile, Ezh2$^{flox/flox}$ and LyM-Cre$^{Cre/Cre}$ mice were obtained from RIKEN BRC Experimental Animal Division (Ibaraki, Japan) and cross-bred until having *Ezh2* littermate control (Ezh$^{fl/fl}$; LysM-Cre$^{-/-}$) or *Ezh2* null (Ezh$^{fl/fl}$; LysM-Cre$^{cre/-}$) in F3 generation of the breeding protocol. As such, the Ezh2$^{flox/flox}$ mice had loxP sites upstream and downstream of the 2.7 kb SET domain, while bred with LysM-Cre$^{Cre/Cre}$ mice, the mice with a Cre recombinase under the control of lysozyme M to target *Ezh2* for deletion in myeloid cells (macrophages and neutrophils). Mice with Ezh2$^{flox/flox}$ without LysM-Cre (Ezh$^{fl/fl}$; LysM-Cre$^{-/-}$) were used as littermate controls (Ezh2 control). To genotype these mice on the loxP sites' insertion, the following primers were used for *Ezh2*: reverse 1: 3′ of loxp: 5′-AGG GCA TCA GCC TGG CTGTA-3′; forward 2: 5′ of loxp: 5′-TTA TTC ATA GAG CCA CCTGG-3′; forward 3: left loxp: 5-ACG AAA CAG CTC CAG ATTCAG GG-3′ according to a previous publication [83]. The mice homozygous for the flox were selected and genotyped for the expression of LysM-Cre using the primers; forward: 5′-CTTGGGCTGCCAGAATTCTC-3′; Reverse: 5′-CCCAGAAATGCCAGATTACG-3′.

4.2. Animal Models

Cecal ligation and puncture (CLP) surgery was used to induce sepsis, following previous publications, under isoflurane anesthesia [98–100]. Briefly, a median abdominal incision was performed and the cecum was ligated at 10 cm from the cecal tip, punctured twice with a 21-gauge needle, and gently squeezed to express a small amount of fecal material before closing the abdominal wall layer by layer with sutures. Then, tramadol (25 mg/kg/dose) in 0.25 mL prewarmed normal saline solution (NSS) and imipenem/cilastatin (14 mg/kg/dose) in 0.2 mL NSS were subcutaneously administered in abdominal areas after surgery, and at 6 and 18 h post-CLP [9]. In sham-operated mice, the cecum was isolated and closed by suturing without ligation or puncture. In parallel, a sham operation was performed with only cecal identification before closing abdomen layer by layer. Because lipopolysaccharide (LPS) tolerance inhibits macrophage cell respiration and induces global proteomic changes in macrophages [35], sepsis during LPS tolerance might be different from the regular condition. Then, CLP after LPS tolerance using twice-administered LPS injection (LPS-CLP) was conducted. Hence, the mice were divided into 3 groups. First, for CLP in LPS tolerance (LPS-CLP), intraperitoneal injection of 0.8 mg/kg LPS (*Escherichia coli* 026:B6) (Sigma-Aldrich, St. Louis, MO, USA) with another dose of 4 mg/kg LPS at 5 days later and followed by CLP surgery at 1 days after the 2nd dose of LPS was performed. Second, for CLP alone (CLP), the experiments started with intraperitoneal injection of phosphate buffer solution (PBS) at 0 and 5 days followed by CLP surgery. Third, in sham control mice (Sham), 2 doses of PBS at 0 and 5th days of experiments followed by sham surgery was conducted. Of note, the lower 1st dose (0.8 mg/kg) followed by the higher 2nd LPS dose (4 mg/kg) for LPS tolerance induction was performed according to a previous protocol [25]. Mice were sacrificed with cardiac puncture under isoflurane anesthesia with sample collection at 24 h or 96 h post-surgery for blood biomarkers and survival analysis, respectively. On the other hand, these mouse protocols, including CLP, LPS-CLP, and sham, were also used for the test of Ezh2 inhibitor using WT mice in all groups. The Ezh2 inhibitor (GSK343; Medchemexpress, Monmouth, NJ, USA) at 4 mM/25 g mice in 3% dimethyl sulfoxide (DMSO) or DMSO alone (vehicle control) was subcutaneously administered 15 min before surgery and 6 h later (15 min before tramadol and the antibiotics). These mice were sacrificed with the same protocol of experiments in the transgenic mice.

4.3. Mouse Sample Analysis

Serum creatinine and alanine transaminase [40] were measured by colorimetric method (QuantiChrom™ Creatinine Assay Kit, BioAssay System, Hayward, CA, USA) and EnzyChrom Alanine Transaminase assay (EALT-100, BioAssay), respectively. Serum cell-free DNA and LPS (endotoxin) were detected by Quanti PicoGreen assay (Sigma-Aldrich) and HEK-Blue LPS Detection Kit 2 (InvivoGen™, San Diego, CA, USA), while ELISA (Invitrogen, Carlsbad, CA, USA) was used for detection of cytokines (TNF-α, IL-6, and IL-10). In parallel, blood bacterial abundance (bacteremia) was evaluated using the direct spread of mouse blood onto blood agar plates (Oxoid, Hampshire, UK) in serial dilutions and incubating at 37 °C for 24 h before colony enumeration. For the kidney injury score, the injury score was semi-quantitatively evaluated on hematoxylin and eosin (H&E) staining in 4 mm thick paraffin-embedded slides at 200× magnification by the area of injury (tubular epithelial swelling, loss of brush border, vacuolar degeneration, necrotic tubules, cast formation, and desquamation) using the following score: 0, area < 5%; 1, area 5–10%; 2, area 10–25%; 3, area 25–50%; 4, area > 50% [23]. In parallel, for spleen apoptosis, spleens with 10% formalin fixation were stained by anti-active caspase 3 antibody (Cell Signaling Technology, Beverly, MA, USA), using immunohistochemistry, and expressed in positive cells per high-power field (200× magnification) as previously published [23].

4.4. Bone Marrow-Derived Macrophages and the In Vitro Experiments

Bone marrow-derived macrophages from mouse femurs using supplemented Dulbecco's Modified Eagle's Medium (DMEM) with conditioned medium of the L929 cells (ATCC CCL-1) were derived as previously described [65,101–103]. The macrophages at 5×10^4 cells/well in supplemented DMEM (Thermo Fisher Scientific, Waltham, MA, USA) were incubated in 5% carbon dioxide (CO_2) at 37 °C for 24 h before being treated by 3 different protocols, including (i) a single LPS stimulation; started with DMEM followed by LPS (100 ng/mL) 24 h later (N/LPS), or (ii) LPS tolerance; using the twice stimulations by 100 ng/mL of LPS (LPS/LPS), or control (N/N) using the twice DMEM incubation, before the sample collection (supernatant and cells). Notably, the supernatant of the stimulated cells in all groups was gently removed and washed with DMEM before the re-administration of LPS or DMEM, as previously mentioned [39,87,104]. Supernatant interleukin (IL)-1β was evaluated by ELISA (Invitrogen, Carlsbad, CA, USA) and the gene expression was evaluated by quantitative real-time polymerase chain reaction (PCR), as previously described [102,105–107]. Briefly, the RNA was extracted from the cells with TRIzol Reagent (Invitrogen, Carlsbad, CA, USA) together with RNeasy Mini Kit (Qiagen, Hilden, Germany) as 1 mg of total RNA was used for cDNA synthesis with iScript reverse transcription supermix (Bio-Rad, Hercules, CA, USA). Quantitative real-time PCR was performed on a QuantStudio 6 real-time PCR system (Thermo Fisher Scientific, Waltham, MA, USA) using SsoAdvance Universal SYBR Green Supermix (Bio-Rad, Hercules, CA, USA). Expression values were normalized to beta-actin (*β-actin*) as an endogenous housekeeping gene and the fold change was calculated by the ΔΔCt method. The primers used in this study are listed in Table 1.

Table 1. List of primers used in the study.

Name	Forward	Reverse
Inducible nitric oxide synthase (*iNOS*); Gene ID: 18126	5′-ACCCACATCTGGCAGAATGAG-3′	5′-AGCCATGACCTTTCGCATTAG-3′
Interleukin-1β (*IL-1β*) Gene ID: 16176	5′-GAAATGCCACCTTTTGACAGTG-3′	5′-TGGATGCTCTCATCAGGACAG-3′
Tumor necrosis factor α (*TNF-α*) Gene ID: 21926	5′-CCTCACACTCAGATCATCTTCTC-3′	5′-AGATCCATGCCGTTGGCCAG-3′
Interleukin-6 (*IL-6*) Gene ID: 16193	5′-TACCACTTCACAAGTCGGAGGC-3′	5′-CTGCAAGTGCA TCA TCGTTGTTC-3′
Interleukin-10 (*IL-10*) Gene ID: 16153	5′-GCTCTTACTGACTGGCATGAG-3′	5′-CGCAGCTCTAGGAGCATGTG-3′
Arginase-1 (*Arg-1*) Gene ID: 11846	5′-CTTGGCTTGCTTCGGAACTC-3′	5′-GGAGAAGGCGTTTGCTTAGTT-3′
Resistin-like molecule-α1 (*FIZZ-1*) Gene ID: 57262	5′-GCCAGGTCCTGGAACCTTTC-3′	5′-GGAGCAGGGAGATGCAGATGA-3′
Transforming growth factor-β (*TGF-β*) Gene ID: 21813	5′-CAGAGCTGCGCTTGCAGAG-3′	5′-GTCAGCAGCCGGTTACCAAG-3′
Nuclear factor kappa B (*NFκB*) Gene ID: 18033	5′-CTTCCTCAGCCATGGTACCTCT-3′	5′-CAAGTCTTCATCAGCATCAAACTG-3′
β-actin Gene ID: 11461	5′-CGGTTCCGATGCCCTGAGGCTCTT-3′	5′-CGTCACACTTCATGATGGAATTGA-3′

4.5. Mass Spectrometry Proteomic and Secretome Analysis

The proteomic and secretome analyses were performed, using the cells and supernatant media, respectively, according to previous publications [35,56,104]. Briefly, for proteome analysis, the activated macrophages (1×10^6 cells/well) with 3 protocols, N/N, N/LPS, and LPS/LPS as mentioned above, were processed for in-solution digestion. For secretome analysis, an equal volume of culture medium from 3 conditions was centrifuged to remove intact cells, concentrated by centrifugation in an Amicon Ultracel—3K (EMD Millipore, Billerica, MA, USA), and the buffer exchanged using 8 M urea lysis buffer. The concentrated proteins were also subjected to in-solution digestion. Then, the peptides from N/N, N/LPS, and LPS/LPS from the cells and culture media samples, for proteome and secretome, respectively, were labeled with light reagents (CH2O and NaBH3CN), medium reagents (CD2O and NaBH3CN), and heavy reagents (13CD2O and NaBD3CN), respectively. The pooled peptides were fractionated using a high pH reversed-phase peptide fractionation kit (Thermo Fisher Scientific, San Jose, CA, USA). Liquid chromatography–tandem mass spectrometry (LC-MS/MS) analysis of samples was performed on an EASY-nLC1000 system coupled to a Q-Exactive Orbitrap Plus mass spectrometer equipped with a nanoelectrospray ion source (Thermo Fisher Scientific, San Jose, CA, USA). The mass spectrometry (MS) raw files were searched against the Mouse Swiss-Prot Database

(17,138 proteins, November 2022) with a list of common protein contaminants. The search parameters were set for the following fixed modifications: carbamidomethylation of cysteine (+57.02146 Da), as well as light, medium, and heavy dimethylation of N termini and lysine (+28.031300, +32.056407, and +36.075670 Da) and variable modification: oxidation of methionine (15.99491 Da). The false positive discovery rate of the identified peptides based on Q-values using The Proteome Discoverer decoy database together with the Percolator algorithm was set to 1%. The relative MS signal intensities of dimethyl labeled peptides were quantified and presented as ratios of single LPS/no stimulation and LPS tolerance/no stimulation. Log 2 of the ratios in triplicate was used to calculate the p-values using Student's t-test. The proteins with a p-value < 0.05 were considered significant proteins, and these proteins were subjected to the online DAVID Bioinformatics Resources 6.8 to investigate the enriched biological processes. The mass spectrometry proteomics data have been deposited to the ProteomeXchange Consortium via the PRIDE partner repository with the dataset identifier PXD041265. Then, the data visualization was performed using R packages. Volcano plots were generated by ggplot2 version 3.4.2. KEGG pathway analyses were generated by PathfindR. Go enrichment analysis was performed using Shiny 0.77 (http://bioinformatics.sdstate.edu/go/) accessed on 20 March 2023.

4.6. Statistical Analysis

The results are shown as mean ± S.E.M. All data were analyzed with GraphPad Prism6. Student's t-test or one-way analysis of variance [41] with Tukey's comparison test was used for the analysis of experiments with two and more than two groups, respectively. The survival analysis was determined by the log-rank test. For all datasets, a p-value less than 0.05 was considered significant.

5. Conclusions

The Ezh2-deleted macrophages induced fewer activities (proteomic and secretome analyses) after LPS stimulation compared with the control states, supporting the less severe sepsis in *Ezh2* null ($Ezh^{fl/fl}$; *LysM-Cre*$^{cre/-}$) over the control ($Ezh^{fl/fl}$; *LysM-Cre*$^{-/-}$) mice. The more severe sepsis in CLP after LPS tolerance over CLP alone supported the less effective microbial control during LPS tolerance. The Ezh2 inhibitor was more effective in the CLP model than the CLP after LPS tolerance, perhaps due to the more profound sepsis severity in the latter condition. More studies on the use of Ezh2 blockage in sepsis are warranted.

Supplementary Materials: The following supporting information can be downloaded at: https://www.mdpi.com/article/10.3390/ijms24108517/s1.

Author Contributions: Conceptualization, N.H., T.P. and A.L.; methodology, P.P., J.M. and A.L.; formal analysis, P.P. and A.L.; investigation, P.P., P.K., J.M., J.I.-A., A.B., S.B., P.R. and A.N.-L.; resources, A.N.-L., N.H., T.P. and A.L.; data curation, P.P., J.M., J.I.-A. and A.L.; writing—original draft preparation, P.P. and A.L.; writing—review and editing, N.H., T.P. and A.L.; supervision, A.L.; funding acquisition, N.H. All authors have read and agreed to the published version of the manuscript.

Funding: This research was funded from NSRF via the Program Management Unit for Human Resources, Institutional Development, Research, and Innovation (B05F640144), (B16F640175), and (B36G660003) with Rachadapisek Sompote Matching Fund (RA-MF-04/66), and the Thailand Science research and Innovation Fund, Chulalongkorn University (CU_FRB65_hea (6)_012_32_07). This research was supported in part by the Intramural Research Program of National Institute of Allergy and Infectious Diseases (NIAID), National Institutes of Health (NIH). T.P. is funded by National Research Council of Thailand (811/2563). P.P. was supported by the Second Century Fund (C2F) for Postdoctoral Fellowship, Chulalongkorn University. P.K. was supported by the Second Century Fund (C2F) for high-efficiency Ph.D. candidates, Chulalongkorn University.

Institutional Review Board Statement: The study was conducted in accordance with the National Institutes of Health (NIH) criteria, and the animal study protocol was approved by the Institutional Animal Care and Use Committee of the Faculty of Medicine, Chulalongkorn University, Bangkok, Thailand (CU-ACUP No. 021/2562).

Informed Consent Statement: Not applicable.

Data Availability Statement: Not applicable.

Conflicts of Interest: The authors declare no conflict of interest.

References

1. Amornphimoltham, P.; Yuen, P.S.T.; Star, R.A.; Leelahavanichkul, A. Gut Leakage of Fungal-Derived Inflammatory Mediators: Part of a Gut-Liver-Kidney Axis in Bacterial Sepsis. *Dig. Dis. Sci.* **2019**, *64*, 2416–2428. [CrossRef]
2. Lacal, I.; Ventura, R. Epigenetic Inheritance: Concepts, Mechanisms and Perspectives. *Front. Mol. Neurosci.* **2018**, *11*, 292. [CrossRef] [PubMed]
3. Han, M.; Jia, L.; Lv, W.; Wang, L.; Cui, W. Epigenetic Enzyme Mutations: Role in Tumorigenesis and Molecular Inhibitors. *Front. Oncol.* **2019**, *9*, 194. [CrossRef] [PubMed]
4. Hotchkiss, R.S.; Moldawer, L.L.; Opal, S.M.; Reinhart, K.; Turnbull, I.R.; Vincent, J.L. Sepsis and septic shock. *Nat. Rev. Dis. Primers* **2016**, *2*, 16045. [CrossRef] [PubMed]
5. Liu, D.; Huang, S.-Y.; Sun, J.-H.; Zhang, H.-C.; Cai, Q.-L.; Gao, C.; Li, L.; Cao, J.; Xu, F.; Zhou, Y.; et al. Sepsis-induced immunosuppression: Mechanisms, diagnosis and current treatment options. *Mil. Med. Res.* **2022**, *9*, 56. [CrossRef] [PubMed]
6. Brady, J.; Horie, S.; Laffey, J.G. Role of the adaptive immune response in sepsis. *Intensive Care Med. Exp.* **2020**, *8*, 20. [CrossRef] [PubMed]
7. Delano, M.J.; Ward, P.A. Sepsis-induced immune dysfunction: Can immune therapies reduce mortality? *J. Clin. Investig.* **2016**, *126*, 23–31. [CrossRef] [PubMed]
8. Németh, K.; Leelahavanichkul, A.; Yuen, P.S.; Mayer, B.; Parmelee, A.; Doi, K.; Robey, P.G.; Leelahavanichkul, K.; Koller, B.H.; Brown, J.M.; et al. Bone marrow stromal cells attenuate sepsis via prostaglandin E(2)-dependent reprogramming of host macrophages to increase their interleukin-10 production. *Nat. Med.* **2009**, *15*, 42–49. [CrossRef]
9. Leelahavanichkul, A.; Yasuda, H.; Doi, K.; Hu, X.; Zhou, H.; Yuen, P.S.; Star, R.A. Methyl-2-acetamidoacrylate, an ethyl pyruvate analog, decreases sepsis-induced acute kidney injury in mice. *Am. J. Physiol. Renal Physiol.* **2008**, *295*, F1825–F1835. [CrossRef]
10. Taratummarat, S.; Sangphech, N.; Vu, C.T.B.; Palaga, T.; Ondee, T.; Surawut, S.; Sereemaspun, A.; Ritprajak, P.; Leelahavanichkul, A. Gold nanoparticles attenuates bacterial sepsis in cecal ligation and puncture mouse model through the induction of M2 macrophage polarization. *BMC Microbiol.* **2018**, *18*, 85. [CrossRef]
11. Panpetch, W.; Chancharoenthana, W.; Bootdee, K.; Nilgate, S.; Finkelman, M.; Tumwasorn, S.; Leelahavanichkul, A. *Lactobacillus rhamnosus* L34 Attenuates Gut Translocation-Induced Bacterial Sepsis in Murine Models of Leaky Gut. *Infect. Immun.* **2018**, *86*, e00700-17. [CrossRef]
12. Issara-Amphorn, J.; Chancharoenthana, W.; Visitchanakun, P.; Leelahavanichkul, A. Syk Inhibitor Attenuates Polymicrobial Sepsis in FcgRIIb-Deficient Lupus Mouse Model, the Impact of Lupus Characteristics in Sepsis. *J. Innate Immun.* **2020**, *12*, 461–479. [CrossRef] [PubMed]
13. Dang, C.P.; Leelahavanichkul, A. Over-expression of miR-223 induces M2 macrophage through glycolysis alteration and attenuates LPS-induced sepsis mouse model, the cell-based therapy in sepsis. *PLoS ONE* **2020**, *15*, e0236038. [CrossRef] [PubMed]
14. Chancharoenthana, W.; Udompronpitak, K.; Manochantr, Y.; Kantagowit, P.; Kaewkanha, P.; Issara-Amphorn, J.; Leelahavanichkul, A. Repurposing of High-Dose Erythropoietin as a Potential Drug Attenuates Sepsis in Preconditioning Renal Injury. *Cells* **2021**, *10*, 3133. [CrossRef] [PubMed]
15. Dang, C.P.; Issara-Amphorn, J.; Charoensappakit, A.; Udompornpitak, K.; Bhunyakarnjanarat, T.; Saisorn, W.; Sae-Khow, K.; Leelahavanichkul, A. BAM15, a Mitochondrial Uncoupling Agent, Attenuates Inflammation in the LPS Injection Mouse Model: An Adjunctive Anti-Inflammation on Macrophages and Hepatocytes. *J. Innate Immun.* **2021**, *13*, 359–375. [CrossRef]
16. Perner, A.; Rhodes, A.; Venkatesh, B.; Angus, D.C.; Martin-Loeches, I.; Preiser, J.C.; Vincent, J.L.; Marshall, J.; Reinhart, K.; Joannidis, M.; et al. Sepsis: Frontiers in supportive care, organisation and research. *Intensive Care Med.* **2017**, *43*, 496–508. [CrossRef]
17. Mithal, L.B.; Arshad, M.; Swigart, L.R.; Khanolkar, A.; Ahmed, A.; Coates, B.M. Mechanisms and modulation of sepsis-induced immune dysfunction in children. *Pediatr. Res.* **2022**, *91*, 447–453. [CrossRef]
18. Schrijver, I.T.; Théroude, C.; Roger, T. Myeloid-Derived Suppressor Cells in Sepsis. *Front. Immunol.* **2019**, *10*, 327. [CrossRef]
19. Cao, C.; Ma, T.; Chai, Y.F.; Shou, S.T. The role of regulatory T cells in immune dysfunction during sepsis. *World J. Emerg. Med.* **2015**, *6*, 5–9. [CrossRef]
20. Vergadi, E.; Vaporidi, K.; Tsatsanis, C. Regulation of Endotoxin Tolerance and Compensatory Anti-inflammatory Response Syndrome by Non-coding RNAs. *Front. Immunol.* **2018**, *9*, 2705. [CrossRef]
21. Chancharoenthana, W.; Sutnu, N.; Visitchanakun, P.; Sawaswong, V.; Chitcharoen, S.; Payungporn, S.; Schuetz, A.; Schultz, M.J.; Leelahavanichkul, A. Critical roles of sepsis-reshaped fecal virota in attenuating sepsis severity. *Front. Immunol.* **2022**, *13*, 940935. [CrossRef]
22. Chancharoenthana, W.; Kamolratanakul, S.; Ariyanon, W.; Thanachartwet, V.; Phumratanaprapin, W.; Wilairatana, P.; Leelahavanichkul, A. Abnormal Blood Bacteriome, Gut Dysbiosis, and Progression to Severe Dengue Disease. *Front. Cell. Infect. Microbiol.* **2022**, *12*, 890817. [CrossRef]

23. Hiengrach, P.; Visitchanakun, P.; Tongchairawewat, P.; Tangsirisatian, P.; Jungteerapanich, T.; Ritprajak, P.; Wannigama, D.L.; Tangtanatakul, P.; Leelahavanichkul, A. Sepsis Encephalopathy Is Partly Mediated by miR370-3p-Induced Mitochondrial Injury but Attenuated by BAM15 in Cecal Ligation and Puncture Sepsis Male Mice. *Int. J. Mol. Sci.* **2022**, *23*, 5445. [CrossRef]
24. Ondee, T.; Pongpirul, K.; Udompornpitak, K.; Sukkummee, W.; Lertmongkolaksorn, T.; Senaprom, S.; Leelahavanichkul, A. High Fructose Causes More Prominent Liver Steatohepatitis with Leaky Gut Similar to High Glucose Administration in Mice and Attenuation by Lactiplantibacillus plantarum dfa1. *Nutrients.* **2023**, *17*, 1462. [CrossRef]
25. Ondee, T.; Surawut, S.; Taratummarat, S.; Hirankarn, N.; Palaga, T.; Pisitkun, P.; Pisitkun, T.; Leelahavanichkul, A. Fc Gamma Receptor IIB Deficient Mice: A Lupus Model with Increased Endotoxin Tolerance-Related Sepsis Susceptibility. *Shock* **2017**, *47*, 743–752. [CrossRef] [PubMed]
26. Hiengrach, P.; Panpetch, W.; Chindamporn, A.; Leelahavanichkul, A. Macrophage depletion alters bacterial gut microbiota partly through fungal overgrowth in feces that worsens cecal ligation and puncture sepsis mice. *Sci. Rep.* **2022**, *12*, 9345. [CrossRef]
27. Makjaroen, J.; Thim-Uam, A.; Dang, C.P.; Pisitkun, T.; Somparn, P.; Leelahavanichkul, A. A Comparison Between 1 Day versus 7 Days of Sepsis in Mice with the Experiments on LPS-Activated Macrophages Support the Use of Intravenous Immunoglobulin for Sepsis Attenuation. *J. Inflamm. Res.* **2021**, *14*, 7243–7263. [CrossRef] [PubMed]
28. Kaewduangduen, W.; Visitchanakun, P.; Saisorn, W.; Phawadee, A.; Manonitnantawat, C.; Chutimaskul, C.; Susantitaphong, P.; Ritprajak, P.; Somboonna, N.; Cheibchalard, T.; et al. Blood Bacteria-Free DNA in Septic Mice Enhances LPS-Induced Inflammation in Mice through Macrophage Response. *Int. J. Mol. Sci.* **2022**, *23*, 1907. [CrossRef]
29. Visitchanakun, P.; Kaewduangduen, W.; Chareonsappakit, A.; Susantitaphong, P.; Pisitkun, P.; Ritprajak, P.; Townamchai, N.; Leelahavanichkul, A. Interference on Cytosolic DNA Activation Attenuates Sepsis Severity: Experiments on Cyclic GMP–AMP Synthase (cGAS) Deficient Mice. *Int. J. Mol. Sci.* **2021**, *22*, 11450. [CrossRef] [PubMed]
30. Doi, K.; Leelahavanichkul, A.; Yuen, P.S.; Star, R.A. Animal models of sepsis and sepsis-induced kidney injury. *J. Clin. Investig.* **2009**, *119*, 2868–2878. [CrossRef] [PubMed]
31. Vu, C.T.B.; Thammahong, A.; Yagita, H.; Azuma, M.; Hirankarn, N.; Ritprajak, P.; Leelahavanichkul, A. Blockade Of PD-1 Attenuated Postsepsis Aspergillosis Via The Activation of IFN-gamma and The Dampening of IL-10. *Shock* **2020**, *53*, 514–524. [CrossRef] [PubMed]
32. Walton, A.H.; Muenzer, J.T.; Rasche, D.; Boomer, J.S.; Sato, B.; Brownstein, B.H.; Pachot, A.; Brooks, T.L.; Deych, E.; Shannon, W.D.; et al. Reactivation of multiple viruses in patients with sepsis. *PLoS ONE* **2014**, *9*, e98819. [CrossRef] [PubMed]
33. Wang, T.; Derhovanessian, A.; De Cruz, S.; Belperio, J.A.; Deng, J.C.; Hoo, G.S. Subsequent infections in survivors of sepsis: Epidemiology and outcomes. *J. Intensive Care Med.* **2014**, *29*, 87–95. [CrossRef] [PubMed]
34. Seeley, J.J.; Ghosh, S. Molecular mechanisms of innate memory and tolerance to LPS. *J. Leukoc. Biol.* **2017**, *101*, 107–119. [CrossRef] [PubMed]
35. Gillen, J.; Ondee, T.; Gurusamy, D.; Issara-Amphorn, J.; Manes, N.P.; Yoon, S.H.; Leelahavanichkul, A.; Nita-Lazar, A. LPS Tolerance Inhibits Cellular Respiration and Induces Global Changes in the Macrophage Secretome. *Biomolecules* **2021**, *11*, 164. [CrossRef] [PubMed]
36. López-Collazo, E.; del Fresno, C. Pathophysiology of endotoxin tolerance: Mechanisms and clinical consequences. *Crit. Care* **2013**, *17*, 242. [CrossRef]
37. Naler, L.B.; Hsieh, Y.P.; Geng, S.; Zhou, Z.; Li, L.; Lu, C. Epigenomic and transcriptomic analyses reveal differences between low-grade inflammation and severe exhaustion in LPS-challenged murine monocytes. *Commun. Biol.* **2022**, *5*, 102. [CrossRef] [PubMed]
38. Koos, B.; Moderegger, E.L.; Rump, K.; Nowak, H.; Willemsen, K.; Holtkamp, C.; Thon, P.; Adamzik, M.; Rahmel, T. LPS-Induced Endotoxemia Evokes Epigenetic Alterations in Mitochondrial DNA That Impacts Inflammatory Response. *Cells* **2020**, *9*, 2282. [CrossRef]
39. Kunanopparat, A.; Leelahavanichkul, A.; Visitchanakun, P.; Kueanjinda, P.; Phuengmaung, P.; Sae-Khow, K.; Boonmee, A.; Benjaskulluecha, S.; Palaga, T.; Hirankarn, N. The Regulatory Roles of Ezh2 in Response to Lipopolysaccharide (LPS) in Macrophages and Mice with Conditional Ezh2 Deletion with LysM-Cre System. *Int. J. Mol. Sci.* **2023**, *24*, 5363. [CrossRef]
40. Falcao-Holanda, R.B.; Brunialti, M.K.C.; Jasiulionis, M.G.; Salomao, R. Epigenetic Regulation in Sepsis, Role in Pathophysiology and Therapeutic Perspective. *Front. Med.* **2021**, *8*, 685333. [CrossRef]
41. De Santa, F.; Narang, V.; Yap, Z.H.; Tusi, B.K.; Burgold, T.; Austenaa, L.; Bucci, G.; Caganova, M.; Notarbartolo, S.; Casola, S.; et al. Jmjd3 contributes to the control of gene expression in LPS-activated macrophages. *EMBO J.* **2009**, *28*, 3341–3352. [CrossRef]
42. De Santa, F.; Totaro, M.G.; Prosperini, E.; Notarbartolo, S.; Testa, G.; Natoli, G. The histone H3 lysine-27 demethylase Jmjd3 links inflammation to inhibition of polycomb-mediated gene silencing. *Cell* **2007**, *130*, 1083–1094. [CrossRef]
43. Kondo, T.; Ito, S.; Koseki, H. Polycomb in Transcriptional Phase Transition of Developmental Genes. *Trends Biochem. Sci.* **2016**, *41*, 9–19. [CrossRef] [PubMed]
44. Neele, A.E.; de Winther, M.P.J. Repressing the repressor: Ezh2 mediates macrophage activation. *J. Exp. Med.* **2018**, *215*, 1269–1271. [CrossRef]
45. Laugesen, A.; Hojfeldt, J.W.; Helin, K. Role of the Polycomb Repressive Complex 2 (PRC2) in Transcriptional Regulation and Cancer. *Cold Spring Harb. Perspect. Med.* **2016**, *6*, a026575. [CrossRef] [PubMed]
46. Tan, J.Z.; Yan, Y.; Wang, X.X.; Jiang, Y.; Xu, H.E. EZH2: Biology, disease, and structure-based drug discovery. *Acta Pharmacol. Sin.* **2014**, *35*, 161–174. [CrossRef] [PubMed]

47. Nakagawa, M.; Kitabayashi, I. Oncogenic roles of enhancer of zeste homolog 1/2 in hematological malignancies. *Cancer Sci.* **2018**, *109*, 2342–2348. [CrossRef]
48. Garber, K. Histone-writer cancer drugs enter center stage. *Nat. Biotechnol.* **2020**, *38*, 909–912. [CrossRef] [PubMed]
49. Liu, Y.; Peng, J.; Sun, T.; Li, N.; Zhang, L.; Ren, J.; Yuan, H.; Kan, S.; Pan, Q.; Li, X.; et al. Epithelial EZH2 serves as an epigenetic determinant in experimental colitis by inhibiting TNFalpha-mediated inflammation and apoptosis. *Proc. Natl. Acad. Sci. USA* **2017**, *114*, E3796–E3805. [CrossRef] [PubMed]
50. Neele, A.E.; Chen, H.J.; Gijbels, M.J.J.; van der Velden, S.; Hoeksema, M.A.; Boshuizen, M.C.S.; Van den Bossche, J.; Tool, A.T.; Matlung, H.L.; van den Berg, T.K.; et al. Myeloid Ezh2 Deficiency Limits Atherosclerosis Development. *Front. Immunol.* **2020**, *11*, 594603. [CrossRef]
51. Qin, H.; Holdbrooks, A.T.; Liu, Y.; Reynolds, S.L.; Yanagisawa, L.L.; Benveniste, E.N. SOCS3 deficiency promotes M1 macrophage polarization and inflammation. *J. Immunol.* **2012**, *189*, 3439–3448. [CrossRef]
52. Ruenjaiman, V.; Butta, P.; Leu, Y.W.; Pongpanich, M.; Leelahavanichkul, A.; Kueanjinda, P.; Palaga, T. Profile of Histone H3 Lysine 4 Trimethylation and the Effect of Lipopolysaccharide/Immune Complex-Activated Macrophages on Endotoxemia. *Front. Immunol.* **2019**, *10*, 2956. [CrossRef]
53. Benjaskulluecha, S.; Boonmee, A.; Pattarakankul, T.; Wongprom, B.; Klomsing, J.; Palaga, T. Screening of compounds to identify novel epigenetic regulatory factors that affect innate immune memory in macrophages. *Sci. Rep.* **2022**, *12*, 1912. [CrossRef] [PubMed]
54. Cheng, Y.; He, C.; Wang, M.; Ma, X.; Mo, F.; Yang, S.; Han, J.; Wei, X. Targeting epigenetic regulators for cancer therapy: Mechanisms and advances in clinical trials. *Signal. Transduct. Target. Ther.* **2019**, *4*, 62. [CrossRef]
55. Jaroonwitchawan, T.; Visitchanakun, P.; Dang, P.C.; Ritprajak, P.; Palaga, T.; Leelahavanichkul, A. Dysregulation of Lipid Metabolism in Macrophages Is Responsible for Severe Endotoxin Tolerance in FcgRIIB-Deficient Lupus Mice. *Front. Immunol.* **2020**, *11*, 959. [CrossRef] [PubMed]
56. Ondee, T.; Jaroonwitchawan, T.; Pisitkun, T.; Gillen, J.; Nita-Lazar, A.; Leelahavanichkul, A.; Somparn, P. Decreased Protein Kinase C-beta Type II Associated with the Prominent Endotoxin Exhaustion in the Macrophage of FcGRIIb$^{-/-}$ Lupus Prone Mice is Revealed by Phosphoproteomic Analysis. *Int. J. Mol. Sci.* **2019**, *20*, 1354. [CrossRef]
57. Thim-Uam, A.; Makjaroen, J.; Issara-Amphorn, J.; Saisorn, W.; Wannigama, D.L.; Chancharoenthana, W.; Leelahavanichkul, A. Enhanced Bacteremia in Dextran Sulfate-Induced Colitis in Splenectomy Mice Correlates with Gut Dysbiosis and LPS Tolerance. *Int. J. Mol. Sci.* **2022**, *23*, 1676. [CrossRef] [PubMed]
58. Duan, R.; Du, W.; Guo, W. EZH2: A novel target for cancer treatment. *J. Hematol. Oncol.* **2020**, *13*, 104. [CrossRef] [PubMed]
59. Carow, B.; Rottenberg, M.E. SOCS3, a Major Regulator of Infection and Inflammation. *Front. Immunol.* **2014**, *5*, 58. [CrossRef]
60. Zhang, X.; Wang, Y.; Yuan, J.; Li, N.; Pei, S.; Xu, J.; Luo, X.; Mao, C.; Liu, J.; Yu, T.; et al. Macrophage/microglial Ezh2 facilitates autoimmune inflammation through inhibition of Socs3. *J. Exp. Med.* **2018**, *215*, 1365–1382. [CrossRef]
61. Hu, G.; Guo, M.; Xu, J.; Wu, F.; Fan, J.; Huang, Q.; Yang, G.; Lv, Z.; Wang, X.; Jin, Y. Nanoparticles Targeting Macrophages as Potential Clinical Therapeutic Agents Against Cancer and Inflammation. *Front. Immunol.* **2019**, *10*, 1998. [CrossRef]
62. Tungsanga, S.; Panpetch, W.; Bhunyakarnjanarat, T.; Udompornpitak, K.; Katavetin, P.; Chancharoenthana, W.; Chatthanathon, P.; Somboonna, N.; Tungsanga, K.; Tumwasorn, S.; et al. Uremia-Induced Gut Barrier Defect in 5/6 Nephrectomized Mice Is Worsened by Candida Administration through a Synergy of Uremic Toxin, Lipopolysaccharide, and β-D-Glucan, but Is Attenuated by *Lacticaseibacillus rhamnosus* L34. *Int. J. Mol. Sci.* **2022**, *23*, 2511. [CrossRef] [PubMed]
63. Visitchanakun, P.; Panpetch, W.; Saisorn, W.; Chatthanathon, P.; Wannigama, D.L.; Thim-uam, A.; Svasti, S.; Fucharoen, S.; Somboonna, N.; Leelahavanichkul, A. Increased susceptibility to dextran sulfate-induced mucositis of iron-overload β-thalassemia mice, another endogenous cause of septicemia in thalassemia. *Clin. Sci.* **2021**, *135*, 1467–1486. [CrossRef]
64. Boonhai, S.; Bootdee, K.; Saisorn, W.; Takkavatakarn, K.; Sitticharoenchai, P.; Tungsanga, S.; Tiranathanagul, K.; Leelahavanichkul, A. TMAO reductase, a biomarker for gut permeability defect induced inflammation, in mouse model of chronic kidney disease and dextran sulfate solution-induced mucositis. *Asian Pac. J. Allergy Immunol.* **2021**. [CrossRef]
65. Binmama, S.; Dang, C.P.; Visitchanakun, P.; Hiengrach, P.; Somboonna, N.; Cheibchalard, T.; Pisitkun, P.; Chindamporn, A.; Leelahavanichkul, A. Beta-Glucan from *S. cerevisiae* Protected AOM-Induced Colon Cancer in cGAS-Deficient Mice Partly through Dectin-1-Manipulated Macrophage Cell Energy. *Int. J. Mol. Sci.* **2022**, *23*, 10951. [CrossRef]
66. Charoensappakit, A.; Sae-Khow, K.; Leelahavanichkul, A. Gut Barrier Damage and Gut Translocation of Pathogen Molecules in Lupus, an Impact of Innate Immunity (Macrophages and Neutrophils) in Autoimmune Disease. *Int. J. Mol. Sci.* **2022**, *23*, 8223. [CrossRef]
67. Ezponda, T.; Licht, J.D. Molecular pathways: Deregulation of histone h3 lysine 27 methylation in cancer-different paths, same destination. *Clin. Cancer Res.* **2014**, *20*, 5001–5008. [CrossRef]
68. Nichol, J.N.; Dupere-Richer, D.; Ezponda, T.; Licht, J.D.; Miller, W.H., Jr. H3K27 Methylation: A Focal Point of Epigenetic Deregulation in Cancer. *Adv. Cancer Res.* **2016**, *131*, 59–95. [CrossRef]
69. Yue, D.; Wang, Z.; Yang, Y.; Hu, Z.; Luo, G.; Wang, F. EZH2 inhibitor GSK343 inhibits sepsis-induced intestinal disorders. *Exp. Ther. Med.* **2021**, *21*, 437. [CrossRef] [PubMed]
70. Zhang, Q.; Sun, H.; Zhuang, S.; Liu, N.; Bao, X.; Liu, X.; Ren, H.; Lv, D.; Li, Z.; Bai, J.; et al. Novel pharmacological inhibition of EZH2 attenuates septic shock by altering innate inflammatory responses to sepsis. *Int. Immunopharmacol.* **2019**, *76*, 105899. [CrossRef] [PubMed]

71. Zhao, D.; Li, Z.; Liu, X.; Liu, N.; Bao, X.; Sun, H.; Meng, Q.; Ren, H.; Bai, J.; Zhou, X.; et al. Lymphocyte expression of EZH2 is associated with mortality and secondary infectious complications in sepsis. *Int. Immunopharmacol.* **2020**, *89*, 107042. [CrossRef]
72. Adema, V.; Colla, S. EZH2 Inhibitors: The Unpacking Revolution. *Cancer Res.* **2022**, *82*, 359–361. [CrossRef]
73. Wang, Q.; Xu, L.; Zhang, X.; Liu, D.; Wang, R. GSK343, an inhibitor of EZH2, mitigates fibrosis and inflammation mediated by HIF-1alpha in human peritoneal mesothelial cells treated with high glucose. *Eur. J. Pharmacol.* **2020**, *880*, 173076. [CrossRef]
74. Bamidele, A.O.; Svingen, P.A.; Sagstetter, M.R.; Sarmento, O.F.; Gonzalez, M.; Braga Neto, M.B.; Kugathasan, S.; Lomberk, G.; Urrutia, R.A.; Faubion, W.A., Jr. Disruption of FOXP3-EZH2 Interaction Represents a Pathobiological Mechanism in Intestinal Inflammation. *Cell. Mol. Gastroenterol. Hepatol.* **2019**, *7*, 55–71. [CrossRef] [PubMed]
75. He, J.; Song, Y.; Li, G.; Xiao, P.; Liu, Y.; Xue, Y.; Cao, Q.; Tu, X.; Pan, T.; Jiang, Z.; et al. Fbxw7 increases CCL2/7 in CX3CR1hi macrophages to promote intestinal inflammation. *J. Clin. Investig.* **2019**, *129*, 3877–3893. [CrossRef]
76. Yong, H.; Wu, G.; Chen, J.; Liu, X.; Bai, Y.; Tang, N.; Liu, L.; Wei, J. lncRNA MALAT1 Accelerates Skeletal Muscle Cell Apoptosis and Inflammatory Response in Sepsis by Decreasing BRCA1 Expression by Recruiting EZH2. *Mol. Ther. Nucleic Acids* **2020**, *21*, 1120–1121. [CrossRef]
77. Alexander, W.S. Suppressors of cytokine signalling (SOCS) in the immune system. *Nat. Rev. Immunol.* **2002**, *2*, 410–416. [CrossRef] [PubMed]
78. Cassatella, M.A.; Gasperini, S.; Bovolenta, C.; Calzetti, F.; Vollebregt, M.; Scapini, P.; Marchi, M.; Suzuki, R.; Suzuki, A.; Yoshimura, A. Interleukin-10 (IL-10) selectively enhances CIS3/SOCS3 mRNA expression in human neutrophils: Evidence for an IL-10-induced pathway that is independent of STAT protein activation. *Blood* **1999**, *94*, 2880–2889. [CrossRef] [PubMed]
79. Prele, C.M.; Keith-Magee, A.L.; Yerkovich, S.T.; Murcha, M.; Hart, P.H. Suppressor of cytokine signalling-3 at pathological levels does not regulate lipopolysaccharide or interleukin-10 control of tumour necrosis factor-alpha production by human monocytes. *Immunology* **2006**, *119*, 8–17. [CrossRef]
80. Berlato, C.; Cassatella, M.A.; Kinjyo, I.; Gatto, L.; Yoshimura, A.; Bazzoni, F. Involvement of suppressor of cytokine signaling-3 as a mediator of the inhibitory effects of IL-10 on lipopolysaccharide-induced macrophage activation. *J. Immunol.* **2002**, *168*, 6404–6411. [CrossRef]
81. Qasimi, P.; Ming-Lum, A.; Ghanipour, A.; Ong, C.J.; Cox, M.E.; Ihle, J.; Cacalano, N.; Yoshimura, A.; Mui, A.L. Divergent mechanisms utilized by SOCS3 to mediate interleukin-10 inhibition of tumor necrosis factor alpha and nitric oxide production by macrophages. *J. Biol. Chem.* **2006**, *281*, 6316–6324. [CrossRef]
82. Mola, S.; Pinton, G.; Erreni, M.; Corazzari, M.; De Andrea, M.; Grolla, A.A.; Martini, V.; Moro, L.; Porta, C. Inhibition of the Histone Methyltransferase EZH2 Enhances Protumor Monocyte Recruitment in Human Mesothelioma Spheroids. *Int. J. Mol. Sci.* **2021**, *22*, 4391. [CrossRef] [PubMed]
83. Kitchen, G.B.; Hopwood, T.; Gali Ramamoorthy, T.; Downton, P.; Begley, N.; Hussell, T.; Dockrell, D.H.; Gibbs, J.E.; Ray, D.W.; Loudon, A.S.I. The histone methyltransferase Ezh2 restrains macrophage inflammatory responses. *FASEB J.* **2021**, *35*, e21843. [CrossRef]
84. Arango Duque, G.; Descoteaux, A. Macrophage cytokines: Involvement in immunity and infectious diseases. *Front. Immunol.* **2014**, *5*, 491. [CrossRef]
85. Wang, Y.; Wang, Q.; Wang, B.; Gu, Y.; Yu, H.; Yang, W.; Ren, X.; Qian, F.; Zhao, X.; Xiao, Y.; et al. Inhibition of EZH2 ameliorates bacteria-induced liver injury by repressing RUNX1 in dendritic cells. *Cell Death Dis.* **2020**, *11*, 1024. [CrossRef] [PubMed]
86. Kang, N.; Eccleston, M.; Clermont, P.L.; Latarani, M.; Male, D.K.; Wang, Y.; Crea, F. EZH2 inhibition: A promising strategy to prevent cancer immune editing. *Epigenomics* **2020**, *12*, 1457–1476. [CrossRef]
87. Makjaroen, J.; Phuengmaung, P.; Saisorn, W.; Udomkarnjananun, S.; Pisitkun, T.; Leelahavanichkul, A. Lipopolysaccharide Tolerance Enhances Murine Norovirus Reactivation: An Impact of Macrophages Mainly Evaluated by Proteomic Analysis. *Int. J. Mol. Sci.* **2023**, *24*, 1829. [CrossRef]
88. Leelahavanichkul, A.; Somparn, P.; Bootprapan, T.; Tu, H.; Tangtanatakul, P.; Nuengjumnong, R.; Worasilchai, N.; Tiranathanagul, K.; Eiam-ong, S.; Levine, M.; et al. High-dose ascorbate with low-dose amphotericin B attenuates severity of disease in a model of the reappearance of candidemia during sepsis in the mouse. *Am. J. Physiol. Regul. Integr. Comp. Physiol.* **2015**, *309*, R223–R234. [CrossRef]
89. Vu, C.T.B.; Thammahong, A.; Leelahavanichkul, A.; Ritprajak, P. Alteration of macrophage immune phenotype in a murine sepsis model is associated with susceptibility to secondary fungal infection. *Asian Pac. J. Allergy Immunol.* **2022**, *40*, 162–171. [CrossRef] [PubMed]
90. Ou, S.M.; Lee, K.H.; Tsai, M.T.; Tseng, W.C.; Chu, Y.C.; Tarng, D.C. Sepsis and the Risks of Long-Term Renal Adverse Outcomes in Patients With Chronic Kidney Disease. *Front. Med.* **2022**, *9*, 809292. [CrossRef]
91. Kalani, C.; Venigalla, T.; Bailey, J.; Udeani, G.; Surani, S. Sepsis Patients in Critical Care Units with Obesity: Is Obesity Protective? *Cureus* **2020**, *12*, e6929. [CrossRef] [PubMed]
92. Camilleri, M. Leaky gut: Mechanisms, measurement and clinical implications in humans. *Gut* **2019**, *68*, 1516–1526. [CrossRef]
93. de Kort, S.; Keszthelyi, D.; Masclee, A.A. Leaky gut and diabetes mellitus: What is the link? *Obes. Rev.* **2011**, *12*, 449–458. [CrossRef]
94. Costantini, E.; Carlin, M.; Porta, M.; Brizzi, M.F. Type 2 diabetes mellitus and sepsis: State of the art, certainties and missing evidence. *Acta Diabetol.* **2021**, *58*, 1139–1151. [CrossRef] [PubMed]

95. Colbert, J.F.; Schmidt, E.P.; Faubel, S.; Ginde, A.A. Severe Sepsis Outcomes Among Hospitalizations With Inflammatory Bowe Disease. *Shock* **2017**, *47*, 128–131. [CrossRef]
96. Arbuckle, J.H.; Gardina, P.J.; Gordon, D.N.; Hickman, H.D.; Yewdell, J.W.; Pierson, T.C.; Myers, T.G.; Kristie, T.M. Inhibitors of th Histone Methyltransferases EZH2/1 Induce a Potent Antiviral State and Suppress Infection by Diverse Viral Pathogens. *mBi* **2017**, *8*, e01141-17. [CrossRef]
97. Martinez, M.L.; Plata-Menchaca, E.P.; Ruiz-Rodriguez, J.C.; Ferrer, R. An approach to antibiotic treatment in patients with sepsis *J. Thorac. Dis.* **2020**, *12*, 1007–1021. [CrossRef]
98. Visitchanakun, P.; Tangtanatakul, P.; Trithiphen, O.; Soonthornchai, W.; Wongphoom, J.; Tachaboon, S.; Srisawat, N.; Leelaha vanichkul, A. Plasma miR-370-3P as a Biomarker of Sepsis-Associated Encephalopathy, the Transcriptomic Profiling Analysis o Microrna-Arrays From Mouse Brains. *Shock* **2020**, *54*, 347–357. [CrossRef]
99. Issara-Amphorn, J.; Surawut, S.; Worasilchai, N.; Thim-Uam, A.; Finkelman, M.; Chindamporn, A.; Palaga, T.; Hirankarn, N. Pisitkun, P.; Leelahavanichkul, A. The Synergy of Endotoxin and (1–>3)-β-D-Glucan, from Gut Translocation, Worsens Sepsis Severity in a Lupus Model of Fc Gamma Receptor IIb-Deficient Mice. *J. Innate Immun.* **2018**, *10*, 189–201. [CrossRef]
100. Panpetch, W.; Somboonna, N.; Bulan, D.E.; Issara-Amphorn, J.; Finkelman, M.; Worasilchai, N.; Chindamporn, A.; Palaga T.; Tumwasorn, S.; Leelahavanichkul, A. Oral administration of live- or heat-killed *Candida albicans* worsened cecal ligation and puncture sepsis in a murine model possibly due to an increased serum (1–>3)-β-D-glucan. *PLoS ONE* **2017**, *12*, e0181439 [CrossRef] [PubMed]
101. Hiengrach, P.; Visitchanakun, P.; Finkelman, M.A.; Chancharoenthana, W.; Leelahavanichkul, A. More Prominent Inflammatory Response to Pachyman than to Whole-Glucan Particle and Oat-β-Glucans in Dextran Sulfate-Induced Mucositis Mice and Mouse Injection through Proinflammatory Macrophages. *Int. J. Mol. Sci.* **2022**, *23*, 4026. [CrossRef]
102. Singkham-In, U.; Phuengmaung, P.; Makjaroen, J.; Saisorn, W.; Bhunyakarnjanarat, T.; Chatsuwan, T.; Chirathaworn, C. Chancharoenthana, W.; Leelahavanichkul, A. Chlorhexidine Promotes Psl Expression in *Pseudomonas aeruginosa* That Enhances Cell Aggregation with Preserved Pathogenicity Demonstrates an Adaptation against Antiseptic. *Int. J. Mol. Sci.* **2022**, *23*, 8308 [CrossRef]
103. Udompornpitak, K.; Charoensappakit, A.; Sae-Khow, K.; Bhunyakarnjanarat, T.; Dang, C.P.; Saisorn, W.; Visitchanakun, P. Phuengmaung, P.; Palaga, T.; Ritprajak, P.; et al. Obesity Exacerbates Lupus Activity in Fc Gamma Receptor IIb Deficient Lupus Mice Partly through Saturated Fatty Acid-Induced Gut Barrier Defect and Systemic Inflammation. *J. Innate Immun.* **2022**, *15* 240–261. [CrossRef]
104. Ondee, T.; Gillen, J.; Visitchanakun, P.; Somparn, P.; Issara-Amphorn, J.; Dang Phi, C.; Chancharoenthana, W.; Gurusamy D.; Nita-Lazar, A.; Leelahavanichkul, A. Lipocalin-2 (Lcn-2) Attenuates Polymicrobial Sepsis with LPS Preconditioning (LPS Tolerance) in FcGRIIb Deficient Lupus Mice. *Cells* **2019**, *8*, 1064. [CrossRef] [PubMed]
105. Hiengrach, P.; Panpetch, W.; Chindamporn, A.; Leelahavanichkul, A. *Helicobacter pylori*, Protected from Antibiotics and Stresses Inside *Candida albicans* Vacuoles, Cause Gastritis in Mice. *Int. J. Mol. Sci.* **2022**, *23*, 8568. [CrossRef] [PubMed]
106. Panpetch, W.; Visitchanakun, P.; Saisorn, W.; Sawatpanich, A.; Chatthanathon, P.; Somboonna, N.; Tumwasorn, S.; Leelaha-vanichkul, A. *Lactobacillus rhamnosus* attenuates Thai chili extracts induced gut inflammation and dysbiosis despite capsaicin bactericidal effect against the probiotics, a possible toxicity of high dose capsaicin. *PLoS ONE* **2021**, *16*, e0261189. [CrossRef]
107. Phuengmaung, P.; Mekjaroen, J.; Saisorn, W.; Chatsuwan, T.; Somparn, P.; Leelahavanichkul, A. Rapid Synergistic Biofilm Production of *Pseudomonas* and *Candida* on the Pulmonary Cell Surface and in Mice, a Possible Cause of Chronic Mixed Organismal Lung Lesions. *Int. J. Mol. Sci.* **2022**, *23*, 9202. [CrossRef] [PubMed]

International Journal of
Molecular Sciences

MDPI

Communication

Proteomic Profiling of Mouse Brain Pyruvate Kinase Binding Proteins: A Hint for Moonlighting Functions of PKM1?

Olga Buneeva [1], Arthur Kopylov [1], Oksana Gnedenko [1], Marina Medvedeva [2], Alexander Veselovsky [1], Alexis Ivanov [1], Victor Zgoda [1] and Alexei Medvedev [1,*]

1 Institute of Biomedical Chemistry, 10 Pogodinskaya Street, Moscow 119121, Russia
2 Department of Biochemistry, School of Biology, Moscow State University, Moscow 119991, Russia
* Correspondence: professor57@yandex.ru

Abstract: Affinity-based proteomic profiling is widely used for the identification of proteins involved in the formation of various interactomes. Since protein–protein interactions (PPIs) reflect the role of particular proteins in the cell, identification of interaction partners for a protein of interest can reveal its function. The latter is especially important for the characterization of multifunctional proteins, which can play different roles in the cell. Pyruvate kinase (PK), a classical glycolytic enzyme catalyzing the last step of glycolysis, exists in four isoforms: PKM1, PKM2, PKL, and PKR. The enzyme isoform expressed in actively dividing cells, PKM2, exhibits many moonlighting (noncanonical) functions. In contrast to PKM2, *PKM1*, predominantly expressed in adult differentiated tissues, lacks well-documented moonlighting functions. However, certain evidence exists that it can also perform some functions unrelated to glycolysis. In order to evaluate protein partners, bound to PKM1, in this study we have combined affinity-based separation of mouse brain proteins with mass spectrometry identification. The highly purified PKM1 and a 32-mer synthetic peptide (PK peptide), sharing high sequence homology with the interface contact region of all PK isoforms, were used as the affinity ligands. This proteomic profiling resulted in the identification of specific and common proteins bound to both affinity ligands. Quantitative affinity binding to the affinity ligands of selected identified proteins was validated using a surface plasmon resonance (SPR) biosensor. Bioinformatic analysis has shown that the identified proteins, bound to both full-length PKM1 and the PK peptide, form a protein network (interactome). Some of these interactions are relevant for the moonlighting functions of PKM1. The proteomic dataset is available via ProteomeXchange with the identifier PXD041321.

Keywords: pyruvate kinase; pyruvate kinase isoforms; PKM; moonlighting functions; PKM binding proteins; interactome

Citation: Buneeva, O.; Kopylov, A.; Gnedenko, O.; Medvedeva, M.; Veselovsky, A.; Ivanov, A.; Zgoda, V.; Medvedev, A. Proteomic Profiling of Mouse Brain Pyruvate Kinase Binding Proteins: A Hint for Moonlighting Functions of PKM1? *Int. J. Mol. Sci.* **2023**, *24*, 7634. https://doi.org/10.3390/ijms24087634

Academic Editors: Paolo Iadarola and Firas Kobeissy

Received: 20 March 2023
Revised: 18 April 2023
Accepted: 19 April 2023
Published: 21 April 2023

1. Introduction

Affinity-based proteomic profiling is a powerful approach used in proteomic studies for separation of various groups of proteins [1]. The combination of affinity chromatography with mass spectrometry represents an effective tool for the generation of a broad range of data for protein–protein interactions (PPIs), mapping post-translational modifications or recovering a certain target from the whole proteome, characterization of intracellular signaling networks and particular interactomes, etc. [2–6].

In the context of PPIs, so-called moonlighting proteins attract much interest because these proteins exhibit more than one function in the cell [7]. The repertoire of protein partners interacting with such moonlighting proteins significantly differs in the dependence of particular biochemical/physiological processes in which the moonlighting protein is involved. Currently, several hundred moonlighting proteins have been recognized [7–10].

Glycolytic enzymes are one of the most abundant groups of moonlighting proteins [10,11]. Pyruvate kinase (PK, EC 2.7.1.40) is a highly conserved enzyme catalyzing the irreversible conversion of ADP and phosphoenolpyruvate to ATP and pyruvate (substrate-level phosphorylation reaction). Mammalian PK exists in four isoforms differed by expression patterns and regulatory properties: PKM1, PKM2, PKL, and PKR [12,13]. PK is most active as a tetramer, and dissociation into dimers and monomers results in its inactivation [13]. The isoform PKM2 is typical of embryonic and adult dividing cells [14]. Besides its role in glycolysis, PKM2 performs many moonlighting functions and participates in different cellular processes, such as: transcription, translation, signaling, and cytoskeletal dynamics [10,13–20]. In contrast to PKM2, PKM1 lacks well-documented moonlighting functions [14]. It is predominantly expressed in adult differentiated tissues including brain and musculature [12], and skeletal muscles are frequently used as the source for purification of this enzyme [21,22]. In the mouse brain, PKM1 was found in gray matter neurons and white matter bundles [23], but no functions of PKM1 unrelated to glycolysis have been reported so far.

Certain evidence exists that PKM1 can perform some functions unrelated to its role in glycolysis. For example, a PKM1 monomer is known as a thyroid hormone (T3) binding protein [24]. PKM1 can bind to ribosomes [20,25]. In a rabbit reticulocyte in vitro translation system, exogenously added PKM1-GST decreased the translation of some mRNAs [25]. In addition, PKM1 can form hetero-oligomers with other PK isoforms [13]. This suggests an important role of oligomer-to-monomer transition and the interface contact region that could be potentially involved in PKM1 functioning unrelated to glycolysis.

Since classical and moonlighting functions obviously require different sets of protein partners, proteomic profiling with the protein of interest as an affinity ligand may give a hint for further search and research of its moonlighting. The latter is especially interesting in the context of PKM1, which shares high sequence similarity with PKM2 [26] and therefore could be considered "as a substitute player in the PKM2 game". Thus, the aim of this study was to investigate profiles of mouse brain proteins bound to the highly purified PKM1 or a synthetic 32-mer linear PK peptide. This PK peptide corresponds to the interface of PKL/R isoforms and exhibits high homology with PKM isoforms. Results of this study suggest that the profiles of proteins bound to either purified PKM1 or the PK peptide significantly differ. However, common proteins, bound to both full-length PKM1 and the PK peptide, form a protein network (interactome) that may reflect some moonlighting functions of PKM1.

2. Results

2.1. Proteomic Profiling of Mouse Brain Proteins Using PKM1 as the Affinity Ligand

Proteomic profiling of cleared lysates of mouse brain homogenates performed using immobilized PKM1 as the affinity ligand resulted in confident identification of 44 individual proteins including brain PKM (Table 1, Supplementary Materials Table S1). This suggests that the immobilized affinity ligand, purified rabbit muscle PKM1, was functionally competent to form PKM oligomers. All the identified brain proteins belonged to the following functional groups: (1) metabolic enzymes; (2) proteins involved in cytoskeleton formation and trafficking; (3) proteins involved in signal transduction and enzyme activity regulation; (4) protective proteins and components of the ubiquitin–proteasome system; and (5) protein regulators of gene expression, cell division, and differentiation.

2.2. Biosensor Validation of PKM1 Interaction with Selected Identified PKM Binding Proteins

Interaction of some of the identified PKM binding proteins with PKM1 has been validated in the optical biosensor experiments using available purified rabbit muscle enzymes: PKM1, aldolase, glyceraldehyde 3-phosphate dehydrogenase (GAPDH), lactate dehydrogenase (LDH), and creatine phosphokinase (CK). Correctness of extrapolation of the results obtained using the rabbit muscle enzyme for validation of PKM1 interaction is determined by high similarity between mouse and rabbit glycolytic enzymes [27]. All the selected proteins demonstrated quantitative binding to the PKM1 (Figure 1), with the K_d values ranging from 10^{-8} M (CK) to about $5{\cdot}10^{-6}$ M (LDH) (Table 2).

Table 1. Distribution of the identified mouse brain proteins among functional groups.

Group No.	Protein Functions	Number of Identified Proteins Bound to Immobilized Full-Length PKM1	Number of Identified Proteins Bound to Immobilized PK Peptide
1	metabolic enzymes	13	8
2	proteins involved in cytoskeleton formation and trafficking	10	29
3	proteins involved in signal transduction and enzyme activity regulation	6	19
4	protective proteins and components of the ubiquitin–proteasome system	6	14
5	protein regulators of gene expression, cell division, and differentiation	9	22
	total	44	92

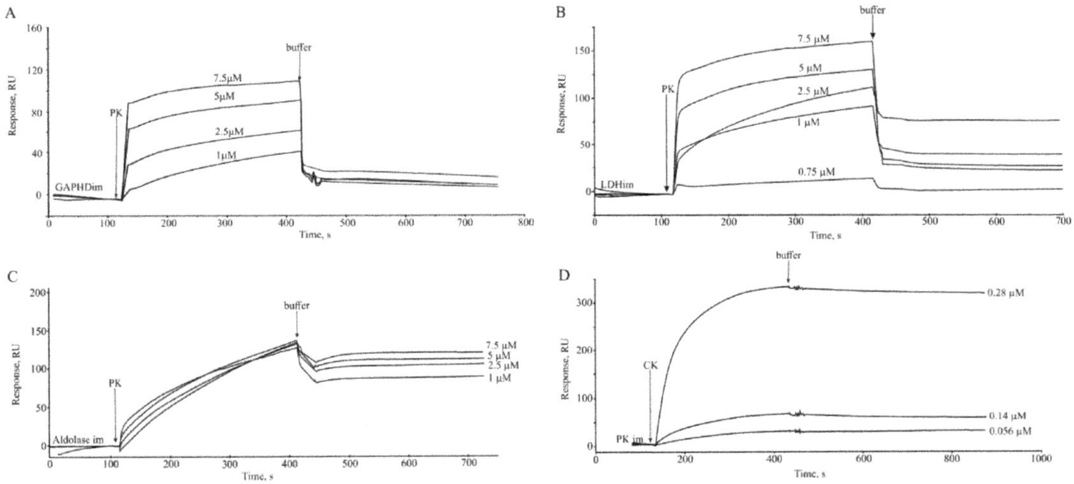

Figure 1. Biosensor validation of PKM1 interaction with selected identified PKM binding proteins. In the case of A, B, and C, purified rabbit muscle PK1 was used as an analyte, and purified rabbit muscle GAPDH (**A**), LDH (**B**), and aldolase A (**C**) were immobilized onto a chip surface; in (**D**) PKM1 was immobilized on the chip surface and CK was used as the analyte.

Table 2. Biosensor validation of the interaction of selected proteins with pyruvate kinase.

Immobilized Ligand	Analyte	ka, $M^{-1} \cdot s^{-1}$	kd, s^{-1}	K_d, M
Pyruvate kinase	Creatine kinase B-type	$(8.64 \pm 0.03) \times 10^4$	$(9.24 \pm 0.18) \times 10^{-4}$	1.07×10^{-8}
Glyceraldehyde-3-phosphate dehydrogenase	Pyruvate kinase	$(1.53 \pm 0.17) \times 10^3$	$(1.62 \pm 0.05) \times 10^{-3}$	1.06×10^{-6}
Lactate dehydrogenase	Pyruvate kinase	$(1.37 \pm 0.08) \times 10^3$	$(6.57 \pm 0.13) \times 10^{-3}$	4.8×10^{-6}
Aldolase A	Pyruvate kinase	$(1.69 \pm 0.06) \times 10^3$	$(2.38 \pm 0.1) \times 10^{-3}$	1.41×10^{-6}

2.3. *Proteomic Profiling of Mouse Brain Proteins Using the PK Peptide as the Affinity Ligand*

Identification of PKM as the protein bound to the immobilized PKM1 (PKM1-binding protein) suggested involvement of a PKM oligomer-forming interface in this interaction. In order to investigate the role of the PK interface in the PKM proteome formation we have compared profiles of mouse brain proteins bound to PKM1 and the PK peptide. The PK peptide corresponds to the sequence fragment (residues 406–437) of PKR/L and shares high homology with PKM1 and PKM2.

Proteomic profiling of cleared lysates of mouse brain homogenates performed using the immobilized PK peptide as the affinity ligand resulted in confident identification of more than 75 individual proteins (Supplementary Table S2). These proteins fell into the same functional groups as the mouse proteins bound to the full-length PKM1. Although the total number of brain proteins bound to the PK peptide was significantly higher than in the case of PKM1, 15 proteins were common for both affinity ligands (Table 3). At least 10 of 15 common proteins are known to exhibit some moonlighting functions (Table 3).

Table 3. Common mouse brain proteins interacting with the full-length PKM1 and the synthetic PK peptide *.

no.	Accession Number	Gene	Protein Name (Uniprot)	Functional Group
1	Q61937	*NPM*	Nucleophosmin	5
2	**P52480**	***KPYM***	**Pyruvate kinase PKM**	1
3	**P61979**	***HNRPK***	**Heterogeneous nuclear ribonucleoprotein K**	5
4	**P63017**	***HSP7C***	**Heat shock cognate 71 kDa protein**	4
5	**P63101**	***1433Z***	**14-3-3 protein zeta/delta**	3
6	**P63038**	***CH60***	**60 kDa heat shock protein, mitochondrial**	4
7	**P0DP26**	***CALM1***	**Calmodulin-1**	3
8	**P16858**	***G3P***	**Glyceraldehyde-3-phosphate dehydrogenase**	1
9	**P17182**	***ENOA***	**Alpha-enolase**	1
10	P63242	*IF5A1*	Eukaryotic translation initiation factor 5A-1	5
11	P57780	*ACTN4*	Alpha-actinin-4	2
12	Q9QUM9	*PSA6*	Proteasome subunit alpha type-6	4
13	**P11499**	***HS90B***	**Heat shock protein HSP 90-beta**	4
14	**P05064**	***ALDOA***	**Fructose-bisphosphate aldolase A**	1
15	P68372	*TBB4B*	Tubulin beta-4B chain	2

* Moonlighting proteins are shown in bold. Numbers in the functional group column designate the following protein functions: 1. metabolic enzymes; 2. proteins involved in cytoskeleton formation and trafficking; 3. proteins involved in signal transduction and enzyme activity regulation; 4. protective proteins and components of the ubiquitin–proteasome system; 5. protein regulators of gene expression, cell division, and differentiation.

The commonly identified proteins bound to both the PK peptide and full-length PKM1 were submitted for plotting an interactome map. The resultant map contained 15 nodes and 60 edges in the interactome (Figure 2). The core of the interactions was formed by glycolytic proteins exhibiting moonlighting functions (ALDOA, GAPDH, ENO1, and PKM) and a group of proteins involved in protein folding and chaperoning (HSP90AB1, HSPA8, and NPM1). Results of our study indicate direct interactions between PKM1 and the mouse brain proteins with documented moonlighting functions. These proteins may be relevant for the performance of moonlighting functions of PKM1.

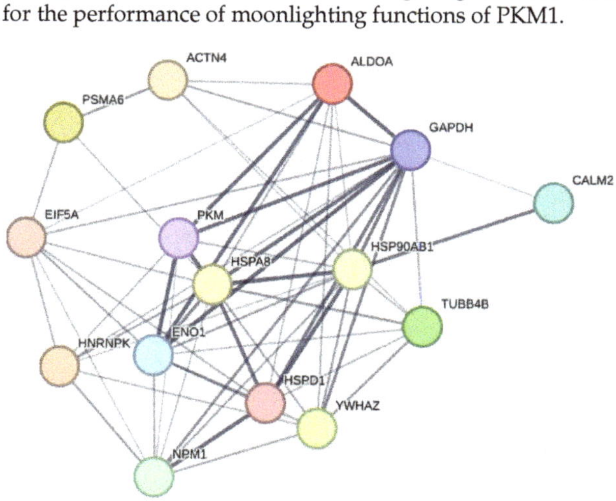

Figure 2. The interactome map reflecting the interaction between mouse brain proteins bound to the full-length PKM1 and the PK peptide. Explanations are given in the text.

3. Discussion

Performance of canonical and moonlighting functions by multifunctional proteins obviously involves different sets of proteins and PPIs. In the case of unknown moonlighting functions of a protein of interest, certain information could be obtained by analyzing its protein partners and PPIs.

In the context of putative moonlighting functions, PKM1, an enzyme of adult non-proliferating tissues (skeletal and cardiac musculature, and the brain), is always "in the shadow" of PKM2, expressed by embryonic and adult actively proliferating tissues. Being encoded by a single gene (*PKM*), PKM1 and PKM2 are formed due to alternative splicing: the alternative exons encode peptide products of the same length, but their identity is about 50% [28]. Moonlighting functions of PKM2 are associated with various cellular processes. Mitogenic and oncogenic stimulation of proliferating cells induces PKM2 translocation into the nucleus [29,30], where it functions as a transcriptional coactivator and a protein kinase phosphorylating histone [31,32].

The moonlighting role of PKM1 is poorly investigated and there are a few examples suggesting that PKM1 could have moonlighting functions in some cells. In this case proteomic profiling with PKM1 as the affinity ligand may help to find putative protein partners for such moonlighting functions.

Using highly purified skeletal rabbit PKM1 and the synthetic 32-mer PK peptide as the affinity ligands, we have identified 44 and 92 individual mouse brain proteins, respectively (Tables 1 and 3, and Supplementary Tables S1 and S2). PKM has been identified among common mouse brain proteins bound to both affinity sorbents (PKM1 and the PK peptide, Table 3). This suggests functional importance of the interface contact region for interaction of PKM1 with potential protein partners. The higher repertoire of proteins bound to the PK peptide may be explained by the higher conformational accessibility of the peptide interaction sites. Nevertheless, there were common proteins bound to both affinity ligands and formed several groups of functional interactions (Figure 2).

PKM1 participates in interactions with proteasome subunits (PSMA6), known glycolytic proteins exhibiting moonlighting functions (ALDOA, GAPDH, and ENO1), and also with proteins involved in protein folding and chaperoning (HSP90AB1, HSPA8, and NPM1). Some of these interactions obviously reflect canonical functions of the glycolytic enzymes supporting proteasomes with additional energy resources and promoting maintenance of nucleotide-dependent functions of proteasomes [33]. The group of proteins involved in protein folding and chaperoning is particularly interesting. It includes several chaperones, HSP90AB1, HSPA8, HSPD1, and NPM1, directly interacting with PKM1. Nucleophosmin (NPM1) is a multifunctional protein, which shuttles between nucleoli, nucleoplasm, and cytoplasm and performs its multifaceted roles (see for review [34]). It exhibits histone- and protein-chaperone activity, participates in DNA replication and repair, ribosome assembly and export, and centrosome duplication and cell cycle control [34–41]. In nucleoli NPM1 acts as a hub protein, which contributes to nucleolar organization through multiple (heterotypic and homotypic) interactions [34]. The number of NPM1-binding proteins exceeds several dozens, and their list is constantly growing [34]. It is possible that PKM1 interaction with NPM1 and other proteins has a regulatory importance in various cell compartments including the nucleus.

Convincing evidence exists in the literature that PKM2, but not PKM1, can move into the nucleus due to the exclusive presence of exon 10, encoding the nuclear localization signal [31,32]. Since the nuclear localization signal is located in the interface contact region, it appears that realization of this scenario requires dissociation of PKM2 tetramers. However, recently it has been demonstrated that under certain conditions PKM1 can also be translocated to the nucleus [42] and therefore interact with potential nuclear targets.

Thus, results of this proteomic profiling have shown that, in addition to proteins related to the classical functioning of PKM1, there are other PKM1-binding proteins. These proteins may be relevant to some activities unrelated to the role of PKM1 as an important glycolytic enzyme. Since binding capacities obviously involve the PKM1 interface contact region, this suggests an important role of oligomer-to-monomer transition for PKM1 moonlighting.

4. Materials and Methods

4.1. Reagents

CNBr-activated Sepharose 4B, creatine kinase from rabbit muscle, dithiothreitol, iodoacetamide, Tris (hydroxymethyl) aminomethane, urea, guanidine hydrochloride, glycerol, a cocktail of protease inhibitors, 4-vinylpyridine, sodium deoxycholate, ammonium bicarbonate, sodium chloride, Triton X-100, and Coomassie brilliant Blue were from Merck (Branchburg, NJ, USA); acetonitrile was from Fisher Chemical (Loughborough, Leicestershire, UK); sodium acetate, boric acid, formic acid, sodium tetraborate, and sodium hydroxide were from Acros Organics (Morris Planes, NJ, USA); modified trypsin (sequencing grade) was purchased from Promega (Madison, WI, USA); 10 kDa membrane filters were from Sartorius Stedium Biotech (Goettingen, Germany); Amicon Ultracel-10K centrifugal concentration filters were from Millipore (Burlington, MA, USA); and Acclaim PepMap® RSLC C18 column (150 mm × 75 μm, particle size 2 μm, pore size 100 Å) were from Dionex (Sunnyvale, CA, USA). The remaining reagents of the highest degree of purity were obtained from local suppliers.

Reagents for the Biacore biosensor were obtained from Cytiva (Marlborough, MA, USA). These included HBS-EP buffer (150 mM NaCl, 3 mM EDTA, 0.005% surfactant P20, 10 mM HEPES, pH 7.4); 10 mM acetate buffer, pH 4.0, pH 5.0, and pH 5.5; and amine coupling reagents kit, containing 1–ethyl–3–(3–dimethylaminopropyl) carbodiimide hydrochloride (EDC), N–hydroxysuccinimide (NHS), and 1 M ethanolamine–HCl, pH 8.5.

Electrophoretically homogeneous PKM1, GAPDH, aldolase A, and LDH were isolated from skeletal muscles of adult (five months old) rabbits according to [22]. The specific activity of the enzyme preparations was 295 μmol/min per mg of protein (PKM1), 170 μmol/min per 1 mg of protein (evaluated in the reaction of 3-phosphoglycerate reduction by NADH; (GAPDH), 10.4 μmol/min per 1 mg of protein (aldolase A), 28 μmol/min per 1 mg of protein (LDH). Before use, the purified enzymes were kept as an ammonium sulfate suspension at 4 °C for not more than two months.

4.2. Selection of the PK Peptide

The PK peptide synthesized by Immunotex (Stavropol, Russia; custom made order) corresponds to the sequence fragment (residues 406–437) of PKR/L. It shares high homology with PKM1 and PKM2 (See Supplementary Materials Figure S1).

4.3. Animals

Adult (three months old) male C57BL/6 mice (weighing 20–25 g) obtained from the Stolbovaya nursery (Moscow region), were used in this study. Experiments were performed one week after their arrival from the nursery. Animals were maintained at natural illumination and had free access to standard laboratory chow and water. All procedures conform to the Russian version of the Guide for the Care and Use of Laboratory Animals (Washington, DC, USA, 1996) and have been approved by the Animal Care and Use Committee at the Institute of Biomedical Chemistry.

4.4. Preparation of Brain Homogenate Lysates

After decapitation of mice under light ether anesthesia, the brain tissue was homogenized using a SilentCrusher S homogenizer (Heidolph, Wood Dale, IL, USA) at 50,000 rpm in 0.05 M potassium phosphate buffer, pH 7.4 (buffer A), to obtain a 30% homogenate. After addition of Triton X-100 (final concentration 3%), incubation for 60 min at 4 °C, and subsequent three-fold dilution with buffer A, the samples were centrifuged at 16,000× g for 30 min. The resultant supernatant (lysate) was used for affinity chromatography on Sepharose with the immobilized PKM1 for subsequent mass spectrometric analysis.

4.5. Immobilization of PKM1 and the PK Peptide on Cyanogen Bromide-Activated Sepharose 4B, Affinity Chromatography and Sample Preparation for Mass Spectrometric Analysis

Immobilization of PKM1 and the PK peptide onto Cyanogen Bromide-Activated Sepharose 4B (CNBr-Sepharose) was carried out according to the previously described protocol [43]. For determination of proteins nonspecifically bound to the sorbent, a control CNBr-Sepharose was used. It was subjected to the same procedures but without the addition of PKM1.

Lysates of brain homogenates (protein concentration of 6 mg/mL) were added to the affinity resin washed with buffer A. The suspension (1:1), containing a cocktail of protease inhibitors added at a concentration recommended by the manufacturer, was incubated overnight at 4 °C and gentle stirring. The affinity resin was then washed with 100 volumes of buffer A to remove nonspecifically bound proteins (protein content in the washings was controlled by OD_{280}). The remaining proteins were eluted at room temperature with 0.1 M glycine buffer, pH 2.8, containing 3 M NaCl (at a flow rate of 0.5 mL/min), using a column 1 cm × 2 cm. The eluate (30 mL) was concentrated to 0.25 mL using an Amicon Ultra centrifuge device. Proteins were extracted with a mixture of chloroform–methanol [44]. The reduction of disulfide bonds, the alkylation of sulfhydryl groups, and trypsinolysis were performed on Vivaspin 500 centrifuge filters with a 10,000 Da membrane, as described in [44]. Samples were evaporated using a 5301 vacuum concentrator (Eppendorf, Hamburg, Germany), dissolved in 0.1% formic acid and analyzed using liquid chromatography tandem mass spectrometry (LC-MS/MS).

4.6. The Mass Spectrometric Analysis (LC-MS/MS)

The mass spectrometric analysis was performed using an Ultimate 3000 RSLCnano (Thermo Scientific, Waltham, MA, USA) integrated system for high-performance liquid separation of peptides in the nanoflow mode. Chromatographic separation of peptides was carried out on an analytical reverse phase column Acclaim Pepmap® C18 (75 μm × 150 mm, 2 μm particle size, Thermo Scientific, USA) in a linear elution gradient of mobile phase A (0.1% aqueous solution of formic acid) and mobile phase B (80% acetonitrile, 0.1% formic acid) from 2% to 40% at a flow rate of 0.3 μL/min for 60 min, followed by equilibration of the chromatographic system in the initial conditions of the gradient (A:B = 2:98) for 5 min.

A Thermo Scientific Q Exactive HF-X mass spectrometer equipped with a nanoelectrospray ionization source (nESI) was operated in the positive ionization mode with a resolution of 120,000 at m/z 200, the ion accumulation volume in the trap was set to 1×10^6, the ion accumulation time in the trap was maximum 50 ms. The dominant charge state of precursor ions was set as 2^+, charge states above 4^+ and below 2^+ were excluded from further analysis. Scanning of tandem spectra was carried out in the mode of automatic selection of 20 dominant peaks of precursor ions recorded at m/z = 350–1400. The resolution for detecting fragment ions was set to 15,000 at m/z 200, the ion accumulation volume in the trap was 1×10^5, and the ion accumulation time in the trap was a maximum of 50 ms. Parent ions were isolated in a window of 2.0 m/z offset by 0.5 m/z for better isotope capture. Measured precursor ions were excluded from subsequent analysis for 20 s after scanning. The obtained mass spectrometric data with the *.raw extension were processed by the MaxQuant software (v 1.6.3.4) with the built-in Andromeda search algorithm. Mouse (*Mus musculus*) Swiss Prot/Uniprot complete proteome protein sequence database was downloaded from the Uniprot database with the addition of reversed sequences and commonly encountered contaminating sequences to apply the target decoy approach. The method was used to calculate the FDR (False Discovery Rate) parameter; the FDR parameter of 1% was taken as a cutoff for protein registration. The following parameters were used for signal extraction and its subsequent processing: the proteolytic cleavage enzyme was trypsin; the maximum allowable amount of intrapeptide residues of lysine or arginine was not more than 1; and the allowable error in measuring the monoisotopic mass of the peptide was ±0.01 Da, and the allowable error in measuring the fragment ion was ±0.05 Da.

Carbamide methylation of cysteine residues was chosen as a fixed chemical modification and methionine oxidation was chosen as a variable modification [20].

The relative abundance (quantitation) of the proteins defined as an emPAI value (Exponentially Modified Protein Abundance index) was based on the protein coverage by the peptide matches in a database search result. The obtained proteome was percolated to extract proteins for bioinformatic analysis. The confidence score for these proteins was at least 95%.

Each protein presented in the tables was identified in at least three independent experiments with a Bonferroni-adjusted *p*-value cut-off of 0.001 (raw *p*-value cut-off 0.01). Protein function identification was performed using UniProt functional information.

4.7. Biosensor Analysis

4.7.1. Immobilization of Proteins

Measurements were performed with a Biacore T200 instrument (Cytiva, Marlborough, MA, USA), thermostated at 25 °C. The surface of the CM5 chip was activated by a mixture 0.2 M 1-ethyl-3(3-dimethylaminopropyl)-carbodiimide hydrochloride/0.05 M N-hydroxysuccinimide for 7 min at 5 μL/min. The 50 μg/mL samples of pyruvate kinase, LDH, aldolase, or GAPD in 10 mM acetate buffer, pH 5.0, pH 4.0, or pH 5.5, respectively, were injected over an activated chip surface for 10 min at 5 μL/min followed by a 2 min injection of HBS-EP buffer to remove excess ligand, and a 2 min injection of 1 M ethanolamine, pH 8.5, to inactivate residual active groups.

4.7.2. Binding Measurements

Protein–protein interactions were monitored by injecting proteins dissolved at various concentrations in buffer A (running buffer) at the flow rate of 10 μL/min for 5 min. The sensor surface was regenerated between sample injections by washing with 1 M NaCl in running buffer for 0.5 min at the flow rate of 50 μL/min. Interactions were estimated by subtracting the response in a blank flow cell from the response in a cell with immobilized ligands.

Data analysis was performed using the BIAevaluation v.4.1 software. Kinetic rate constants were calculated from the sensorgrams by globally fitting response curves obtained at various ligand concentrations to the 1:1 binding model.

4.8. Bioinformatic Analysis of Protein–Protein Interactions

Analysis of protein interactions was performed using the STRING (version 11.5) tool. Common proteins interacting with both full-length PKM1 and the PK peptide were submitted for plotting an interactome map with at least 0.7 confidence interaction score. Pathways and processes were picked up from the KEGG and Reactome databases provided that FDR was $<1.0 \times 10^{-5}$, the interaction strength coefficient was more than 1.0, and a subset of at least four different members of the network fell into the certain pathway or process. We selected the most preferable signaling pathways and biological processes that fit the proposed network of interactions. The resultant map contained 15 nodes and 60 edges in the interactome (with the confidence level at least 0.40 and higher, and the average local clustering coefficient of 0.698).

5. Conclusions

The affinity-based proteomic profiling of mouse brain proteins, performed by using highly purified PKM1 as the affinity ligand, resulted in the identification of 44 PKM1-binding proteins. Since one of the identified proteins was PKM, the immobilized affinity ligand, purified rabbit muscle PKM1, was functionally competent to form PKM oligomers. This suggests involvement of the interface contact region in the interaction with PKM binding proteins. The use of the 32-mer peptide (PK peptide) significantly increased the repertoire of brain proteins bound to this affinity ligand. Nevertheless, there were 15 common proteins bound to both PKM1 and the PK peptide. These common proteins form a protein network (interactome) and some interactions especially with proteins involved in protein folding and chaperoning are relevant for moonlighting functions of PKM1.

Supplementary Materials: The following supporting information can be downloaded at: https://www.mdpi.com/article/10.3390/ijms24087634/s1.

Author Contributions: Conceptualization, A.M. and O.B.; proteomic profiling O.B., A.K., and V.Z.; enzyme purification and characterization, M.M.; bioinformatic analysis, A.K. and A.V.; SPR-based validation, O.G., A.I., and O.B., data analysis, O.B., A.K., and A.M.; writing—original draft preparation, A.M.; writing—review and editing, A.M. All authors have read and agreed to the published version of the manuscript.

Funding: The work was performed within the framework of the Program for Basic Research in the Russian Federation for a long-term period (2021–2030) (no. 122030100170-5).

Institutional Review Board Statement: All procedures conform to the Russian version of the Guide for the Care and Use of Laboratory Animals (Washington, 1996) and have been approved by the Animal Care and Use Committee at the Institute of Biomedical Chemistry (approval code: AE1433RU60-BMC dated 9 August 2022).

Informed Consent Statement: Not applicable.

Data Availability Statement: The mass spectrometry proteomics data have been deposited in the ProteomeXchange Consortium via the PRIDE partner repository with the dataset identifier PXD041321.

Acknowledgments: Mass spectrometry analysis of proteins and SPR analysis of protein–protein interactions were performed using equipment and resources of the Center for Collective Use at the Institute of Biomedical Chemistry (Moscow, Russia).

Conflicts of Interest: The authors declare no conflict of interest.

References

1. Medvedev, A.; Kopylov, A.; Buneeva, O.; Zgoda, V.; Archakov, A. Affinity-based proteomic profiling: Problems and achievements. *Proteomics* **2012**, *12*, 621–637. [CrossRef] [PubMed]
2. Zhang, X.; Wang, T.; Zhang, H.; Han, B.; Wang, L.; Kang, J. Profiling of drug binding proteins by monolithic affinity chromatography in combination with liquid chromatography-tandem mass spectrometry. *J. Chromatogr. A* **2014**, *1359*, 84–90. [CrossRef] [PubMed]
3. Shan, M.; Gregory, B.D. Using RNA Affinity Purification Followed by Mass Spectrometry to Identify RNA-Binding Proteins (RBPs). *Methods Mol. Biol.* **2020**, *2166*, 241–253. [PubMed]
4. Hwang, H.Y.; Kim, T.Y.; Szász, M.A.; Dome, B.; Malm, J.; Marko-Varga, G.; Kwon, H.J. Profiling the Protein Targets of Unmodified Bio-Active Molecules with Drug Affinity Responsive Target Stability and Liquid Chromatography/Tandem Mass Spectrometry. *Proteomics* **2020**, *20*, e1900325. [CrossRef]
5. Low, T.Y.; Syafruddin, S.E.; Mohtar, M.A.; Vellaichamy, A.A.; Rahman, N.S.; Pung, Y.F.; Tan, C.S.H. Recent progress in mass spectrometry-based strategies for elucidating protein-protein interactions. *Cell Mol. Life Sci.* **2021**, *78*, 5325–5339. [CrossRef]
6. Iacobucci, I.; Monaco, V.; Cozzolino, F.; Monti, M. From classical to new generation approaches: An excursus of -omics methods for investigation of protein-protein interaction networks. *J. Proteom.* **2021**, *230*, 103990. [CrossRef]
7. Jeffery, C.J. Protein moonlighting: What is it, and why is it important? *Philos. Trans. R. Soc. B Biol. Sci.* **2018**, *373*, 20160523. [CrossRef]
8. Jeffery, C.J. Moonlighting Proteins—Nature's Swiss Army Knives. *Sci. Prog.* **2017**, *100*, 363–373. [CrossRef]
9. Jeffery, C.J. Enzymes, pseudoenzymes, and moonlighting proteins: Diversity of function in protein superfamilies. *FEBS J.* **2020**, *287*, 4141–4149. [CrossRef]
10. Snaebjornsson, M.T.; Schulze, A. Non-canonical functions of enzymes facilitate cross-talk between cell metabolic and regulatory pathways. *Exp. Mol. Med.* **2018**, *50*, 1–16. [CrossRef]
11. Dhar-Chowdhury, P.; Harrell, M.D.; Han, S.Y.; Jankowska, D.; Parachuru, L.; Morrissey, A.; Srivastava, S.; Liu, W.; Malester, B.; Yoshida, H.; et al. The glycolytic enzymes, glyceraldehyde-3-phosphate dehydrogenase, triose-phosphate isomerase, and pyruvate kinase are components of the K(ATP) channel macromolecular complex and regulate its function. *J. Biol. Chem.* **2005**, *1280*, 38464–38470. [CrossRef] [PubMed]
12. Imamura, K.; Tanaka, T. Multimolecular forms of pyruvate kinase from and other mammalian tissue. *J. Biochem.* **1972**, *71*, 1043–1051. [CrossRef] [PubMed]
13. Israelsen, W.J.; Vander Heiden, M.G. Pyruvate kinase: Function, regulation and role in cancer. *Semin. Cell Dev. Biol.* **2015**, *43*, 43–51. [CrossRef] [PubMed]

14. Alquraishi, M.; Puckett, D.L.; Alani, D.S.; Humidat, A.S.; Frankel, V.D.; Donohoe, D.R.; Whelan, J.; Bettaieb, A. Pyruvate kinase M2: A simple molecule with complex functions. *Free Radic. Biol. Med.* **2019**, *143*, 176–192. [CrossRef] [PubMed]
15. Gupta, V.; Bamezai, R.N. Human pyruvate kinase M2: A multifunctional protein. *Protein Sci.* **2010**, *19*, 2031–2044. [CrossRef]
16. Zhang, Z.; Deng, X.; Liu, Y.; Liu, Y.; Sun, L.; Chen, F. PKM2, function and expression and regulation. *Cell Biosci.* **2019**, *9*, 52. [CrossRef]
17. Zahra, K.; Dey, T.; Mishra, S.P.; Pandey, U. Pyruvate Kinase M2 and Cancer: The Role of PKM2 in Promoting Tumorigenesis. *Front. Oncol.* **2020**, *10*, 159. [CrossRef]
18. Wiese, E.K.; Hitosugi, T. Tyrosine Kinase Signaling in Cancer Metabolism: PKM2 Paradox in the Warburg Effect. *Front. Cell Dev. Biol.* **2018**, *6*, 79. [CrossRef]
19. Lee, J.; Kim, H.K.; Han, Y.M.; Kim, J. Pyruvate kinase isozyme type M2 (PKM2) interacts and cooperates with Oct-4 in regulating transcription. *Int. J. Biochem. Cell Biol.* **2008**, *40*, 1043–1054. [CrossRef]
20. Simsek, D.; Tiu, G.C.; Flynn, R.A.; Byeon, G.W.; Leppek, K.; Xu, A.F.; Chang, H.Y.; Barna, M. The Mammalian Ribo-interactome Reveals Ribosome Functional Diversity and Heterogeneity. *Cell* **2017**, *169*, 1051–1065.e18. [CrossRef]
21. Hall, E.R.; Cottam, G.L. Isozymes of pyruvate kinase in vertebrates: Their physical, chemical, kinetic and immunological properties. *Int. J. Biochem.* **1978**, *9*, 785–793. [CrossRef] [PubMed]
22. Scopes, R.K.; Stoter, A. Purification of all glycolytic enzymes from one muscle extract. *Methods Enzymol.* **1982**, *90 Pt E*, 479–490.
23. Casson, R.J.; Wood, J.P.M.; Han, G.; Kittipassorn, T.; Peet, D.J.; Chidlow, G. M type pyruvate kinase isoforms and lactate dehydrogenase A in the mammalian retina: Metabolic implications. *Investig. Ophthalmol. Vis. Sci.* **2016**, *57*, 66–80. [CrossRef] [PubMed]
24. Cheng, S.-Y.; Leonard, J.L.; Davis, P.J. Molecular Aspects of Thyroid Hormone Actions. *Endocr. Rev.* **2010**, *31*, 139–170. [CrossRef] [PubMed]
25. Kejiou, N.S.; Ilan, L.; Aigner, S.; Luo, E.; Rabano, I.; Rajakulendran, N.; Najafabadi, H.S.; Angers, S.; Yeo, G.W.; Palazzo, A.F. Pyruvate Kinase M Links Glucose Availability to Protein Synthesis. *bioRxiv* **2019**. [CrossRef]
26. Li, X.; Kim, W.; Arif, M.; Gao, C.; Hober, A.; Kotol, D.; Strandberg, L.; Forsström, B.; Sivertsson, Å.; Oksvold, P.; et al. Discovery of Functional Alternatively Spliced *PKM* Transcripts in Human Cancers. *Cancers* **2021**, *13*, 348. [CrossRef]
27. Buneeva, O.; Gnedenko, O.; Zgoda, V.; Kopylov, A.; Glover, V.; Ivanov, A.; Medvedev, A.; Archakov, A. Isatin binding proteins of rat and mouse brain: Proteomic identification and optical biosensor validation. *Proteomics* **2010**, *10*, 23–37. [CrossRef] [PubMed]
28. Chen, M.; Zhang, J.; Manley, J.L. Turning on a Fuel Switch of Cancer: hnRNP Proteins Regulate Alternative Splicing of Pyruvate Kinase mRNA hnRNP Proteins Regulate PKM Splicing. *Cancer Res.* **2010**, *70*, 8977–8980. [CrossRef]
29. Lv, L.; Xu, Y.-P.; Zhao, D.; Li, F.; Wang, W.; Sasaki, N.; Jiang, Y.; Zhou, X.; Li, T.-T.; Guan, K.-L.; et al. Mitogenic and oncogenic stimulation of K433 acetylation promotes PKM2 protein kinase activity and nuclear localization. *Mol. Cell.* **2013**, *52*, 340–352. [CrossRef]
30. Yang, W.; Zheng, Y.; Xia, Y.; Ji, H.; Chen, X.; Guo, F.; Lyssiotis, C.A.; Aldape, K.; Cantley, L.C.; Lu, Z. ERK1/2-dependent phosphorylation and nuclear translocation of PKM2 promotes the Warburg effect. *Nat. Cell. Biol.* **2012**, *14*, 1295–1304. [CrossRef]
31. Yang, W.; Xia, Y.; Hawke, D.; Li, X.; Liang, J.; Xing, D.; Aldape, K.; Humter, T.; Alfred Yung, W.K.; Lu, Z. PKM2 phosphorylates histone H3 and promotes gene transcription and tumorigenesis. *Cell* **2012**, *150*, 685–696. [CrossRef] [PubMed]
32. Yang, W.; Xia, Y.; Ji, H.; Zheng, Y.; Liang, J.; Huang, W.; Gao, X.; Aldape, K.; Lu, Z. Nuclear PKM2 regulates beta-catenin transactivation upon EGFR activation. *Nature* **2011**, *480*, 118–122. [CrossRef] [PubMed]
33. Buneeva, O.A.; Kopylov, A.T.; Medvedev, A.E. Proteasome Interactome and Its Role in the Mechanisms of Brain Plasticity. *Biochemistry* **2023**, *88*, 319–336. [CrossRef]
34. Cela, I.; Di Matteo, A.; Federici, L. Nucleophosmin in Its Interaction with Ligands. *Int. J. Mol. Sci.* **2020**, *21*, 4885. [CrossRef] [PubMed]
35. Maggi, L.B.; Kuchenruether, M.; Dadey, D.Y.A.; Schwope, R.M.; Grisendi, S.; Townsend, R.R.; Pandolfi, P.P.; Weber, J.D. Nucleophosmin Serves as a Rate-Limiting Nuclear Export Chaperone for the Mammalian Ribosome. *Mol. Cell. Biol.* **2008**, *28*, 7050–7065. [CrossRef]
36. Murano, K.; Okuwaki, M.; Hisaoka, M.; Nagata, K. Transcription Regulation of the rRNA Gene by a Multifunctional Nucleolar Protein, B23/Nucleophosmin, through Its Histone Chaperone Activity. *Mol. Cell. Biol.* **2008**, *28*, 3114–3126. [CrossRef]
37. Okuda, M. The role of nucleophosmin in centrosome duplication. *Oncogene* **2002**, *21*, 6170–6174. [CrossRef]
38. Ziv, O.; Zeisel, A.; Mirlas-Neisberg, N.; Swain, U.; Nevo, R.; Ben-Chetrit, N.; Martelli, M.P.; Rossi, R.; Schiesser, S.; Canman, C.E.; et al. Identification of novel DNA-damage tolerance genes reveals regulation of translesion DNA synthesis by nucleophosmin. *Nat. Commun.* **2014**, *5*, 5437. [CrossRef]
39. Scott, D.D.; Oeffinger, M. Nucleolin and nucleophosmin: Nucleolar proteins with multiple functions in DNA repair. *Biochem. Cell Biol.* **2016**, *94*, 419–432. [CrossRef]
40. Szebeni, A.; Olson, M.O.J. Nucleolar protein B23 has molecular chaperone activities. *Protein Sci.* **1999**, *8*, 905–912. [CrossRef]
41. Okuwaki, M.; Matsumoto, K.; Tsujimoto, M.; Nagata, K. Function of nucleophosmin/B23, a nucleolar acidic protein, as a histone chaperone. *FEBS Lett.* **2001**, *506*, 272–276. [CrossRef] [PubMed]
42. Kim, S.H.; Wi, J.H.; Gwak, H.; Yang, E.G.; Kim, S.Y. Single-Cell FISH Analysis Reveals Distinct Shifts in PKM Isoform Populations during Drug Resistance Acquisition. *Biomolecules* **2022**, *12*, 1082. [CrossRef] [PubMed]

3. Buneeva, O.A.; Kopylov, A.T.; Gnedenko, O.V.; Medvedeva, M.V.; Kapitsa, I.G.; Ivanova, E.A.; Ivanov, A.S.; Medvedev, A.E. Changes in the mitochondrial subproteome of mouse brain Rpn13-binding proteins induced by the neurotoxin MPTP and the neuroprotector isatin. *Biomed. Khim.* **2021**, *67*, 51–65. [CrossRef] [PubMed]
4. Medvedev, A.; Buneeva, O.; Kopylov, A.; Gnedenko, O.; Ivanov, A.; Zgoda, V.; Makarov, A.A. Amyloid-binding proteins: Affinity-based separation, proteomic identification, and optical biosensor validation. *Methods Mol. Biol.* **2015**, *1295*, 465–477. [PubMed]

International Journal of
Molecular Sciences

Article

Deciphering the Kidney Matrisome: Identification and Quantification of Renal Extracellular Matrix Proteins in Healthy Mice

Umut Rende [1,2], Seong Beom Ahn [2], Subash Adhikari [2,3,4], Edward S. X. Moh [5], Carol A. Pollock [6], Sonia Saad [6] and Anna Guller [1,2,*]

1 ARC Centre of Excellence in Nanoscale Biophotonics, The Graduate School of Biomedical Engineering, University of New South Wales, Sydney, NSW 2052, Australia
2 Macquarie Medical School, Macquarie University, Macquarie Park, NSW 2109, Australia
3 Advanced Technology and Biology Division, The Walter and Eliza Hall Institute of Medical Research, Melbourne, VIC 3052, Australia
4 Department of Medical Biology, University of Melbourne, Melbourne, VIC 3052, Australia
5 ARC Centre of Excellence for Nanoscale BioPhotonics, Macquarie University, Sydney, NSW 2109, Australia
6 Department of Medicine, Kolling Institute of Medical Research, University of Sydney, St. Leonards, NSW 2065, Australia
* Correspondence: anna.guller@mq.edu.au

Abstract: Precise characterization of a tissue's extracellular matrix (ECM) protein composition (matrisome) is essential for biomedicine. However, ECM protein extraction that requires organ-specific optimization is still a major limiting factor in matrisome studies. In particular, the matrisome of mouse kidneys is still understudied, despite mouse models being crucial for renal research. Here, we comprehensively characterized the matrisome of kidneys in healthy C57BL/6 mice using two ECM extraction methods in combination with liquid chromatography tandem mass spectrometry (LC-MS/MS), protein identification, and label-free quantification (LFQ) using MaxQuant. We identified 113 matrisome proteins, including 22 proteins that have not been previously listed in the Matrisome Database. Depending on the extraction approach, the core matrisome (structural proteins) comprised 45% or 73% of kidney ECM proteins, and was dominated by glycoproteins, followed by collagens and proteoglycans. Among matrisome-associated proteins, ECM regulators had the highest LFQ intensities, followed by ECM-affiliated proteins and secreted factors. The identified kidney ECM proteins were primarily involved in cellular, developmental and metabolic processes, as well as in molecular binding and regulation of catalytic and structural molecules' activity. We also performed in silico comparative analysis of the kidney matrisome composition in humans and mice based on publicly available data. These results contribute to the first reference database for the mouse renal matrisome.

Keywords: extracellular matrix; matrisome; kidneys; proteomics; mass spectrometry; mouse; tissue extraction; protein identification; label-free quantification (LFQ) of proteins

Citation: Rende, U.; Ahn, S.B.; Adhikari, S.; Moh, E.S.X.; Pollock, C.A.; Saad, S.; Guller, A. Deciphering the Kidney Matrisome: Identification and Quantification of Renal Extracellular Matrix Proteins in Healthy Mice. *Int. J. Mol. Sci.* **2023**, *24*, 2827. https://doi.org/10.3390/ijms24032827

Academic Editor: Paolo Iadarola

Received: 20 December 2022
Revised: 23 January 2023
Accepted: 24 January 2023
Published: 1 February 2023

1. Introduction

The extracellular matrix (ECM) is a complex macromolecular network that surrounds the cells of all tissues and organs [1]. The ECM is a product of the cells; thus, it is organ- and tissue-specific. By providing adhesion and anchorage to the cells [2], the ECM ensures tissue integrity. The composition and architecture of the ECM change with modifications of cellular phenotypes. In turn, the differentiation and biological activity of the cells are reciprocally controlled via signaling from the ECM [1,3,4]. Proteins of the ECM (collectively termed the "matrisome" [5]) play an essential role in tissue and organ development [6] and cellular metabolism [7]. According to the classification proposed by Naba et al. [5], the matrisome consists of the "core matrisome" and "matrisome-associated proteins". The core matrisome is formed by structural ECM components (collagens, ECM glycoproteins and proteoglycans) [8,9]. The term "matrisome-associated proteins" covers the proteins which are found in the ECM but can not be classified as core matrisome components. These proteins are further categorized as (i) ECM-affiliated proteins, which have a similar architecture to ECM proteins and/or are known to be associated with ECM proteins, (ii) ECM regulators, and (iii) secreted factors, which are shown to interact with the ECM [5,10]. Identified matrisome proteins of different human and mouse tissues and tumors are deposited in the "Matrisome Database" (MD) (http://matrisomeproject.mit.edu/, accessed on 23 January 2023).

Pathological changes of the matrisome (e.g., excessive accumulation, destruction or dysfunctionality of certain proteins or protein groups) orchestrate many diseases of diverse etiologies, with particular involvement in inflammation, wound healing, cancers and fibrosis [11–13]. A deep analysis of the matrisome is an invaluable instrument for understanding tissues' and organs' functions, as well as the mechanisms and biomarkers of various diseases.

In the last decade, mass spectrometry (MS) has become the leading analytical approach in matrisome studies. It is intensively used for the discovery of biomarkers and assessment of the ECM [14]. Rapid technological developments in proteomics, such as enhanced sample preparation protocols, database searching, and bioinformatics analysis now allow unbiased protein quantification, identification and characterization, including the discovery of post-translational modifications (PTMs) [15]. However, despite these impressive advancements, standard protein extraction techniques, which use whole tissue lysates, need to be better adapted to matrisome studies. The continuing challenges in ECM proteomics are the enrichment of matrisome proteins in the samples by separating low-abundance ECM proteins from intracellular proteins, and solubilizing heavily cross-linked [9] extracted proteins. Additionally, identification and quantification of the matrisome proteins in the presence of abundant PTMs is a complex task [16].

To tackle the difficulty of matrisome enrichment, several variations of decellularization methodologies that gradually reduce the amount of cellular components in the sample while preserving the ECM proteins have been proposed. The most widely used extraction method applicable for proteomic studies of the matrisome [5,17–20] employs a commercially available kit from Millipore for tissue protein fractionation by sequential incubations of the sample in buffers of different pH, salt and detergent concentrations. This procedure results in the biochemical separation and removal of the proteins of the cells' cytosolic, nuclear, membrane and cytoskeletal compartments. It then allows the enrichment of ECM proteins in the residual insoluble product. This method is further referred to as "Compartmental Matrix Enrichment" (CME). Another method is termed "Sequential Matrix Enrichment" (SME). It uses guanidine hydrochloride (Gu-HCl) to further solubilize insoluble matrisome proteins after a decellularization step [20–23]. This extraction method is performed with an ionic (NaCl) buffer to additionally extract loosely bound ECM proteins (enzymes, secreted factors, ECM-associated and newly deposited proteins), as well as a low-concentration detergent (sodium dodecyl sulphate, SDS) with a shorter incubation time to eliminate intracellular proteins, and then uses a Gu-HCl buffer to enhance the obtaining of heavily cross-linked ECM proteins.

In the current study, we applied these two ECM extraction methods (Figure 1a) to explore the matrisome of kidneys in healthy adult mice with an overall goal of creating a reference database for future renal research. This work was particularly motivated by the fact that, now, information on the matrisome composition of mouse kidneys is very limited and incomplete. While mouse models are crucial for studies of renal morphogenesis and diseases [24], chronic kidney disease (CKD) [25], as well as for drug and biomarker discovery [26,27], only two recent studies are currently available on the identification of ECM proteins in the insoluble tissue fractions of mouse kidney extracts [20,28]. Importantly, the methodology applied in these studies did not allow the identification of loosely bound

ECM proteins, which suggests incomplete characterization of the matrisome. Moreover, quantification of ECM proteins in mouse kidneys has yet to be achieved.

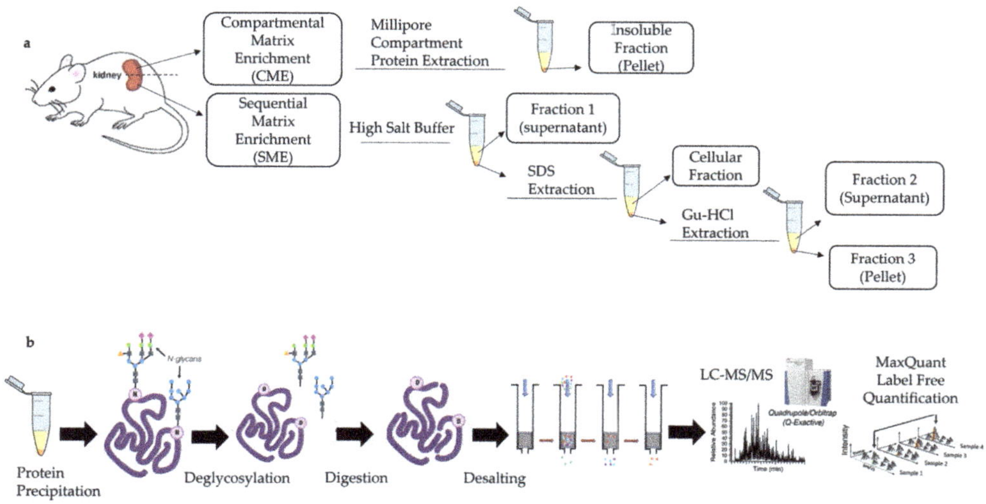

Figure 1. Schematic illustration of the methods applied in the current study. (**a**) Matrisome protein extraction from healthy mouse kidneys by CME and SME methods. (**b**) Further processing and analysis of proteins after obtaining samples from methods CME and SME. Yellow color in the extraction products indicates the supernatant, and the orange color shows insoluble pellets. Abbreviations: SDS (Sodium Dodecyl Sulfate) and Gu-HCl (Guanidine Hydrochloride).

The matrisome proteins extracted using CME and SME methods were further studied with liquid chromatography using a tandem mass spectrometry (LC-MS/MS) system (Figure 1b). Firstly, the identified proteins were explored with the Matrisome Database (MD) and UniProt databases. To compare the abundance of identified matrisome proteins from the two extraction methods, MaxQuant label-free quantification (LFQ) was employed, and the data was then analyzed using LFQ-Analyst [29].

2. Results

2.1. *Qualitative Analysis of Mouse Kidney Matrisome*

2.1.1. Identification of the Proteins of the Mouse Kidney Matrisome

The raw data results of the proteomic study are available in Table S1 in Supplementary Materials.

Based on unique peptides, a total of 2442 proteins were identified by the CME and SME methods. Collectively, this resulted in the identification of a total of 113 matrisome proteins in healthy mouse kidneys. The summary on the mouse kidney matrisome composition identified in the current study is shown in Figure 2. The detailed and referenced lists of the matrisome proteins identified in the current study by using CME and SME approaches and classified by the MD divisions and categories are presented in Table A1 (collagens), Table A2 (ECM glycoproteins), Table A3 (ECM proteoglycans), Table A4 (ECM regulators), Table A5 (ECM-affiliated proteins), and Table A6 (ECM secreted factors), in Appendix A.

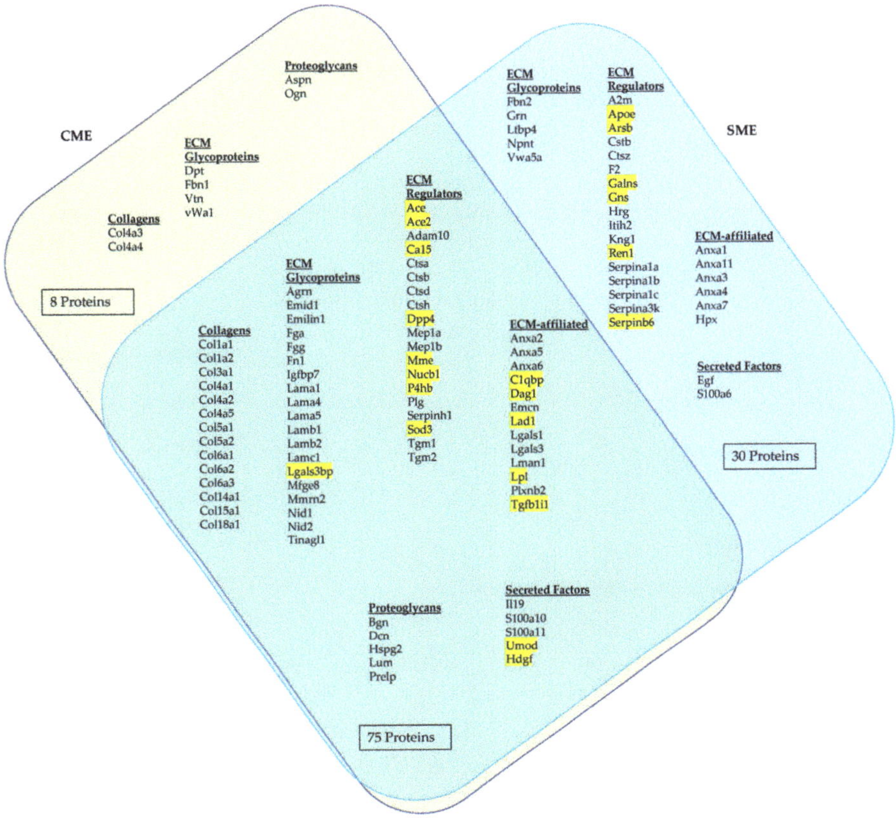

Figure 2. Identified matrisome proteins with unique peptides obtained from the samples processed via CME and SME methods. Venn diagram illustrates the common and unique matrisome proteins identified using each method. The proteins highlighted proteins in yellow have not been previously listed in the MD.

As follows from Figure 2, among the proteins identified by combination of the two ECM extraction and enrichment methods, 51 (45%) were classified as core matrisome proteins. By frequency of observation, the core matrisome of healthy mouse kidneys was dominated by ECM glycoproteins (28 identified proteins), followed by sixteen collagens and seven ECM proteoglycans. The remaining 62 identified proteins (55%) were regarded as matrisome-associated proteins. Most of the identified matrisome-associated proteins belonged to the category of ECM regulators (36 proteins). This division of the matrisome also included nineteen ECM-affiliated proteins and seven secreted factors.

Notably, 22 proteins among the 113 identified ECM proteins have not been previously classified in the MD (see Figure 2, the yellow highlights; refer to Table A7, Appendix A for the more detailed descriptions). These 22 proteins were attributed as matrisome components based on their extracellular location, functions, and, also, on their interactions with extracellular proteins—as reported in the literature and in the UniProt database. Then, they were classified into the Matrisome categories of ECM glycoproteins, ECM regulators, ECM-affiliated proteins, and secreted factors. The majority of the newly classified mouse kidney matrisome proteins belonged to the division of the matrisome-associated proteins, including 14 ECM regulators, five ECM-affiliated proteins, and two secreted factors, and only one was classified as a core matrisome component (an ECM glycoprotein: Galectin-3-binding protein). Notably, six of these matrisome proteins identified in mouse kidneys for the first time were revealed only by SME.

2.1.2. Gene Ontology Analysis of Identified Mouse Matrisome Proteins

Gene Ontology (GO)-enriched terms analysis based on the top five biological processes revealed that the majority of the identified mouse kidney matrisome proteins primarily belonged to the classes of "cellular process" (GO:0009987), "developmental process" (GO:0032502), "response to stimulus" (GO:0048583) and "metabolic processes (GO:0019538) and other processes, including multicellular organismal processes, biological adhesion localization, immune system processes, locomotion, interspecies biological processes, reproductive processes, behavior, biomineral tissue development, growth, viral processes, trans-synaptic signaling, removal of superoxide radicals and estrous cycle (Figure 3a). These attributions indicate a key involvement of the matrisome in the regulation of the cellular physiological activity in the kidneys.

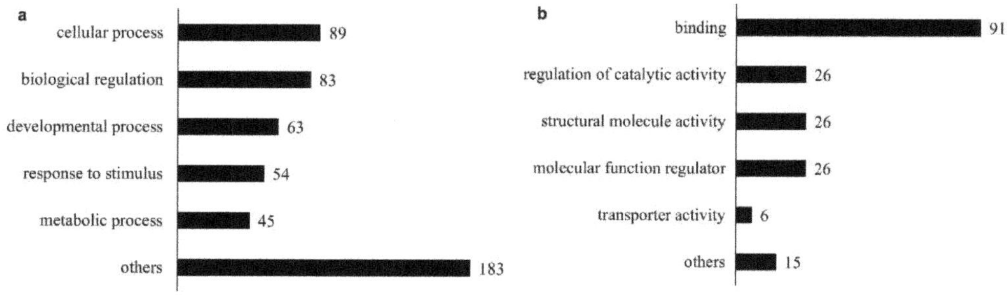

Figure 3. Mapping of the identified healthy mouse kidney matrisome proteins against GO terms using Uniprot database: (**a**) GO terms for the Biological Process and (**b**) GO terms for the Molecular Function.

The top five GO-enriched terms for molecular function of identified proteins revealed the "binding (GO:0005515)" proteins, which interact with other molecules via a specific site, as a leading category. This was followed by "regulation of catalytic activity (GO:0050790)", "structural molecule activity" (GO:0005198), "molecular function regulator" (GO:0098772), "transporter activity" (GO:0005215) and others including signaling receptor activity, transcription coregulators, scavenger receptor activity, antioxidant activity, ATPase activity, translation activator activity and protein–macromolecule adaptors (Figure 3b).

2.1.3. Comparative Efficiency of ECM Extraction Methods in Mouse Kidneys Matrisome Identification

Table A8 in Appendix A shows the outcomes of the CME and SME methods in detection of matrisome composition in terms of the number of proteins belonging to various categories of the matrisome. As follows from Figure 2 and detailed in Table A8, the SME method allowed identification of more matrisome proteins than CME (105 vs. 83, respectively).

There were no statistically significant differences in identification of proteins belonging to matrisome divisions and categories compared to the total number of proteins revealed by each method individually or both methods in total. However, in comparing the 113 mouse kidney matrisome proteins identified in total in the current study to the combination of two ECM extraction methods, SME contributed to the observed outcomes more than CME (105/113, or 93% vs. 83/113, or 73%, with the $CI_{95\%}$ [87; 99]% vs. [64; 82]%, respectively). The relative efficiency of CME and SME in the detection of core matrisome proteins was statistically similar (41%, $CI_{95\%}$ [31; 51]% vs. 38% $CI_{95\%}$ [28; 48]%, respectively). At the same time, SME identified more matrisome-associated proteins than CME in the totally detected matrisome (62/113, or 55%, $CI_{95\%}$ [45; 65]% vs. 37/113, or 33%, $CI_{95\%}$ [23; 42]%, respectively). Both of the ECM enrichment methods used in this study demonstrated similar efficiency in identification of the proteins belonging to the subordinal categories of the matrisome.

2.1.4. Comparative Analysis of the Composition of the Mouse and Human Kidney Matrisome

Next, we compared our results regarding matrisome protein identification with the published data on the healthy mouse kidney ECM proteins identified in recent studies [20,28,30] and the published data on the matrisome proteins from healthy human kidneys [31,32]. It is important to note that across the studies included in this comparison, different protein extraction and analysis methods were used (Figure A1 in Appendix A). There was also a variation in the sample size, age and gender of mice and humans (Table A9 in Appendix A).

By combining our data with other published data obtained in mouse kidney matrisome studies [20,28,30], in total 229 different proteins have been identified. Prior studies [31,32] together identified 178 matrisome proteins in adult human kidneys. The comparison of mouse and human data revealed that 134 matrisome proteins are shared between mouse and human matrisomes, while 95 of these proteins were found only in mouse kidneys and 44 were identified only in human kidneys (Tables A10–A15 in Appendix A).

A summary of the contribution of the identified proteins of different categories to mouse and human matrisomes based on the data reported in the current study and the cited references is shown in Figure A2 in Appendix A. This figure indicates that the qualitative composition of mouse and human matrisome is similar. The average numbers of identified matrisome proteins in mouse and human kidneys are 126 and 124, respectively. A similar number of proteins were identified in the matrisome categories as well (eighteen and seventeen collagens, forty-one and forty-three ECM glycoproteins, seven and nine proteoglycans, thirty-seven and twenty-nine ECM regulators, seventeen and nineteen ECM-affiliated proteins, and seven and nine secreted factors, respectively).

2.2. Mouse Kidney Matrisome Protein Quantification

The quantitative characterization of the healthy mouse kidney matrisome was performed by analyzing the MaxQuant LFQ protein intensities data. The analysis revealed a total of 87 distinct matrisome proteins that were able to be quantitatively examined (Figure 4).

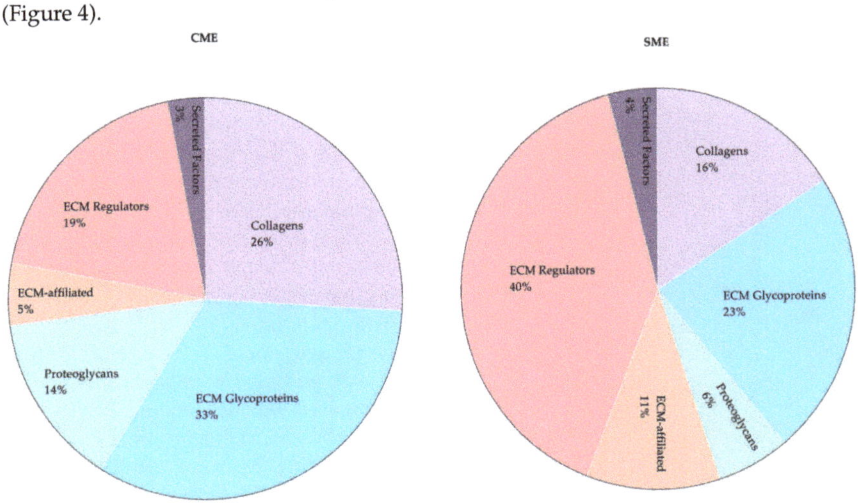

Figure 4. Relative abundance of matrisome proteins in healthy mouse kidneys according to MaxQuant LFQ protein intensities. The results are presented separately for the products of CME and SME methods.

From the total LFQ intensities, the core matrisome of the healthy mouse kidneys is predominantly composed of ECM glycoproteins (45% or 51%, as quantified following the CME and SME sample preparation methods, respectively). Collagens comprise 36% of the core mouse kidney matrisome according to both extraction methods. Proteoglycans form

19% or 13% of the core matrisome, according to the quantifications of the samples prepared by CME and SME, respectively. The matrisome-associated proteins in mouse kidneys are represented mostly by the ECM regulators, followed by the ECM-affiliated proteins and secreted factors.

The detailed presentation of the results of quantitative analysis of the mouse kidney matrisome is visualized in Figure 5. As can be seen from this figure, using both ECM enrichment methods, collagen type IV (Col4a1) is defined as the most abundant collagen, with collagen type VI (Col6a1, Col6a2, Col6a3) and collagen type XVIII (Col18a1) also being expressed at above the average level of LFQ intensities observed for collagens. The most highly expressed ECM glycoproteins quantified in the products of both CME and SME include nidogen (Nid1, Nid 2), agrin (Agrn), laminin (Lama5, Lamb2, Lamc1), and fibronectin (Fn1). The top-expressed proteoglycan is perlecan (Hspg2). Among ECM regulators, the top-expressed proteins include meprin A (Mep1a, Mep1b), protein-glutamine gamma-glutamyltransferase 2 (Tgm2), serpin H1 (Serpinh1), and Dipeptidyl peptidase 4 (Dpp4). The category of ECM-affiliated proteins is dominated by the complement component 1 Q subcomponent-binding protein (C1qbp), protein ERGIC-53 (Lman1), and annexin A2 (Anxa2; plus, Anxa5 and Anxa6 that were only overexpressed in the product of SME). Overexpressed matrisome-associated secreted factors included uromodulin (Umod) and hepatoma-derived growth factor (Hdgf).

Some of the quantifiable proteins were more abundant either in CME or SME, while other proteins were detected in similar intensities in the products of both extraction methods. In general, CME allowed quantification of more intensities of core matrisome proteins, but fewer ECM-associated proteins, compared to SME. On the other hand, SME processing detected a higher abundance of ECM-affiliated proteins, ECM regulators and secreted factors than did CME. The complete list of protein FC data between SME fractions and CME is presented in Table S2 in the Supplementary Information.

The comparative analysis of the LFQ quantification of ECM proteins extracted by CME and SME methods also revealed the following. Fifty one matrisome proteins (ten collagens, sixteen ECM glycoproteins, three proteoglycans, eleven ECM regulators, seven ECM-affiliated and four secreted factors) were shared between the products of both extraction methods. CME additionally allowed quantification of six matrisome proteins (two collagens, one ECM glycoprotein, two proteoglycans and one ECM regulator). SME added another 30 quantifiable matrisome proteins to the list (two ECM glycoprotein, fifteen ECM regulator and twelve ECM-affiliated and one secreted Factor). The comparative analysis of the quantification efficiency of the two studied protein extraction methods shows that there were no significant differences in abundances of shared matrisome proteins. However, higher FCs were observed in the majority of the proteins in CME, with the exception of one collagen (Col14a1), four ECM regulators (Ace, P4hb, Plg and Nucb1), four ECM-affiliated (Anxa2, Anxa5, Anxa6 and Lgals3) and two secreted factors (S100a10 and S100a11) (Figure A3 in Appendix A).

Furthermore, the comparison of CME with SME fractions showed that Ace, Mep1b and Ctsa (ECM regulators, and matrisome-associated proteins) were significantly higher in SME-F1 fraction, compared to CME, whereas CME obtained more abundances of core matrisome proteins including ECM glycoproteins such as Nid1, Nid2, Tinagl1 and Agrn. There were significant FC differences for some matrisome proteins in the comparison between SME-F2 fraction and CME. Serpina1a and A2m are matrisome-associated ECM regulators and were 49.8 (p-value < 0.0001) and 20.1 (p-value < 0.001) folds higher in SME-F2, respectively. Nid1 and Lamb1 are core matrisome ECM glycoproteins and were 64.4 (p-value < 0.001) and 33.3 (p-value < 0.001) folds higher in CME, respectively, compared to SME-F2. These observations indicate that specific extraction methods may be required for the enrichment of specific matrisome proteins. The SME-F3 and CME comparison also showed that the majority of highly abundant proteins are matrisome-associated (e.g., ECM-affiliated proteins and ECM regulators), whereas core matrisome proteins (e.g., ECM glycoproteins and collagens) were higher in the CME.

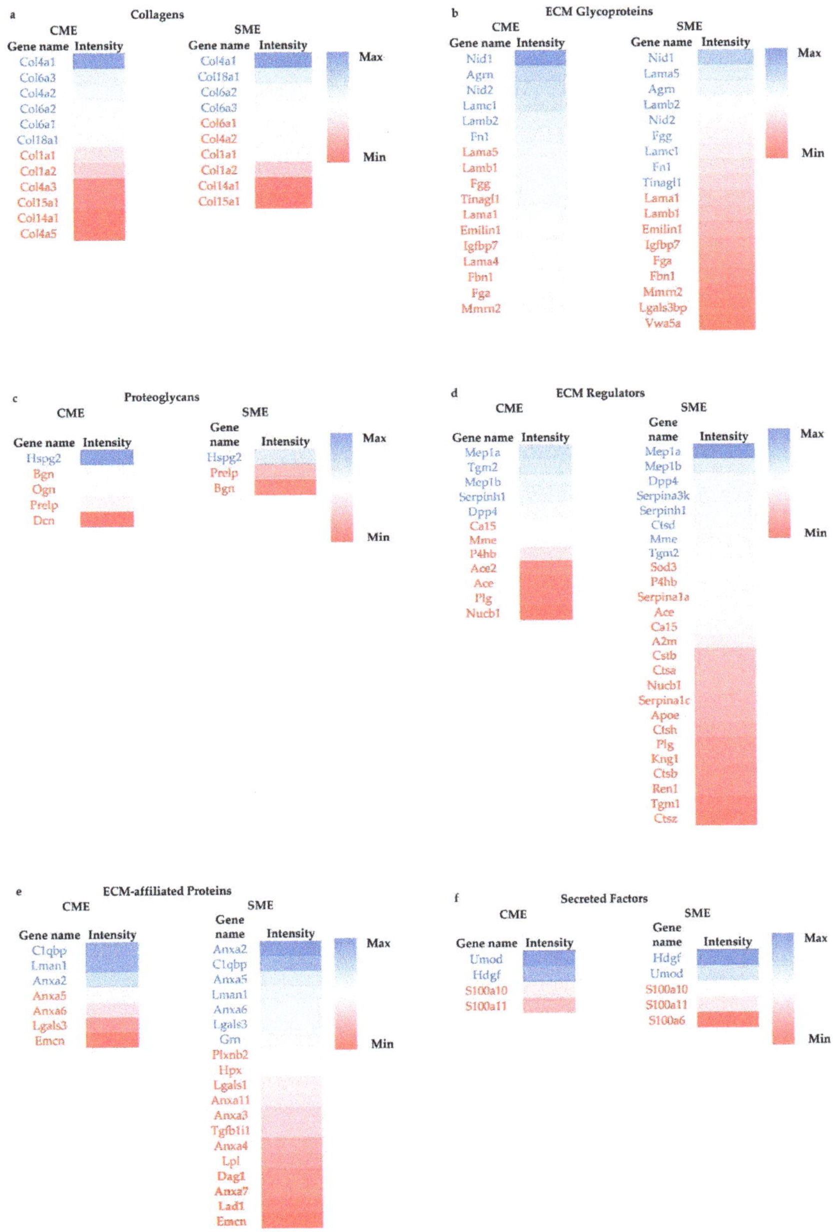

Figure 5. Heatmap of abundance of quantifiable matrisome proteins in the products of the CME and SME methods of sample preparation. Color coding in the heatmap depicts the variation between the maximum (coded in blue tones) to minimum (coded in red tones) observed LFQ intensity for each matrisome category and protein extraction method. The protein names shown by blue and red fonts are expressed above and below the average LFQ values in each matrisome category, respectively. The color coding scales (from blue for maximum LFQ intensity to red for minimum intensity per category) are provided on side of heatmaps (**a**–**f**).

In addition, we compared the composition of the quantifiable matrisome proteins in SME fractions (Figure A4 in Appendix A). This analysis showed that 48, 42 and 56 matrisome proteins were quantitatively detected in SME-F1, SME-F2, and SME-F3, respectively. Among core matrisome proteins, collagens and glycoproteins were mostly detected (LFQ intensities) in SME-F1 and SME-F3, while proteoglycans were predominantly detected in SME-F3. In parallel, different ECM-associated proteins were detected in different fractions. However, SME-F2 allowed for the detection of more ECM regulators, ECM-affiliated proteins and secreted factors alone compared to other fractions.

3. Discussion

The current understanding of matrisome composition and how it is regulated during pathophysiological processes remains very limited. One of the major obstacles is the need for optimization of ECM protein extraction and characterization methods. Many previous studies which examined kidney tissues of different species mainly used the decellularization approach that was developed for tissue engineering applications and involved removal of cellular proteins followed by matrisome protein examination in the decellularized scaffolds [33–35]. However, long detergent incubations in tissue engineering-focused decellularization methods may cause degradation and elimination of some matrisome proteins [36].

In the current study, we aimed to comprehensively characterize the mouse renal matrisome using proteomic technologies. In order to overcome the limitations imposed by the tissue engineering decellularization technologies, we first examined two ECM extraction and enrichment methods (CME and SME) in healthy mouse kidney tissues (see Figure 1).

The first method, CME, relied on using a commercially available Millipore Compartment Protein fractionation kit. This method biochemically separates subcellular proteins and enriches matrisome proteins in the insoluble pellet at the final step of the extraction series. The method is well documented in previous studies, and many ECM proteins have been identified by using this extraction method. For example, Naba et al. [5] identified 100 matrisome proteins in mouse lungs and colon, Schiller et al. [18] identified 435 matrisome proteins in healthy mouse lungs and Gocheva et al. [19] identified 113 matrisome proteins in human lungs. Just recently, this method has been applied for the first time by Lipp et al. [28] to identify 79 matrisome proteins of mouse kidneys. In our study using Millipore Compartment Fractionation (here termed the CME method), we detected 83 matrisome proteins of which 16 proteins were not previously listed in the MD. Although our study and Lipp et al. [28] used the same matrisome enrichment method, the protein precipitation, deglycosylation and LC-MS/MS set-up were different. This could explain the higher total number of matrisome proteins identified in our work, compared to the reference [28]. It is important to note that the Millipore Compartment Fractionation only analyses one fraction, while the loosely bound and soluble matrisome proteins potentially remain in the "cellular" fractions that are not used in matrisome studies.

In an attempt to improve the ECM protein isolation on the second half of the same kidney sample, we applied another matrisome enrichment method, a sequential matrix extraction (SME). This method allowed the extraction of loosely bound or soluble matrisome proteins using a high-salt buffer before removal of cellular components by SDS and enriched the matrisome fractions by solubilizing the insoluble pellet with Gu-HCl [22]. Using a method similar to our SME approach, Massey et al. [23] identified 79 matrisome proteins from all three fractions of mouse liver. This extraction technique is potentially more comprehensive for identifying matrisome proteins, but it has not yet been optimized for kidney tissue. Our study, for the first time, used this sequential approach (SME) to identify matrisome proteins in mouse kidneys. We have identified 105 matrisome proteins, of which 22 proteins were yet to be listed in the currently available MD (http://matrisomeproject.mit.edu/, accessed on 23 January 2023) (see Figure 2). Compared to the results obtained by Lipp et al. [28], our study could distinctly detect 57 matrisome proteins. These 57 proteins included mostly the matrisome division of ECM-associated proteins. We can infer from these findings that SME fractions allow for more efficient detection of ECM-associated proteins compared to CME.

It should be noted that the location of most of the 22 proteins have not previously been exclusively attributed to ECM compartments, but to different "Gene Ontology-Cellular Components" or/and "Subcellular Locations", including ECM space/regions in UniProt. For example, Nucleobindin-1 (Nucb1) was previously shown to be localized in the Golgi apparatus, cytoplasm, or extracellular regions which were inferred from genome-based computational annotations (https://www.uniprot.org/help/gene_ontology). However, it was additionally manually asserted in UniProt as "Secreted" and recently it has been shown that Nucb1 is secreted with metalloproteinase-2 (MMP-2) and regulates ECM remodeling [37]. Hence, we described this protein as an ECM-regulator. Another set of examples includes enzymes with transmembrane domains such as Ace, Ace2, Dpp-4 and Mme. These enzymes are localized at the cell surface and have extracellular, cytoplasmic and transmembrane domains. However, it has been shown that they may also be present in soluble forms after cleavage and play a role in ECM remodeling ([37–42]). Similar to the examples above, we referred to the literature in order to confirm the identified proteins as components of the matrisome (see Table A7 in Appendix A).

We then compared mouse matrisome proteins identified by studies of McCabe et al. [20], Lipp et al. [28] and Liu et al. [30], and human matrisome proteins identified by the studies of Louzao-Martinez et al. [31] and Randles et al. [32]. These studies also differed in matrisome protein enrichment, protein precipitation, deglycosylation, digestion and LC-MS/MS methods (see Figure A1, Table A9 in Appendix A). The comparison was undertaken to identify mouse and human species-specificity of kidney matrisome proteins. The comparison could identify 95 matrisome proteins only identified in mice and 44 only identified in human kidneys (see Tables A10–A15 in Appendix A). Interestingly, five of the matrisome proteins shown in this study—namely Lgals3bp (glycoprotein), Mfge8 (glycoprotein) Emcn (ECM-affiliated), Apoe (ECM regulator) and Sod3 (ECM regulator)—were previously identified only in human kidneys but not in mouse kidneys [32]. However, among mouse kidney studies, only our work identified those proteins in mouse kidneys. Hence, our study shows the necessity of using different approaches to identify matrisome proteins and thus develop a robust matrisome protein list.

In addition to matrisome protein identification, our study for the first time allowed quantification of the ECM proteins extracted from heathy mouse kidneys. We performed a discovery proteomics study known as shotgun proteomics, which uses bottom-up approaches based on unbiased analysis. To compare the quantities of proteins obtained via each extraction method, we applied label-free-based quantification (LFQ) data generated with MaxQuant analysis [29]. This is a widely used method in which the quantitation can be performed either based on chromatographic ion intensities or based on spectral counting of identified proteins [43]. It provides high throughput and does not have the limitations of label-based quantifications such as high complexity of sample preparation, requirement of high concentrations of samples and incomplete labelling [15,43]. In our study using LFQ, we successfully quantified 57 matrisome proteins via CME and 81 proteins via SME (see Figure 5). Some of the quantifiable proteins were more abundant either in CME or SME. This study demonstrated that CME could detect greater intensities of core matrisome proteins, but fewer ECM-associated proteins, compared to SME (see Figure 4). To show the benefit of each extraction method, which can be used as a guideline to study different matrisome proteins in different kidney diseases, we compared SME fractions with CME, separately (see Table S2 in Supplementary Information). For example, IgA nephropathy (IgAN), which is one of the most prevalent chronic glomerular diseases, had significantly more matrisome proteins (Col4a1, Lamb1, Hspg2, Emilin, Fgg and Fbln) in diseased patients compared to healthy patients [34]. These proteins could be better detected using CME, while Col4a1 and Col15a1, which can be detected by both methods and were also elevated in IgAN, could be better detected by CME or in the SME-F3 fraction. In renal fibrosis, the hallmark of CKD, it is known that Col 1, 2, 3, 5, 6, 7 and 15, Fn1, Dcn and Bgn accumulate in the ECM [44]. We detected those matrisome proteins using both extraction methods, although some of them had higher abundances in CME product. Collectively,

CME was powerful at extracting several matrisome proteins in higher abundances, while SME allowed for the quantification of more matrisome proteins.

Subsequently, we quantitatively analyzed matrisome proteins in SME fractions. It has previously been shown in mouse lung [21] and liver [23] that sequential extraction methods, similar to that of the SME used in our study, can extract loosely bound matrisome proteins by using a NaCl buffer which displaces polyanionic interactions and further solubilizes the insoluble proteins using Gu-HCl [21,23]. The SME-F1 fraction allowed detection of mostly secreted, basement membrane proteins and some interstitial matrix proteins (Col1 and Col6). SME-F3 distinctly revealed proteoglycans which could not be found in other fractions, but also detected interstitial matrix and basement membrane proteins. Conversely, the SME-F2 fraction allowed for detection of only a few core matrisome proteins which are located in the interstitial matrix, and also secreted and basement membrane proteins. Although the fractions of the extracted samples revealed matrisome proteins from different ECM locations, each fraction had advantages in extracting different matrisome proteins.

Although many matrisome proteins identified in this study have also been previously identified in other studies, our study was able to detect additional matrisome proteins. The most plausible explanation is the use of different extraction methods. For example, LOX enzymes, which are involved in collagen cross-linking, could not be detected in our extraction methods, but were detected in another mouse kidney proteomic study where different extraction methods were used with a combination of a modified deglycosylation/protein digestion process [20,28]. In addition, the removal of glycosaminoglycans (GAGs) by digestion enzymes such as chondroitinase and heparinase has been suggested to improve peptide identification [22,45,46]. However, McCabe et al. [20] reported that using GAG-digesting enzymes only improved proteoglycan identification, but not for other matrisome proteins. It is important to note that this study was performed on healthy kidney tissue, while it is known that GAG deposition in the ECM is increased during disease, e.g., fibrosis [47,48]. Hence, the opportunities for the improvement of analysis by GAG-digesting enzymes are still not completely clear, and our study is limited as we used only PNGaseF to remove N-glycans. We would suggest performing a pilot study using GAG-digesting enzymes to analyze whether it improves matrisome protein identification before applying it to all samples. Finally, the native biological variation in the proteome needs to be considered. In particular, the role of the age of the animals has been demonstrated in several tissue-specific proteomic studies in mice [49,50].

4. Materials and Methods

4.1. Animal Preparation and Kidney Tissue Collection

Mouse kidney tissues were obtained via the post-mortem animal tissues sharing program encouraged and approved by the UNSW Animal Care and Ethics Committee (ACEC). WT C57BL/6 mice (female, aged 4–10 weeks, Australian Bioresources, Moss Vale NSW, Australia) were housed in a stable environment at 21 ± 2 °C with a 12 h/12 h light-dark cycle. On the day of experimentation, the animals were anesthetized with 4% vaporized isoflurane delivered into an induction chamber and euthanized by cervical dislocation. First, the retinas were collected for the main experiment approved by the UNSW ACEC. After that, the animal bodies were placed on ice and underwent further dissection and kidney extirpation. The obtained kidneys were kept on ice, and one of the kidneys per animal was quickly transversely cut. Then, each half of a kidney (~50 mg) per animal was designated for CME and SME protein extraction (see Figure 1a).

4.2. ECM Proteins Extraction

The dissected kidney tissue samples were processed using the CME and SME methods of protein extraction as described below. For each extraction method, three biological replicates were used.

4.2.1. Compartmental Matrix Enrichment (CME) Method

The dissected kidney halves were grinded using a glass tissue homogenizer, on ice. The homogenized tissue was processed using the CME method in the following way. The Millipore Compartment Protein Extraction Kit (Merck-Millipore, Bayswater VIC 3153, Australia, Cat. #2145) was used to deplete cytosolic, nuclear, membrane and cytoskeletal proteins and to enrich ECM proteins as described in [17]. The obtained ECM enriched pellets were washed with 1× PBS (1.7 mM KH_2PO_4, 5 mM Na_2HPO_4, 150 mM NaCl, 25 mM EDTA, pH 7.4) containing 1:100 (*v:v*) of protease inhibitors (PI) (Halt Protease Inhibitor Cocktail, #78429, Thermo Scientific, North Ryde NSW 2113, Australia) three times to remove detergents, and stored at −20 °C.

4.2.2. Sequential Matrix Enrichment (SME) Method

When using the SME method, enrichment of ECM proteins was performed according to the protocol described elsewhere [22]. Briefly, half-kidney samples were diced into 2–3 mm pieces and washed five times with ice-cold 1x PBS containing 1:100 (*v:v*) of PI (Halt Protease Inhibitor Cocktail, #78429, Thermo Scientific) to minimize blood contamination. To extract ECM-associated, loosely bound ECM proteins and newly synthesized ECM proteins, the washed samples were incubated in NaCl buffer (0.5 M NaCl, 10 mM Tris-HCl and 25 mM EDTA, pH 7.5 and 1:100 (*v:v*) of PI) for 1 h at room temperature (RT) by vortexing at a speed of 65 rpm (Stuart Orbital Shaker). Samples were centrifuged at 16,000× *g* for 10 min at 4 °C and the supernatant was saved as Fraction 1. The pellet was treated with SDS buffer (0.1% SDS, 25 mM EDTA and 1:100 (*v:v*) of PI) by vortexing at RT for 16 h at a speed of 65 rpm and was centrifuged at 16,000× *g* for 10 min at 4 °C to separate intracellular proteins. Then, pellets were treated with GuHCl buffer (4 M guanidine hydrochloride, 50 mM Na acetate, 25 mM EDTA, pH 5.8, and 1:100 (*v:v*) of PI) by vortexing at RT for 72 h at speed 225 rpm to solubilize ECM proteins. Supernatants after centrifugation at 16,000 × *g* for 10 min at 4 °C were saved as Fraction 2 and pellets were saved as Fraction 3 after washing three times with ice-cold 1x PBS containing 1:100 (*v:v*) of PI. All saved fractions were stored at −20 °C.

4.2.3. Protein Precipitation

Pellets obtained from the CME and the three fractions obtained via SME methods (see Figure 1b) were precipitated in EtOH separately to remove detergents/agents. Ten times volume of ice-cold 100% EtOH was added to each fraction and incubated overnight at −20 °C. Protein precipitates were obtained by centrifugation at 16,000× *g* for 10 min at 4 °C and then by drying pellets. Dried pellets were resuspended in the buffer (8 M urea in 100 mM ammonium bicarbonate, pH 8.0) for the following downstream processes.

4.2.4. Deglycosylation and In-Solution Digestion

Deglycosylation and in-solution digestion were performed as described in [17]. Briefly, the obtained pellets for each fraction after protein precipitation were resuspended by adding the appropriate volume (50 µL/5–10 mg dry weight) of 8 M urea and dithiotreitol (DTT) at a final concentration of 10 mM. Samples were incubated with continuous agitation at 150 rpm (Stuart Orbital Shaker) for 2 h at 37 °C.

Alkylation was performed by adding iodoacetamide (IAA) to a final concentration of 25 mM. To complete alkylation, the DTT:IAA ratio was adjusted to 1: 2.5 and incubation was carried out in the dark for 30 min at RT.

Deglycosylation was performed by diluting urea to 2M using 100 mM ammonium bicarbonate, pH 8.0 and adding 2 µL/5–10 mg dry weight (DW) of PNGaseF (Peptide-N-Glycosidase F) (New England Biolabs, #P0704S). Samples were incubated with continuous agitation at 150 rpm (Stuart Horizontal Shaker) for 2 h at 37 °C.

Digestion was continued by diluting samples with 100 mM ammonium bicarbonate pH 8.0 to reduce the concentration of urea in the samples' solution to 1 M. Then, 2 µL/5–10 mg DW of Trypsin/Lys-C (Endoproteinase LysC) Mix (Promega, Cat #V5071) was added to samples and incubated with continuous agitation at 150 rpm overnight a 37 °C. The digestion reaction was inactivated with freshly prepared 8 μL of 50% trifluoro acetic acid (TFA) per 5–10 mg DW of pellet to reach a final concentration of 1% of TFA. The acidified samples were centrifuged at 16,000× *g* for 5 min at RT and the supernatants were saved for desalting steps.

Desalting was performed by using Pierce C18 stage tips (#SP301, Thermo Scientific). Prior to proteomics analysis, desalted peptides were eluted with freshly prepared HPLC-grade water-based solution containing 60% acetonitrile and 0.1% TFA, followed by concentrating in a vacuum concentrator. Samples were then resuspended in freshly prepared 3% acetonitrile and 0.1% TFA water-based solution and analyzed via LC-MS/MS

4.3. LC-MS/MS and Data Analysis

Digested peptides were separated using nanoLC with an Ultimate nanoRSLC UPLC and autosampler system (Dionex, Amsterdam, The Netherlands). Samples (2.5 μL) were concentrated and desalted onto a micro C18 precolumn (300 μm × 5 mm, Dionex) with H_2O:CH_3CN (98:2, 0.1% TFA) at 15 μL/min. After a 4 min wash the pre-column was switched (Valco 10 port UPLC valve, Valco, Houston, TX, USA) into line with a fritless nano column (75 μ × ~20 cm) containing C18AQ media (1.9 μ, 120 Å Dr Maisch, Ammerbuch Entringen, Germany) manufactured according to Gatlin. Peptides were eluted using a linear gradient of H_2O:CH_3CN (98:2, 0.1% formic acid) to H_2O:CH_3CN (64:36, 0.1% formic acid) at 200 nL/min over 30 min. High voltage (2000 V) was applied to low-volume Titanium union (Valco) with the column oven-heated to 45 °C (Sonation, Biberach, Germany) and the tip positioned at ~0.5 cm from the heated capillary (T = 300 °C) of a QExactive Plus (Thermo Electron, Bremen, Germany) mass spectrometer. Positive ions were generated by electrospray and the QExactive operated in data-dependent acquisition mode (DDA).

A survey scan m/z 350–1750 was acquired (resolution = 70,000 at m/z 200, with an accumulation target value of 1,000,000 ions) and lock mass enabled (m/z 445.12003). Up to the 10 most abundant ions (>80,000 counts, underfill ratio 10%) with charge states > +2 and <+7 were sequentially isolated (width m/z 2.5) and fragmented by higher-energy C-trap dissociation (HCD) (normalized collision energy, NCE = 30) with an automatic gain control (AGC) target of 100,000 ions (resolution = 17,500 at m/z 200). M/z ratios selected for MS/MS were dynamically excluded for 30 s.

MS raw files were analyzed by the MaxQuant software [51] (version 2.0.3.0), and peak lists were searched against the mouse UniProt FASTA (reviewed and unreviewed) database (version Aug 2019), and a common contaminants database using the Andromeda search engine [52]. For protein identification and LFQ quantification, fractions of SME were nominated as fractions in MaxQuant [53], and a proprietary Maxquant algorithm was applied to merge and normalize the fractions to give an output as one sample. A fixed modification carbamidomethylation (C) and variable modifications, oxidation (M,P), acetyl (protein N-term), hydroxyproline, and deamidation (N,Q) were used. The false discovery rate was set to 1% for proteins and peptides (minimum length of 7 amino acids) and was determined by searching a reverse database. Enzyme specificity was set as trypsin and lys-C, and a maximum of two missed cleavages were allowed in the database search. Peptide identification was performed with an allowed precursor mass deviation of up to 4.5 ppm after time-dependent mass calibration and an allowed fragment mass deviation of 20 ppm. For LFQ, in MaxQuant, the minimum ratio count was set to two. For matching between runs, the retention time alignment window was set to 30 min and the match time window was 1 min.

For protein identification in CME vs. SME—the first round of matrisome protein identification—unique peptides which were present in at least two of the biological replicates were searched using "Matrisome Annotator" in the MD (http://matrisomeproject.mit.edu/, accessed on 23 January 2023). The matrisome proteins were classified as core matrisome proteins (collagens, ECM glycoproteins, and proteoglycans) and matrisome-associated proteins (ECM-affiliated, ECM Regulators, and Secreted Factors). In the second round, the protein list obtained from LC-MS/MS was matched using the UniProt Database for *Mus musculus*, and possible matrisome proteins were searched based on their tissue locations using "extracellular matrix", "extracellular space", "basement membrane", "secreted", "lysosome" and "exosome". To verify potential candidates, their possible interactions with matrisome proteins were searched in protein interaction databases (BioGRID, STRING). In the final stage, the short list of proteins was searched in the literature to confirm that proteins have been located in the ECM by other studies.

For protein quantification, LFQ intensity values were taken from the MaxQuant protein "Groups" table, which represented the values after inter-experiment normalization [54]. Differential abundant analysis was performed using the LFQ-Analyst online software (https://bioinformatics.erc.monash.edu/apps/LFQ-Analyst/, accessed on 23 January 2023) [29]. Matrisome proteins which had LFQ intensities in at least 2 of 3 biological replicates were included in the comparison of CME and SME or SME-fractions. In LFQ-Analyst, parameters were set as Perseus-type imputation, adjusted p-value cutoff <0.05 and Log2 fold- change (FC) cutoff = 1 (i.e., 2-FC). The comparison was presented with FC and p-values. All p-values were corrected for multiple hypothesis testing using the Benjamini–Hochberg method. Proteins with $p < 0.05$ and up- or downregulated 2-FC were presented as significantly different between the samples obtained by the extraction methods CME and SME.

To compare the fractions of the SME and CME products, LFQ analysis was performed separately for each SME fraction, and the same parameters used to compare CME and SME were applied, although in this fraction comparison analysis, fractions were set as "separate method" instead of as "fractions".

5. Conclusions

In conclusion, a better understanding of the mouse kidney matrisome is likely to inform future studies in the areas of kidney disease modelling, kidney development biology, ageing and tissue engineering. In this study, we have successfully progressed knowledge towards understanding the composition of the mouse kidney matrisome via identification and quantification of matrisome proteins and via identification of a set of proteins which were not previously listed as matrisome components. In addition, this study showed the importance of using different protein extraction steps to fractionate matrisome proteins for a better understanding of the mouse kidney matrisome composition. Hence, we suggest to obtain different fractions of matrisome proteins to discover more ECM proteins for further studies of kidney ECM turnover in normal and pathological conditions, and also to identify novel drug targets and biomarkers for the treatment and diagnosis of chronic kidney disease.

Supplementary Materials: The following supporting information can be downloaded at: https://www.mdpi.com/article/10.3390/ijms24032827/s1.

Author Contributions: Conceptualization, U.R., S.B.A., E.S.X.M. and A.G.; methodology, U.R., S.B.A. and S.A.; validation, U.R., S.B.A. and A.G.; formal analysis, U.R., S.B.A. and S.A.; investigation, U.R., S.B.A. and A.G.; resources, A.G.; data curation, U.R., S.B.A., S.A., C.A.P. and S.S.; writing—original draft preparation, U.R. and A.G.; writing—review and editing, U.R., S.B.A., S.A., E.S.X.M., C.A.P., S.S. and A.G.; visualization, U.R., S.B.A. and A.G.; supervision, C.A.P., S.S. and A.G.; project administration, A.G.; funding acquisition, S.S and A.G. All authors have read and agreed to the published version of the manuscript.

Funding: This research was funded by NSW HEALTH DEPARTMENT, UNSW AND PHARMAXIS LTD., grant for NSW Health PhD Partnership Program Scholarship.

Institutional Review Board Statement: Ethical review and approval were waived for this study due to sharing unused intact animal tissues from an unrelated experiment for which an animal study protocol was approved by the Animal Care and Ethics Committee (ACEC) of University of New South Wales (protocol 21/24A approved on 12 March 2021).

Informed Consent Statement: Not applicable.

Data Availability Statement: The data presented and discussed in this study is available in (Appendix A and Supplementary Material Tables S1 and S2).

Acknowledgments: Authors thank Ewa Goldys, Wolfgang Jarolimek, Long Nguyen, Yimin Yao, and Nicolle Packer for helpful discussions. Mass spectrometric results were obtained at the Bioanalytical Mass Spectrometry Facility within the Mark Wainwright Analytical Centre of the University of New South Wales. We thank Ling Zhong for the support in mass spectrometric analysis. We are grateful to Tianruo Guo and Madhuvanthi Muralidharan (UNSW) for sharing the animal tissues, and Naomi Craig for the professional support in organizing the tissue sharing workflow. U.R. acknowledges the support by NSW Health Department, UNSW and Pharmaxis Ltd. A.G. thanks Macquarie University for supporting her work via the MRF fellowship.

Conflicts of Interest: The authors declare no conflict of interest. The funders had no role in the design of the study; in the collection, analyses, or interpretation of data; in the writing of the manuscript; or in the decision to publish the results.

Appendix A

Table A1. The list of core matrisome proteins belonging to the category of collagens and identified in at least two of the biological replicates.

Protein ID	Protein Name	Gene Symbol	Identified by Method		Ref *
			CME	SME	
P11087	Collagen alpha-1(I) chain	Col1a1	+	+	MD
Q01149	Collagen alpha-2(I) chain	Col1a2	+	+	MD
P08121	Collagen alpha-1(III) chain	Col3a1	+	+	MD
P02463	Collagen alpha-1(IV) chain	Col4a1	+	+	MD
P08122	Collagen alpha-2(IV) chain	Col4a2	+	+	MD
Q9QZS0	Collagen alpha-3(IV) chain	Col4a3	+	−	MD
Q9QZR9	Collagen alpha-4(IV) chain	Col4a4	+	−	MD
Q9CPW5	Collagen alpha-5(IV) chain	Col4a5	+	+	MD
O88207	Collagen alpha-1(V) chain	Col5a1	+	+	MD
Q3U962	Collagen alpha-2(V) chain	Col5a2	+	+	MD
Q04857	Collagen alpha-1(VI) chain	Col6a1	+	+	MD
Q02788	Collagen alpha-2(VI) chain	Col6a2	+	+	MD
Q61001	Collagen alpha-3(VI) chain	Col6a3	+	+	MD
Q80X19	Collagen alpha-1(XIV) chain	Col14a1	+	+	MD
O35206	Collagen alpha-1(XV) chain	Col15a1	+	+	MD
P39061	Collagen alpha-1(XVIII) chain	Col18a1	+	+	MD

Abbreviations: CME—compartmental matrix enrichment; SME—sequential matrix enrichment; MD—Matrisome Database; Reference *—source of information for the provided protein identification; "+" indicates that protein has been identified; "−" indicates that protein has not been identified by the method.

Table A2. The list of core matrisome proteins belonging to the category of glycoproteins and identified in at least two of the biological replicates. A row highlighted in yellow describes an identified matrisome protein which was not previously listed in the MD.

Protein ID	Protein Name	Gene Symbol	Identified by Method		Ref *
			CME	SME	
A2ASQ1	Agrin	Agrn	+	+	MD
Q9QZZ6	Dermatopontin	Dpt	+	−	MD
Q91VF5	EMI domain-containing protein 1	Emid1	+	+	MD
Q99K41	EMILIN-1	Emilin1	+	+	MD
Q61554	Fibrillin-1	Fbn1	+	−	MD
Q61555	Fibrillin-2	Fbn2	−	+	MD
E9PV24	Fibrinogen alpha chain	Fga	+	+	MD

Table A2. *Cont.*

Protein ID	Protein Name	Gene Symbol	Identified by Method		Ref *
			CME	SME	
Q8VCM7	Fibrinogen gamma chain	Fgg	+	+	MD
P11276	Fibronectin	Fn1	+	+	MD
A2ASQ1	Progranulin	Grn	−	+	MD
Q61581	Insulin-like growth factor-binding protein 7	Igfbp7	+	+	MD
P19137	Laminin subunit alpha-1	Lama1	+	+	MD
P97927	Laminin subunit alpha-4	Lama4	+	+	MD
Q61001	Laminin subunit alpha-5	Lama5	+	+	MD
P02469	Laminin subunit beta-1	Lamb1	+	+	MD
Q61292	Laminin subunit beta-2	Lamb2	+	+	MD
P02468	Laminin subunit gamma-1	Lamc1	+	+	MD
Q07797	Galectin-3-binding protein	Lgals3bp	+	+	[55]
Q8K4G1	Latent TGF-b-binding protein 4	Ltbp4	−	+	MD
P21956	Lactadherin	Mfge8	+	+	MD
A6H6E2	Multimerin-2	Mmrn2	+	+	MD
P10493	Nidogen-1	Nid1	+	+	MD
O88322	Nidogen-2	Nid2	+	+	MD
Q91V88	Nephronectin	Npnt	−	+	MD
Q99JR5	Tubulointerstitial nephritis antigen-like	Tinagl1	+	+	MD
P29788	Vitronectin	Vtn	+	−	MD
Q8R2Z5	von Willebrand factor A domain-containing protein 1	Vwa1	+	−	MD
Q99KC8	von Willebrand factor A domain-containing protein 5A	Vwa5a	−	+	MD

Abbreviations: CME—compartmental matrix enrichment; SME—sequential matrix enrichment; MD—Matrisome Database; Ref *—source of information for the provided protein identification; ; "+" indicates that protein has been identified; "−" indicates that protein has not been identified by the method.

Table A3. The list of core matrisome proteins belonging to the category of proteoglycans and identified in at least two of the biological replicates.

Protein ID	Protein Name	Gene Symbol	Identified by Method		Ref *
			CME	SME	
Q99MQ4	Asporin	Aspn	+	−	MD
P28653	Biglycan	Bgn	+	+	MD
P28654	Decorin	Dcn	+	+	MD
Q05793	Heparan sulphate proteoglycan core protein	Hspg2	+	+	MD
P51885	Lumican	Lum	+	+	MD
Q62000	Mimecan	Ogn	+	−	MD
Q9JK53	Prolargin	Prelp	+	+	MD

Abbreviations: CME—compartmental matrix enrichment; SME—sequential matrix enrichment; MD—Matrisome Database; Ref *—source of information for the provided protein identification; ; "+" indicates that protein has been identified; "−" indicates that protein has not been identified by the method.

Table A4. The list of matrisome-associated proteins belonging to the category of ECM regulators and identified in at least two of the biological replicates. Rows highlighted in yellow describe identified matrisome proteins which were not previously listed in the MD.

Protein ID	Protein Name	Gene Symbol	Identified by Method		Ref *
			CME	SME	
Q8JZZ0	Alpha-2-macroglobulin; Alpha-2-macroglobulin 165 kDa subunit; Alpha-2-macroglobulin 35 kDa subunit	A2m	–	+	MD
P09470	Angiotensin-converting enzyme	Ace	+	+	[40]
Q8R0I0	Angiotensin-converting enzyme 2	Ace2	+	+	[38,39]
O35598	Disintegrin and metalloproteinase domain-containing protein 10	Adam10	+	+	MD
P08226	Apolipoprotein E	Apoe	–	+	[56]
P50429	Arylsulfatase B	Arsb	–	+	[57]
Q99N23	Carbonic anhydrase 15	Ca15	+	+	[58]
Q62426	Cystatin-B	Cstb	-	+	MD
P16675	Lysosomal protective protein	Ctsa	+	+	MD
P10605	Cathepsin B	Ctsb	+	+	MD
P18242	Cathepsin D	Ctsd	+	+	MD
P49935	Pro-cathepsin H	Ctsh	+	+	MD
Q9WUU7	Cathepsin Z	Ctsz	–	+	MD
P28843	Dipeptidyl peptidase 4	Dpp4	+	+	[59,60]
P19221	Prothrombin	F2	–	+	MD
Q571E4	N-acetylgalactosamine-6-sulfatase	Galns	–	+	[61]
Q8BFR4	N-acetylglucosamine-6-sulfatase	Gns	–	+	[62]
Q9ESB3	Histidine-rich glycoprotein	Hrg	–	+	MD
Q61703	Inter-alpha-trypsin inhibitor heavy chain H2	Itih2	–	+	MD
O08677	Kininogen-1	Kng1	–	+	MD
P28825	Meprin A subunit alpha	Mep1a	+	+	MD
Q61847	Meprin A subunit beta	Mep1b	+	+	MD
Q61391	Neprilysin	Mme	+	+	[42]
Q02819	Nucleobindin-1	Nucb1	+	+	[37]
P09103	Protein disulfide-isomerase	P4hb	+	+	[63,64]
P20918	Plasminogen	Plg	+	+	MD
P06281	Renin-1	Ren1	–	+	[65]
P07758	Alpha-1-antitrypsin 1-1	Serpina1a	–	+	MD
P22599	Alpha-1-antitrypsin 1-2	Serpina1b	–	+	MD
Q00896	Alpha-1-antitrypsin 1-3	Serpina1c	–	+	MD
P07759	Serine protease inhibitor A3K	Serpina3k	–	+	MD
Q60854	Serpin B6	Serpinb6	–	+	[66]
P19324	Serpin H1	Serpinh1	+	+	MD
O09164	Extracellular superoxide dismutase	Sod3	+	+	[67]
Q9JLF6	Protein-glutamine gamma-glutamyltransferase K	Tgm1	+	+	MD
P21981	Protein-glutamine gamma-glutamyltransferase 2	Tgm2	+	+	MD

Abbreviations: CME—compartmental matrix enrichment; SME—sequential matrix enrichment; MD—Matrisome Database; Ref *—source of information for the provided protein identification; ; "+" indicates that protein has been identified; "–" indicates that protein has not been identified by the method.

Table A5. The list of matrisome-associated proteins belonging to the category of ECM-affiliated proteins and identified in at least two of the biological replicates. Rows highlighted in yellow describe identified matrisome proteins which were not previously listed in the MD.

Protein ID	Protein Name	Gene Symbol	Identified by Method		Ref *
			CME	SME	
P10107	Annexin A1	Anxa1	−	+	MD
P07356	Annexin A2	Anxa2	+	+	MD
O35639	Annexin A3	Anxa3	−	+	MD
P97429	Annexin A4	Anxa4	−	+	MD
P48036	Annexin A5	Anxa5	+	+	MD
P14824	Annexin A6	Anxa6	+	+	MD
Q07076	Annexin A7	Anxa7	−	+	MD
P97384	Annexin A11	Anxa11	−	+	MD
O35658	Complement component 1 Q subcomponent-binding protein	C1qbp	+	+	[68,69]
Q62165	Dystroglycan	Dag1	+	+	[70]
Q9R0H2	Endomucin	Emcn	+	+	MD
Q91X72	Hemopexin	Hpx	−	+	MD
P57016	Ladinin-1	Lad1	+	+	[71,72]
P16045	Galectin-1	Lgals1	+	+	MD
P16110	Galectin-3	Lgals3	+	+	MD
Q9D0F3	Protein ERGIC-53	Lman1	+	+	MD
P11152	Lipoprotein lipase	Lpl	+	+	[73,74]
B2RXS4	Plexin-B2	Plxnb2	+	+	MD
Q62219	TGF-b-1-induced transcript 1 protein	Tgfb1i1	+	+	[75]

Abbreviations: CME—compartmental matrix enrichment; SME—sequential matrix enrichment; MD—Matrisome Database; Ref *—source of information for the provided protein identification; ; "+" indicates that protein has been identified; "−" indicates that protein has not been identified by the method.

Table A6. The list of matrisome-associated proteins belonging to the category of secreted factors and identified in at least two of the biological replicates. Rows highlighted in yellow describe identified matrisome proteins which were not previously listed in the MD.

Protein ID	Protein Name	Gene Symbol	Identified by Method		Ref *
			CME	SME	
P01132	Pro-epidermal growth factor	Egf	−	+	MD
Q8CJ70	Interleukin-19	Il19	+	+	MD
P14069	Protein S100-A6	S100a6	−	+	MD
P08207	Protein S100-A10	S100a10	+	+	MD
P50543	Protein S100-A11	S100a11	+	+	MD
Q91X17	Uromodulin	Umod	+	+	[76,77]
P51859	Hepatoma-derived growth factor	Hdgf	+	+	[78,79]

Abbreviations: CME—compartmental matrix enrichment; SME—sequential matrix enrichment; MD—Matrisome Database; Ref *—source of information for the provided protein identification; ; "+" indicates that protein has been identified; "−" indicates that protein has not been identified by the method.

Table A7. Summary of the newly detected and classified mouse kidney matrisome proteins that are not listed in the current Matrisome Database.

Protein ID	Protein Name	Gene Symbol	Identified by Methods		Role of Protein	Reference
			CME	SME		
Core matrisome: ECM Glycoproteins						
Q07797	Galectin-3-binding protein	Lgals3bp	+	+	ECM modulation, immune response	[55]
Matrisome-associated proteins: ECM Regulators						
P09470	Angiotensin-converting enzyme	Ace	+	+	ECM remodeling	[40]
Q8R0I0	Angiotensin-converting enzyme 2	Ace2	+	+	ECM remodeling	[38,39]
P08226	Apolipoprotein E	Apoe	–	+	Protein binding	[56]
P50429	Arylsulfatase B	Arsb	–	+	Modulation of signaling in cytoskeletal rearrangement and a component of ECM	[57]
Q99N23	Carbonic anhydrase 15	Ca15	+	+	Regulation of acid-base balance	[58]
P28843	Dipeptidyl peptidase 4	Dpp4	+	+	Signaling and endothelial-mesenchymal transition	[59,60]
Q571E4	N-acetylgalactosamine-6-sulfatase	Galns	–	+	Glycosaminoglycan degradation	[61]
Q8BFR4	N-acetylglucosamine-6-sulfatase	Gns	–	+	Glycosaminoglycan degradation	[62]
Q61391	Neprilysin	Mme	+	+	Elastin degradation	[42]
Q02819	Nucleobindin-1	Nucb1	+	+	ECM remodeling and metalloptroteinase-2 transport	[37]
P09103	Protein disulfide-isomerase	P4hb	+	+	Metalloproteinase activation and platelet adhesion and migration	[63,64]
P06281	Renin-1	Ren1	–	+	Protease activity	[65]
Q60854	Serpin B6	Serpinb6	–	+	Protease binding	[66]
O09164	Extracellular superoxide dismutase	Sod3	+	+	Removal of oxygen radicals and response to hypoxia	[67,80]
Matrisome-associated proteins: ECM-affiliated Proteins						
O35658	Complement component 1 Q subcomponent-binding protein	C1qbp	+	+	Regulation of complement activation and cell adhesion	[68,69]
Q62165	Dystroglycan (α-Dystroglycan)	Dag1	+	+	Laminin and dystroglycan binding	[70,81]
P57016	Ladinin-1	Lad1	+	+	Component of basement membrane and integrin signaling	[71,72]
P11152	Lipoprotein lipase	Lpl	+	+	Heparan sulphate binding and lipolytic processing	[73,74]
Q62219	TGF-b-1-induced transcript 1 protein	Tgfb1i1	+	+	ECM remodeling and deposition	[75]
Matrisome-associated proteins: Secreted Factors						
Q91X17	Uromodulin	Umod	+	+	Regulation of kidney electrolyte balance	[76,77]
P51859	Hepatoma-derived growth factor	Hdgf	+	+	Matrix bound growth factor	[78,79]

Abbreviations: CME—compartmental matrix enrichment; SME—sequential matrix enrichment; "+" indicates that protein has been identified; "–" indicates that protein has not been identified by the method.

Table A8. Composition of mouse kidney matrisome proteins identified in this study depending on the ECM enrichment method.

Categories of Identified Proteins	Number of Identified Proteins		
	CME	SME	Total, CEM and SME *
Total proteins	1526	2246	2442
Matrisome proteins	83	105	113
Including:			
Core matrisome, overall	46	43	51
Including:			
Collagens	16	14	16
ECM glycoproteins	23	24	28
Proteoglycans	7	5	7
Matrisome-associated proteins, overall	37	62	62
Including:			
ECM Regulators	19	36	36
ECM-affiliated Proteins	13	19	19
Secreted Factors	5	7	7

* This column shows all different proteins (identified based on unique peptides) obtained by CME and SME methods together, without duplication of the proteins found by both methods.

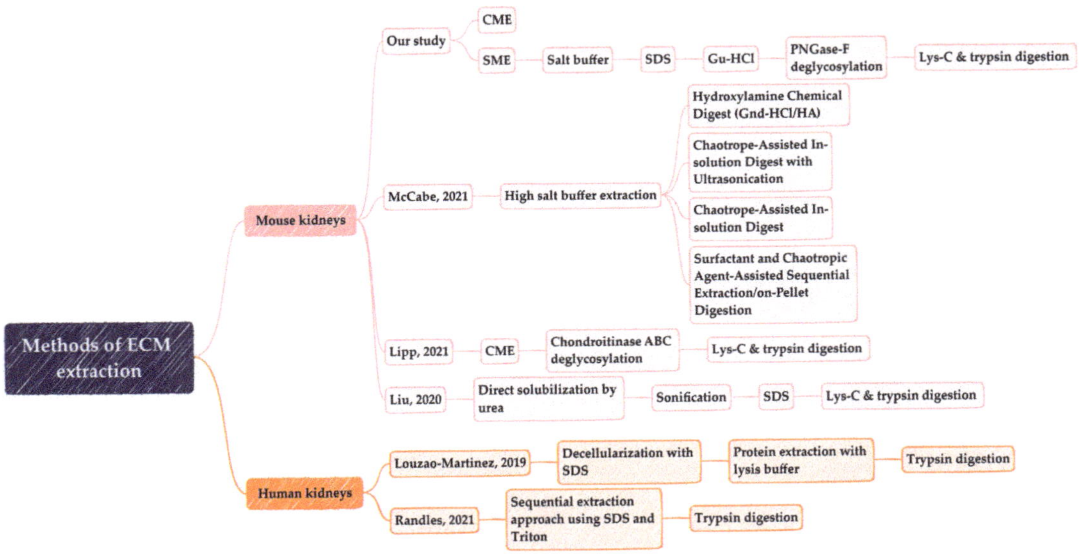

Figure A1. Comparison of the methods of extraction of ECM proteins from mouse and human kidneys tissues.

Table A9. The comparative analysis of the matrisome composition revealed in mouse and human kidney samples by different studies, including the current study, as well as publications by McCabe et al. [20], Lipp et al. [28], Liu et al. [30], Louzao-Martinez et al. [31], and Randles et al. [32].

Characteristics	Identification of Kidney Matrisome Proteins					
Organism	Mouse	Mouse	Mouse	Mouse	Human	Human
Data source	Our data	McCabe et al.	Lipp et al.	Liu et al.	Louzao-Martinez 2019	Randles 2021
# of methods used	2	4	1	1	1	1
Age	4–10 weeks	N/A	adult	15 weeks	55 years	15–37 years
Gender	F	M	N/A	N/A	M/F	M
Strain	C57BL/6J	C67BL/6J	C57BL/6J	C57BL/6J	N/A	N/A
Size	3	3	3	N/A	13	3
# Total proteins	2442	N/A	N/A	5044	N/A	N/A
# Total matri-someproteins	113	173	79	139	76	172
# Collagens	16	22	18	16	12	21
# GPs	28	55	38	41	33	53
# PGs	7	10	6	6	6	11
# ECM-regulators	36	57	9	46	11	47
# ECM-affiliated	19	22	6	20	9	28
# Secreted factors	7	7	2	10	5	12

N/A—not available; F—female; M—male; #—number of; GPs—ECM Glycoproteins; PGs—Proteoglycans.

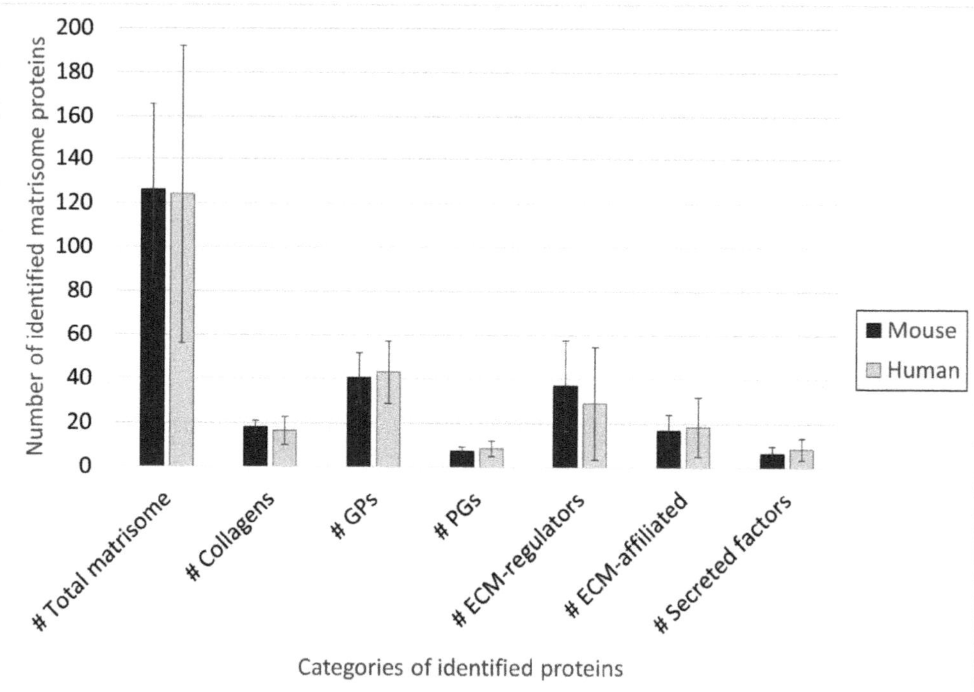

Figure A2. The average composition of the matrisome revealed in mouse and human kidney samples by different studies, including the current study, and the following publications by McCabe et al. [20], Lipp et al. [28], Liu et al. [30], Louzao-Martinez et al. [31], and Randles et al. [32]. The bar graphs show Mean ± Standard Deviation of the number of proteins in each category averaged across the species-specific data reported in Table A9. Abbreviations: #—number of; GPs—ECM Glycoproteins; PGs—Proteoglycans.

Table A10. The combined list of core matrisome proteins belonging to the category of collagens and identified in mouse and human kidneys. The mouse matrisome proteins were obtained from our data and extracted from references [20,28,30]. Human matrisome proteins were obtained from the studies [31,32]. Highlighted "X" refers to identified protein in the corresponding study.

Protein Name	Our data	McCabe 2021	Lipp 2021	Liu 2020	Louzao-Martinez 2019	Randles 2021
	Mouse	Mouse	Mouse	Mouse	Human	Human
COL1A1	X	X	X	X	X	X
COL1A2	X	X	X	X	X	X
COL2A1		X				
COL3A1	X	X	X		X	X
COL4A1	X	X	X	X	X	X
COL4A2	X	X	X	X	X	X
COL4A3	X	X	X	X		X
COL4A4	X	X	X	X		X
COL4A5	X		X	X		X
COL4A6			X			X
COL5A1	X	X	X			X
COL5A2	X	X	X			X
COL6A1	X	X	X	X	X	X
COL6A2	X	X	X	X	X	X
COL6A3	X		X	X	X	X
COL6A5		X		X		
COL6A6		X		X		
COL7A1						X
COL8A1						X
COL11A1		X				
COL11A2		X				
COL12A1		X	X	X	X	X
COL14A1	X	X	X	X	X	X
COL15A1	X	X	X	X	X	X
COL16A1		X				X
COL18A1	X	X	X	X	X	X
COL24A1		X				

Table A11. The combined list of core matrisome proteins belonging to the category of ECM glycoproteins and identified in mouse and human kidneys. The mouse matrisome proteins were obtained from our data and extracted from references [20,28,30]. Human matrisome proteins were obtained from the studies [31,32]. Highlighted "X" refers to identified protein in the corresponding study.

Protein Name	Our data	McCabe 2021	Lipp 2021	Liu 2020	Louzao-Martinez 2019	Randles 2021
	Mouse	Mouse	Mouse	Mouse	Human	Human
ADIPOQ				X		
AEBP1		X				
AGRN	X	X	X	X	X	X
AHSG						X
APOH						X
CRELD1						X
CRELD2				X		
DPT	X	X	X		X	X
ECM1		X	X	X		X

Table A11. *Cont.*

Protein Name	Our data	McCabe 2021	Lipp 2021	Liu 2020	Louzao-Martinez 2019	Randles 2021
	Mouse	Mouse	Mouse	Mouse	Human	Human
EFEMP1		X				X
EFEMP2		X				
ELN		X	X		X	X
EMID1	X	X				
EMILIN1	X	X	X	X	X	X
FBLN1		X		X	X	X
FBLN2		X				
FBLN5	X	X	X		X	X
FBN1	X	X	X	X	X	X
FBN2	X				X	
FGA	X	X	X	X	X	X
FGB		X	X	X	X	X
FGG	X	X	X	X	X	X
FGL1						X
FGL2						X
FN1	X	X	X	X	X	X
FRAS1			X	X	X	X
IGFBP7	X	X	X	X	X	X
KCP		X	X	X		X
LAMA1	X	X	X	X	X	X
LAMA2		X	X	X		X
LAMA3	X	X	X	X		X
LAMA4	X	X	X	X	X	X
LAMA5	X	X	X	X	X	X
LAMB1	X	X	X	X	X	X
LAMB2	X	X	X	X	X	X
LAMB3		X				
LAMC1	X	X	X	X	X	X
LAMC3		X		X		
LGALS3BP	X					X
LTBP1		X	X			
LTBP4	X	X		X		X
MATN2		X	X	X		X
MFAP2		X	X		X	
MFAP4		X		X		X
MFAP5		X		X	X	
MFGE8	X					
MMRN2	X	X	X	X		X
NID1	X	X	X	X	X	X
NID2	X	X	X	X	X	X
NPNT	X	X	X	X	X	
NTN1				X		X
NTN4		X		X		
PAPLN		X		X		X
POSTN	X	X	X	X	X	X
PXDN		X	X	X		X
SBSPON		X				X
SPARC		X				X
TGFBI		X	X	X	X	X
THBS1		X			X	X
THSD4		X				X
TINAG			X		X	X
TINAGL1	X	X	X	X	X	X
TNC		X	X		X	X
TNXB			X			
VTN	X	X	X	X	X	X
VWA1	X	X	X			X
VWA2		X		X		
VWA5A	X	X		X		
VWF		X			X	X

Table A12. The combined list of core matrisome proteins belonging to the category of proteoglycans and identified in mouse and human kidneys. The mouse matrisome proteins were obtained from our data and extracted from references [20,28,30]. Human matrisome proteins were obtained from the studies [31,32]. Highlighted "X" refers to identified protein in the corresponding study.

Protein Name	Our data	McCabe 2021	Lipp 2021	Liu 2020	Louzao-Martinez 2019	Randles 2021
	Mouse	Mouse	Mouse	Mouse	Human	Human
ANGPTL2		X				X
ASPN	X	X	X			X
BGN	X	X	X	X	X	X
DCN	X	X	X	X	X	X
HAPLN1		X				
HSPG2	X	X	X	X	X	X
LUM	X	X	X	X	X	X
OGN	X	X		X		X
PRELP	X	X	X	X	X	X
PRG2		X				X
PRG3						X
VCAN		X			X	X

Table A13. The combined list of matrisome-associated proteins belonging to the category of ECM regulators and identified in mouse and human kidneys. The mouse matrisome proteins were obtained from our data and extracted from references [20,28,30]. Human matrisome proteins were obtained from the studies [31,32]. Highlighted "X" refers to identified protein in the corresponding study.

Protein Name	Our Data	McCabe 2021	Lipp 2021	Liu 2020	Louzao-Martinez 2019	Randles 2021
	Mouse	Mouse	Mouse	Mouse	Human	Human
A2M	X	X			X	X
ACE	X					
ACE2	X					
ADAM8				X		
ADAM9						X
ADAM10	X	X		X		
ADAM17				X		
ADAMTSL1		X				
ADAMTSL4		X				
ADAMTSL5						X
AGT		X				
AMBP		X		X	X	X
APOE	X					X
ARSB	X					
CA15	X					
CASP14						X
CPN2		X		X		
CST3		X				
CSTB	X	X		X		
CTSA	X	X		X		X
CTSB	X	X		X		X
CTSC		X				X
CTSD	X	X		X	X	X
CTSF		X				
CTSH	X	X		X		X
CTSL		X		X		
CTSS				X		X
CTSZ	X	X		X		X
DPP4	X					
ELANE						X
F13A1		X				

Table A13. *Cont.*

Protein Name	Our Data	McCabe 2021	Lipp 2021	Liu 2020	Louzao-Martinez 2019	Randles 2021
	Mouse	Mouse	Mouse	Mouse	Human	Human
F2	X	X		X		
F3						X
F9		X		X		X
FAM20B	X	X		X		
GALNS	X					
GANAB						X
GNS	X					
HRG	X	X		X	X	X
HTRA1		X				X
ITIH1		X	X	X	X	X
ITIH2	X	X		X	X	
ITIH4		X		X		X
ITIH5		X				X
KNG1	X	X		X		X
LOX		X				
LOXL1		X				
LOXL2			X			
LOXL4		X				
MEP1A	X	X	X	X		
MEP1B	X	X	X	X		
MME	X					
MMP9						X
MUG2		X		X		
NGLY1				X		
NUCB1	X					
OGFOD2				X		
P4HA1		X				X
P4HB	X					
PLAU		X				
PLG	X	X	X	X		X
PLOD1						X
PLOD2						X
PLOD3						X
PLSCR1						X
PRDX1						X
PRTN3						X
REN1	X					
SERPINA1					X	
SERPINA1A	X	X	X	X		
SERPINA1B	X	X		X		
SERPINA1C	X	X		X		
SERPINA1D		X		X		
SERPINA1E		X		X		
SERPINA3					X	
SERPINA3G				X		
SERPINA3K	X	X		X		
SERPINA3M		X		X		
SERPINA3N		X				
SERPINA5						X
SERPINA6		X		X		
SERPINB1A		X		X		
SERPINB6	X			X		
SERPINB12						X
SERPINC1		X		X	X	X
SERPIND1		X		X		X
SERPINF1		X				X
SERPINF2		X		X		X
SERPING1		X		X		X
SERPINH1	X	X	X	X		X
SOD1						X
SOD3	X					X
ST14		X				
TGM1	X	X	X	X		X
TGM2	X	X	X	X	X	X
TGM3						X
TIMP3		X		X	X	X

Table A14. The combined list of matrisome-associated proteins belonging to the category of ECM-affiliated proteins and identified in mouse and human kidneys. The mouse matrisome proteins were obtained from our data and extracted from references [20,28,30]. Human matrisome proteins were obtained from the studies [31,32]. Highlighted "X" refers to identified protein in the corresponding study.

Protein Name	Our data	McCabe 2021	Lipp 2021	Liu 2020	Louzao-Martinez 2019	Randles 2021
	Mouse	Mouse	Mouse	Mouse	Human	Human
ANXA1	X	X		X	X	X
ANXA2	X	X	X	X		X
ANXA3	X	X		X		
ANXA4	X	X		X	X	X
ANXA5	X	X		X	X	X
ANXA6	X	X	X	X	X	X
ANXA7	X	X	X	X	X	X
ANXA9		X				
ANXA11	X	X	X	X	X	X
ANXA13				X		X
APCS						X
C1QBP	X					
CCT2						X
CCT6A						X
CD109						X
CSPG4		X		X		X
CXCL14						X
DAG1	X					
EMCN	X					X
FREM1						X
FREM2		X	X	X	X	
GPC4		X		X		
GRN	X					
HPX	X	X		X		X
LAD	X					
LGALS1	X	X		X	X	X
LGALS2						X
LGALS3	X	X				X
LGALS8		X				X
LGALS9		X		X		
LMAN1	X	X	X	X	X	X
LPL	X					
MBL1		X				
MBL2		X				
MGP						X
MUC1						X
PLXNB1						X
PLXNB2	X	X		X		X
PLXND1				X		X
SDC4				X		
SEMA4D		X		X		
TGFB1I1	X					

Table A15. The combined list of matrisome-associated proteins belonging to the category of secreted factors and identified in mouse and human kidneys. The mouse matrisome proteins were obtained from our data and extracted from references [20,28,30]. Human matrisome proteins were obtained from the studies [31,32]. Highlighted "X" refers to identified protein in the corresponding study.

Protein Name	Our data	McCabe 2021	Lipp 2021	Liu 2020	Louzao-Martinez 2019	Randles 2021
	Mouse	Mouse	Mouse	Mouse	Human	Human
CLU						X
CRLF3				X		
DCD						X
DEFA1						X
EGF	X	X		X		
EGFL7		X	X			
FGF1				X		
FGF2						X
HCFC1		X				
HDGF	X					
HRNR					X	X
Il19	X					
INHBE						X
PF4				X		
S100A1	X			X		
S100A6	X					
S100A7					X	
S100A8				X	X	X
S100A9	X				X	X
S100A10	X	X		X		X
S100A11	X	X	X	X	X	X
S100A13	X					
S100A14						X
S100G		X		X		
SFRP1				X		
UMOD	X					

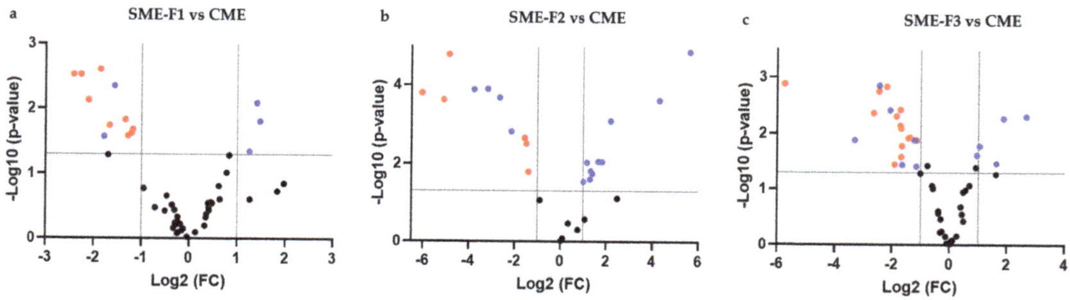

Figure A3. Quantitative comparison of protein extraction levels between SME fractions and CME. Volcano plots represent significant differences in protein abundances (FC > 2, p-value < 0.05) between (**a**) SME (fraction 1) and CME, (**b**) SME (fraction 2) and CME and (**c**) SME (fraction 3) and CME. Top-right-hand sides of volcano plots are proteins that were significantly higher in SME fractions. Top-left-hand sides are proteins that were significantly higher in CME. Blue dots indicate Matrisome-associated proteins and red dots indicate core matrisome proteins that were significantly higher in either SME fractions or CME.

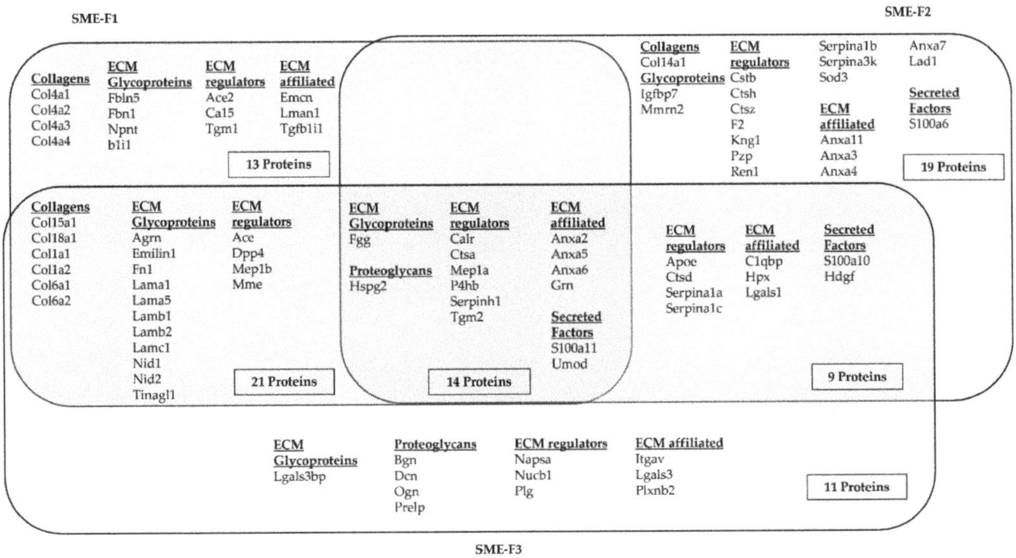

Figure A4. Quantified matrisome proteins in fractions of SME. Venn diagram illustrates the common and unique matrisome proteins in each fraction. Matrisome proteins were classified as collagens, ECM glycoproteins, proteoglycans, ECM-affiliated, ECM regulators and secreted factors.

References

1. Theocharis, A.D.; Skandalis, S.S.; Gialeli, C.; Karamanos, N.K. Extracellular matrix structure. *Adv. Drug Deliv. Rev.* **2016**, *97*, 4–27. [CrossRef]
2. Manninen, A. Epithelial polarity—Generating and integrating signals from the ECM with integrins. *Exp. Cell Res.* **2015**, *334*, 337–349. [CrossRef]
3. Frantz, C.; Stewart, K.M.; Weaver, V.M. The extracellular matrix at a glance. *J. Cell Sci.* **2010**, *123*, 4195–4200. [CrossRef]
4. Karamanos, N.K.; Theocharis, A.D.; Piperigkou, Z.; Manou, D.; Passi, A.; Skandalis, S.S.; Vynios, D.H.; Orian-Rousseau, V.; Ricard-Blum, S.; Schmelzer, C.E.H.; et al. A guide to the composition and functions of the extracellular matrix. *FEBS J.* **2021**, *288*, 6850–6912. [CrossRef]
5. Naba, A.; Clauser, K.R.; Hoersch, S.; Liu, H.; Carr, S.A.; Hynes, R.O. The matrisome: In silico definition and in vivo characterization by proteomics of normal and tumor extracellular matrices. *Mol. Cell Proteom.* **2012**, *11*, M111.014647. [CrossRef]
6. Bonnans, C.; Chou, J.; Werb, Z. Remodelling the extracellular matrix in development and disease. *Nat. Rev. Mol. Cell Biol.* **2014**, *15*, 786–801. [CrossRef]
7. Kim, S.H.; Turnbull, J.; Guimond, S. Extracellular matrix and cell signalling: The dynamic cooperation of integrin, proteoglycan and growth factor receptor. *J. Endocrinol.* **2011**, *209*, 139–151. [CrossRef]
8. Mecham, R.P. Overview of extracellular matrix. *Curr. Protoc. Cell Biol.* **2012**, *57*, 1–10. [CrossRef]
9. Hynes, R.O.; Naba, A. Overview of the matrisome—An inventory of extracellular matrix constituents and functions. *Cold Spring Harb. Perspect. Biol.* **2012**, *4*, a004903. [CrossRef]
10. Naba, A.; Clauser, K.R.; Ding, H.; Whittaker, C.A.; Carr, S.A.; Hynes, R.O. The extracellular matrix: Tools and insights for the "omics" era. *Matrix Biol.* **2016**, *49*, 10–24. [CrossRef]
11. Grassian, A.R.; Coloff, J.L.; Brugge, J.S. Extracellular matrix regulation of metabolism and implications for tumorigenesis. *Cold Spring Harb. Symp. Quant. Biol.* **2011**, *76*, 313–324. [CrossRef]
12. Iozzo, R.V.; Gubbiotti, M.A. Extracellular matrix: The driving force of mammalian diseases. *Matrix Biol.* **2018**, *71–72*, 1–9. [CrossRef]
13. Xue, M.; Jackson, C.J. Extracellular Matrix Reorganization During Wound Healing and Its Impact on Abnormal Scarring. *Adv. Wound Care* **2015**, *4*, 119–136. [CrossRef]
14. Taha, I.N.; Naba, A. Exploring the extracellular matrix in health and disease using proteomics. *Essays Biochem.* **2019**, *63*, 417–432. [CrossRef]
15. Rinschen, M.M.; Saez-Rodriguez, J. The tissue proteome in the multi-omic landscape of kidney disease. *Nat. Rev. Nephrol.* **2021**, *17*, 205–219. [CrossRef]
16. Chang, C.W.; Dalgliesh, A.J.; Lopez, J.E.; Griffiths, L.G. Cardiac extracellular matrix proteomics: Challenges, techniques, and clinical implications. *Proteom. Clin. Appl.* **2016**, *10*, 39–50. [CrossRef]

17. Naba, A.; Clauser, K.R.; Hynes, R.O. Enrichment of Extracellular Matrix Proteins from Tissues and Digestion into Peptides fo Mass Spectrometry Analysis. *JoVE—J. Vis. Exp.* **2015**, *101*, e53057. [CrossRef]
18. Schiller, H.B.; Fernandez, I.E.; Burgstaller, G.; Schaab, C.; Scheltema, R.A.; Schwarzmayr, T.; Strom, T.M.; Eickelberg, O.; Mann, M Time- and compartment-resolved proteome profiling of the extracellular niche in lung injury and repair. *Mol. Syst. Biol.* **2015** *11*, 819. [CrossRef]
19. Gocheva, V.; Naba, A.; Bhutkar, A.; Guardia, T.; Miller, K.M.; Li, C.M.; Dayton, T.L.; Sanchez-Rivera, F.J.; Kim-Kiselak, C. Jailkhani, N.; et al. Quantitative proteomics identify Tenascin-C as a promoter of lung cancer progression and contributor to a signature prognostic of patient survival. *Proc. Natl. Acad. Sci. USA* **2017**, *114*, E5625–E5634. [CrossRef]
20. McCabe, M.C.; Schmitt, L.R.; Hill, R.C.; Dzieciatkowska, M.; Maslanka, M.; Daamen, W.F.; van Kuppevelt, T.H.; Hof, D.J.; Hansen K.C. Evaluation and Refinement of Sample Preparation Methods for Extracellular Matrix Proteome Coverage. *Mol. Cell Proteom* **2021**, *20*, 100079. [CrossRef]
21. Decaris, M.L.; Gatmaitan, M.; FlorCruz, S.; Luo, F.; Li, K.; Holmes, W.E.; Hellerstein, M.K.; Turner, S.M.; Emson, C.L. Proteomic analysis of altered extracellular matrix turnover in bleomycin-induced pulmonary fibrosis. *Mol. Cell Proteom.* **2014**, *13*, 1741–1752 [CrossRef]
22. Barallobre-Barreiro, J.; Baig, F.; Fava, M.; Yin, X.; Mayr, M. Glycoproteomics of the Extracellular Matrix: A Method for Intact Glycopeptide Analysis Using Mass Spectrometry. *JoVE—J. Vis. Exp.* **2017**, *122*, e55674. [CrossRef]
23. Massey, V.L.; Dolin, C.E.; Poole, L.G.; Hudson, S.V.; Siow, D.L.; Brock, G.N.; Merchant, M.L.; Wilkey, D.W.; Arteel, G.E. The hepatic "matrisome" responds dynamically to injury: Characterization of transitional changes to the extracellular matrix in mice *Hepatology* **2017**, *65*, 969–982. [CrossRef]
24. Rabe, M.; Schaefer, F. Non-Transgenic Mouse Models of Kidney Disease. *Nephron* **2016**, *133*, 53–61. [CrossRef]
25. Eddy, A.A.; Lopez-Guisa, J.M.; Okamura, D.M.; Yamaguchi, I. Investigating mechanisms of chronic kidney disease in mouse models. *Pediatr. Nephrol.* **2012**, *27*, 1233–1247. [CrossRef]
26. Clarke, T. Mice make medical history. *Nature* **2002**, *417*, 74–78. [CrossRef]
27. Zuberi, A.; Lutz, C. Mouse Models for Drug Discovery. Can New Tools and Technology Improve Translational Power? *ILAR J.* **2016**, *57*, 178–185. [CrossRef]
28. Lipp, S.N.; Jacobson, K.R.; Hains, D.S.; Schwarderer, A.L.; Calve, S. 3D Mapping Reveals a Complex and Transient Interstitial Matrix During Murine Kidney Development. *J. Am. Soc. Nephrol.* **2021**, *32*, 1649–1665. [CrossRef]
29. Shah, A.D.; Goode, R.J.A.; Huang, C.; Powell, D.R.; Schittenhelm, R.B. LFQ-Analyst: An Easy-To-Use Interactive Web Platform to Analyze and Visualize Label-Free Proteomics Data Preprocessed with MaxQuant. *J. Proteome Res.* **2020**, *19*, 204–211. [CrossRef]
30. Liu, P.; Xie, X.; Jin, J. Isotopic Nitrogen-15 Labeling of Mice Identified Long-lived Proteins of the Renal Basement Membranes. *Sci. Rep.* **2020**, *10*, 5317. [CrossRef]
31. Louzao-Martinez, L.; Van Dijk, C.G.; Xu, Y.J.; Korn, A.; Bekker, N.J.; Brouwhuis, R.; Nicese, M.N.; Demmers, J.A.; Goumans, M.-J.T.; Masereeuw, R. A proteome comparison between human fetal and mature renal extracellular matrix identifies EMILIN1 as a regulator of renal epithelial cell adhesion. *Matrix Biol. Plus* **2019**, *4*, 100011. [CrossRef]
32. Randles, M.J.; Lausecker, F.; Kong, Q.; Suleiman, H.; Reid, G.; Kolatsi-Joannou, M.; Davenport, B.; Tian, P.; Falcone, S.; Potter, P.; et al. Identification of an Altered Matrix Signature in Kidney Aging and Disease. *J. Am. Soc. Nephrol.* **2021**, *32*, 1713–1732. [CrossRef]
33. Nakayama, K.H.; Lee, C.C.; Batchelder, C.A.; Tarantal, A.F. Tissue specificity of decellularized rhesus monkey kidney and lung scaffolds. *PLoS ONE* **2013**, *8*, e64134. [CrossRef]
34. Paunas, F.T.I.; Finne, K.; Leh, S.; Osman, T.A.; Marti, H.P.; Berven, F.; Vikse, B.E. Characterization of glomerular extracellular matrix in IgA nephropathy by proteomic analysis of laser-captured microdissected glomeruli. *BMC Nephrol.* **2019**, *20*, 410. [CrossRef]
35. Sobreiro-Almeida, R.; Melica, M.E.; Lasagni, L.; Osorio, H.; Romagnani, P.; Neves, N.M. Particulate kidney extracellular matrix: Bioactivity and proteomic analysis of a novel scaffold from porcine origin. *Biomater. Sci.* **2021**, *9*, 186–198. [CrossRef]
36. Keane, T.J.; Swinehart, I.T.; Badylak, S.F. Methods of tissue decellularization used for preparation of biologic scaffolds and in vivo relevance. *Methods* **2015**, *84*, 25–34. [CrossRef]
37. Pacheco-Fernandez, N.; Pakdel, M.; Blank, B.; Sanchez-Gonzalez, I.; Weber, K.; Tran, M.L.; Hecht, T.K.; Gautsch, R.; Beck, G.; Perez, F.; et al. Nucleobindin-1 regulates ECM degradation by promoting intra-Golgi trafficking of MMPs. *J. Cell Biol.* **2020**, *219*, e201907058. [CrossRef]
38. Warner, F.J.; Lew, R.A.; Smith, A.I.; Lambert, D.W.; Hooper, N.M.; Turner, A.J. Angiotensin-converting enzyme 2 (ACE2), but not ACE, is preferentially localized to the apical surface of polarized kidney cells. *J. Biol. Chem.* **2005**, *280*, 39353–39362. [CrossRef]
39. Bu, L.; Qu, S.; Gao, X.; Zou, J.J.; Tang, W.; Sun, L.L.; Liu, Z.M. Enhanced angiotensin-converting enzyme 2 attenuates angiotensin II-induced collagen production via AT1 receptor-phosphoinositide 3-kinase-Akt pathway. *Endocrine* **2011**, *39*, 139–147. [CrossRef]
40. Fang, H.; Chen, W.; Gao, Y.; Shen, Y.; Luo, M. Molecular mechanisms associated with Angiotensin-converting enzyme-inhibitory peptide activity on vascular extracellular matrix remodeling. *Cardiology* **2014**, *127*, 247–255. [CrossRef]
41. Schlicht, K.; Rohmann, N.; Geisler, C.; Hollstein, T.; Knappe, C.; Hartmann, K.; Schwarz, J.; Tran, F.; Schunk, D.; Junker, R.; et al. Circulating levels of soluble Dipeptidylpeptidase-4 are reduced in human subjects hospitalized for severe COVID-19 infections. *Int. J. Obes.* **2020**, *44*, 2335–2338. [CrossRef] [PubMed]

42. Mora Huertas, A.C.; Schmelzer, C.E.H.; Luise, C.; Sippl, W.; Pietzsch, M.; Hoehenwarter, W.; Heinz, A. Degradation of tropoelastin and skin elastin by neprilysin. *Biochimie* **2018**, *146*, 73–78. [CrossRef] [PubMed]
43. Zhu, W.; Smith, J.W.; Huang, C.M. Mass spectrometry-based label-free quantitative proteomics. *J. Biomed. Biotechnol.* **2010**, *2010*, 840518. [CrossRef]
44. Bulow, R.D.; Boor, P. Extracellular Matrix in Kidney Fibrosis: More Than Just a Scaffold. *J. Histochem. Cytochem.* **2019**, *67*, 643–661. [CrossRef]
45. Klein, J.A.; Meng, L.; Zaia, J. Deep Sequencing of Complex Proteoglycans: A Novel Strategy for High Coverage and Site-specific Identification of Glycosaminoglycan-linked Peptides. *Mol. Cell Proteom.* **2018**, *17*, 1578–1590. [CrossRef]
46. Raghunathan, R.; Sethi, M.K.; Zaia, J. On-slide tissue digestion for mass spectrometry based glycomic and proteomic profiling. *MethodsX* **2019**, *6*, 2329–2347. [CrossRef]
47. Westergren-Thorsson, G.; Hedstrom, U.; Nybom, A.; Tykesson, E.; Ahrman, E.; Hornfelt, M.; Maccarana, M.; van Kuppevelt, T.H.; Dellgren, G.; Wildt, M.; et al. Increased deposition of glycosaminoglycans and altered structure of heparan sulfate in idiopathic pulmonary fibrosis. *Int. J. Biochem. Cell Biol.* **2017**, *83*, 27–38. [CrossRef]
48. Hu, Y.; Wang, M.; Pan, Y.; Li, Q.; Xu, L. Salvianolic acid B attenuates renal interstitial fibrosis by regulating the HPSE/SDC1 axis. *Mol. Med. Rep.* **2020**, *22*, 1325–1334. [CrossRef]
49. Amelina, H.; Cristobal, S. Proteomic study on gender differences in aging kidney of mice. *Proteome Sci.* **2009**, *7*, 16. [CrossRef]
50. Lofaro, F.D.; Cisterna, B.; Lacavalla, M.A.; Boschi, F.; Malatesta, M.; Quaglino, D.; Zancanaro, C.; Boraldi, F. Age-Related Changes in the Matrisome of the Mouse Skeletal Muscle. *Int. J. Mol. Sci.* **2021**, *22*, 10564. [CrossRef]
51. Cox, J.; Mann, M. MaxQuant enables high peptide identification rates, individualized p.p.b.-range mass accuracies and proteome-wide protein quantification. *Nat. Biotechnol.* **2008**, *26*, 1367–1372. [CrossRef]
52. Cox, J.; Neuhauser, N.; Michalski, A.; Scheltema, R.A.; Olsen, J.V.; Mann, M. Andromeda: A peptide search engine integrated into the MaxQuant environment. *J. Proteome Res.* **2011**, *10*, 1794–1805. [CrossRef]
53. Tyanova, S.; Temu, T.; Cox, J. The MaxQuant computational platform for mass spectrometry-based shotgun proteomics. *Nat. Protoc.* **2016**, *11*, 2301–2319. [CrossRef] [PubMed]
54. Cox, J.; Hein, M.Y.; Luber, C.A.; Paron, I.; Nagaraj, N.; Mann, M. Accurate proteome-wide label-free quantification by delayed normalization and maximal peptide ratio extraction, termed MaxLFQ. *Mol. Cell Proteom.* **2014**, *13*, 2513–2526. [CrossRef] [PubMed]
55. Nangia-Makker, P.; Balan, V.; Raz, A. Galectin-3 binding and metastasis. *Methods Mol. Biol.* **2012**, *878*, 251–266. [CrossRef]
56. Klezovitch, O.; Scanu, A.M. Domains of apolipoprotein E involved in the binding to the protein core of biglycan of the vascular extracellular matrix: Potential relationship between retention and anti-atherogenic properties of this apolipoprotein. *Trends Cardiovasc. Med.* **2001**, *11*, 263–268. [CrossRef] [PubMed]
57. Fujita, K.; Teramura, N.; Hattori, S.; Irie, S.; Mitsunaga-Nakatsubo, K.; Akimoto, Y.; Sakamoto, N.; Yamamoto, T.; Akasaka, K. Mammalian arylsulfatase A functions as a novel component of the extracellular matrix. *Connect. Tissue Res.* **2010**, *51*, 388–396. [CrossRef] [PubMed]
58. Leitans, J.; Kazaks, A.; Balode, A.; Ivanova, J.; Zalubovskis, R.; Supuran, C.T.; Tars, K. Efficient Expression and Crystallization System of Cancer-Associated Carbonic Anhydrase Isoform IX. *J. Med. Chem.* **2015**, *58*, 9004–9009. [CrossRef]
59. Takagaki, Y.; Shi, S.; Katoh, M.; Kitada, M.; Kanasaki, K.; Koya, D. Dipeptidyl peptidase-4 plays a pathogenic role in BSA-induced kidney injury in diabetic mice. *Sci. Rep.* **2019**, *9*, 7519. [CrossRef]
60. Lambeir, A.M.; Durinx, C.; Scharpe, S.; De Meester, I. Dipeptidyl-peptidase IV from bench to bedside: An update on structural properties, functions, and clinical aspects of the enzyme DPP IV. *Crit. Rev. Clin. Lab. Sci.* **2003**, *40*, 209–294. [CrossRef]
61. Bank, R.A.; Groener, J.E.; van Gemund, J.J.; Maaswinkel, P.D.; Hoeben, K.A.; Schut, H.A.; Everts, V. Deficiency in N-acetylgalactosamine-6-sulfate sulfatase results in collagen perturbations in cartilage of Morquio syndrome A patients. *Mol. Genet. Metab.* **2009**, *97*, 196–201. [CrossRef]
62. Valstar, M.J.; Bertoli-Avella, A.M.; Wessels, M.W.; Ruijter, G.J.; de Graaf, B.; Olmer, R.; Elfferich, P.; Neijs, S.; Kariminejad, R.; Suheyl Ezgu, F.; et al. Mucopolysaccharidosis type IIID: 12 new patients and 15 novel mutations. *Hum. Mutat.* **2010**, *31*, E1348–E1360. [CrossRef]
63. Khan, M.M.; Simizu, S.; Suzuki, T.; Masuda, A.; Kawatani, M.; Muroi, M.; Dohmae, N.; Osada, H. Protein disulfide isomerase-mediated disulfide bonds regulate the gelatinolytic activity and secretion of matrix metalloproteinase-9. *Exp. Cell Res.* **2012**, *318*, 904–914. [CrossRef]
64. Bi, S.; Hong, P.W.; Lee, B.; Baum, L.G. Galectin-9 binding to cell surface protein disulfide isomerase regulates the redox environment to enhance T-cell migration and HIV entry. *Proc. Natl. Acad. Sci. USA* **2011**, *108*, 10650–10655. [CrossRef]
65. McKee, C.M.; Xu, D.; Kessler, B.M.; Muschel, R.J. Proteomic analysis reveals a proteolytic feedback loop in murine seminal fluid. *Prostate* **2013**, *73*, 1427–1440. [CrossRef]
66. Booth, A.J.; Hadley, R.; Cornett, A.M.; Dreffs, A.A.; Matthes, S.A.; Tsui, J.L.; Weiss, K.; Horowitz, J.C.; Fiore, V.F.; Barker, T.H.; et al. Acellular normal and fibrotic human lung matrices as a culture system for in vitro investigation. *Am. J. Respir. Crit. Care Med.* **2012**, *186*, 866–876. [CrossRef]
67. Yamada, H.; Adachi, T.; Yamada, Y.; Misao, S.; Suzuki, K.; Watanabe, H.; Kitagawa, W.; Miura, N.; You, C.; Sakuma, M.; et al. Extracellular-superoxide dismutase production in mesangial cell growing in extracellular matrix. *Nihon Jinzo Gakkai Shi* **2005**, *47*, 32–37.

68. Jamaluddin, M.F.B.; Nahar, P.; Tanwar, P.S. Proteomic Characterization of the Extracellular Matrix of Human Uterine Fibroids. *Endocrinology* **2018**, *159*, 2656–2669. [CrossRef]
69. Geuens, T.; Ruiter, F.A.A.; Schumacher, A.; Morgan, F.L.C.; Rademakers, T.; Wiersma, L.E.; van den Berg, C.W.; Rabelink, T.J.; Baker, M.B.; LaPointe, V.L.S. Thiol-ene cross-linked alginate hydrogel encapsulation modulates the extracellular matrix of kidney organoids by reducing abnormal type 1a1 collagen deposition. *Biomaterials* **2021**, *275*, 120976. [CrossRef]
70. Goddeeris, M.M.; Wu, B.; Venzke, D.; Yoshida-Moriguchi, T.; Saito, F.; Matsumura, K.; Moore, S.A.; Campbell, K.P. LARGE glycans on dystroglycan function as a tunable matrix scaffold to prevent dystrophy. *Nature* **2013**, *503*, 136–140. [CrossRef]
71. Klobucar, M.; Sedic, M.; Gehrig, P.; Grossmann, J.; Bilic, M.; Kovac-Bilic, L.; Pavelic, K.; Kraljevic Pavelic, S. Basement membrane protein ladinin-1 and the MIF-CD44-beta1 integrin signaling axis are implicated in laryngeal cancer metastasis. *Biochim. Biophys. Acta* **2016**, *1862*, 1938–1954. [CrossRef]
72. Motoki, K.; Megahed, M.; LaForgia, S.; Uitto, J. Cloning and chromosomal mapping of mouse ladinin, a novel basement membrane zone component. *Genomics* **1997**, *39*, 323–330. [CrossRef]
73. Eisenberg, S.; Sehayek, E.; Olivecrona, T.; Vlodavsky, I. Lipoprotein lipase enhances binding of lipoproteins to heparan sulfate on cell surfaces and extracellular matrix. *J. Clin. Investig.* **1992**, *90*, 2013–2021. [CrossRef]
74. Allan, C.M.; Larsson, M.; Jung, R.S.; Ploug, M.; Bensadoun, A.; Beigneux, A.P.; Fong, L.G.; Young, S.G. Mobility of "HSPG-bound" LPL explains how LPL is able to reach GPIHBP1 on capillaries. *J. Lipid Res.* **2017**, *58*, 216–225. [CrossRef]
75. Goreczny, G.J.; Forsythe, I.J.; Turner, C.E. Hic-5 regulates fibrillar adhesion formation to control tumor extracellular matrix remodeling through interaction with tensin1. *Oncogene* **2018**, *37*, 1699–1713. [CrossRef]
76. Bokhove, M.; Nishimura, K.; Brunati, M.; Han, L.; de Sanctis, D.; Rampoldi, L.; Jovine, L. A structured interdomain linker directs self-polymerization of human uromodulin. *Proc. Natl. Acad. Sci. USA* **2016**, *113*, 1552–1557. [CrossRef]
77. Brunati, M.; Perucca, S.; Han, L.; Cattaneo, A.; Consolato, F.; Andolfo, A.; Schaeffer, C.; Olinger, E.; Peng, J.; Santambrogio, S.; et al. The serine protease hepsin mediates urinary secretion and polymerisation of Zona Pellucida domain protein uromodulin. *Elife* **2015**, *4*, e08887. [CrossRef]
78. Keppie, S.J.; Mansfield, J.C.; Tang, X.; Philp, C.J.; Graham, H.K.; Onnerfjord, P.; Wall, A.; McLean, C.; Winlove, C.P.; Sherratt, M.J.; et al. Matrix-Bound Growth Factors are Released upon Cartilage Compression by an Aggrecan-Dependent Sodium Flux that is Lost in Osteoarthritis. *Function* **2021**, *2*, zqab037. [CrossRef]
79. Thakar, K.; Krocher, T.; Savant, S.; Gollnast, D.; Kelm, S.; Dietz, F. Secretion of hepatoma-derived growth factor is regulated by N-terminal processing. *Biol. Chem.* **2010**, *391*, 1401–1410. [CrossRef]
80. Suliman, H.B.; Ali, M.; Piantadosi, C.A. Superoxide dismutase-3 promotes full expression of the EPO response to hypoxia. *Blood* **2004**, *104*, 43–50. [CrossRef]
81. Durbeej, M.; Talts, J.F.; Henry, M.D.; Yurchenco, P.D.; Campbell, K.P.; Ekblom, P. Dystroglycan binding to laminin alpha1LG4 module influences epithelial morphogenesis of salivary gland and lung in vitro. *Differentiation* **2001**, *69*, 121–134. [CrossRef] [PubMed]

International Journal of
Molecular Sciences

MDPI

Article

Modeling SILAC Data to Assess Protein Turnover in a Cellular Model of Diabetic Nephropathy

Barbara Di Camillo [1,*,†], Lucia Puricelli [2,3,†], Elisabetta Iori [2], Gianna Maria Toffolo [1], Paolo Tessari [2] and Giorgio Arrigoni [3,4,*]

1 Department of Information Engineering, University of Padova, 35131 Padova, Italy
2 Department of Medicine, University of Padova, 35128 Padova, Italy
3 Proteomics Center, University of Padova and Azienda Ospedaliera di Padova, 35128 Padova, Italy
4 Department of Biomedical Sciences, University of Padova, 35131 Padova, Italy
* Correspondence: barbara.dicamillo@unipd.it (B.D.C.); giorgio.arrigoni@unipd.it (G.A.)
† These authors contributed equally to this work.

Abstract: Protein turnover rate is finely regulated through intracellular mechanisms and signals that are still incompletely understood but that are essential for the correct function of cellular processes. Indeed, a dysfunctional proteostasis often impacts the cell's ability to remove unfolded, misfolded, degraded, non-functional, or damaged proteins. Thus, altered cellular mechanisms controlling protein turnover impinge on the pathophysiology of many diseases, making the study of protein synthesis and degradation rates an important step for a more comprehensive understanding of these pathologies. In this manuscript, we describe the application of a dynamic-SILAC approach to study the turnover rate and the abundance of proteins in a cellular model of diabetic nephropathy. We estimated protein half-lives and relative abundance for thousands of proteins, several of which are characterized by either an altered turnover rate or altered abundance between diabetic nephropathic subjects and diabetic controls. Many of these proteins were previously shown to be related to diabetic complications and represent therefore, possible biomarkers or therapeutic targets. Beside the aspects strictly related to the pathological condition, our data also represent a consistent compendium of protein half-lives in human fibroblasts and a rich source of important information related to basic cell biology.

Keywords: proteomics; protein turnover rate; protein half-life; SILAC; diabetes; nephropathy; fibroblasts

Citation: Di Camillo, B.; Puricelli, L.; Iori, E.; Toffolo, G.M.; Tessari, P.; Arrigoni, G. Modeling SILAC Data to Assess Protein Turnover in a Cellular Model of Diabetic Nephropathy. *Int. J. Mol. Sci.* **2023**, *24*, 2811. https://doi.org/10.3390/ijms24032811

Academic Editor: Paolo Iadarola

Received: 14 December 2022
Revised: 27 January 2023
Accepted: 29 January 2023
Published: 1 February 2023

1. Introduction

Modern technological advances have provided investigators with sophisticated methodologies that allow an extensive characterization of the cell proteome. Wide-search proteomics is a powerful tool to identify and quantify an enormous number of proteins and to tackle problems related to all areas of biology and medicine [1]. In the past few years, the study of the protein turnover rate on a large-scale has emerged as a new dimension in proteomics research, and a number of different approaches have been proposed to allow a reliable estimate of protein synthesis and degradation rates (extensively reviewed in [2]). Indeed, protein turnover is tightly regulated by several cellular processes which allow the maintenance of an efficient and functional protein pool and, at the same time, remove from the cellular environment unfolded, misfolded, degraded, non-functional or damaged proteins, as well as insoluble aggregates [3]. Hence, a dysfunctional proteostasis impacts the pathophysiology of many diseases, in particular neurodegenerative disorders but also cancer and metabolic conditions [4–6].

Historically, the first studies investigating the issue of protein synthesis and degradation involved the incorporation into proteins of radiolabeled amino acids and allowed the protein turnover rate to be expressed only as total protein dynamics [7]. Later on, proteomic approaches, mainly based on the exposure to labelled amino acid(s) of either cells cultured in vitro, or of small experimental animals in vivo, were used to estimate the rate of synthesis and/or degradation of individual proteins on a large-scale level [8–11]. The most widely used approach is known as dynamic-SILAC (often referred also as pulsed-SILAC or pSILAC), an evolution of the classical SILAC approach [12] in which cells are first cultured in normal unlabeled medium (the "light" medium) and then switched to a medium containing stable isotope labeled amino acids (usually ^{13}C labelled Lysine and/or Arginine, generally indicated as "heavy" amino acids). Samples are collected at different time points, and the rates of incorporation of the heavy amino acids (which correspond mainly to the rate of synthesis) and of degradation of the pre-existing protein pool (corresponding to the degradation rate) are measured using liquid chromatography coupled to tandem mass spectrometry (LC-MS/MS). The ratio between the signal of the heavy peptide and the signal of the residual light peptide directly reflects protein turnover [2]. By applying this approach, the kinetics of each protein can be associated with its known function(s), thus expanding our knowledge of the relationships between protein expression, function, and turnover.

Primary cultures of cells are a powerful model to investigate several aspects of protein metabolism, among them protein turnover [13]. Cultures of human fibroblasts have been widely exploited as an in vitro system to investigate some pathophysiological mechanisms of disease, such as diabetes mellitus [14–17], particularly those associated with the development of diabetic complications [18–20]. Diabetic nephropathy (DN) is a leading cause of morbidity and mortality in diabetes [21]. Both genetic and environmental factors are associated with the development of diabetic nephropathy, particularly in type 1 diabetes mellitus (T1DM) [22]. As a matter of fact, genetic factors may either convey the risk of, or protect from, diabetic nephropathy [23]. Their expression profiles in skin fibroblasts from type 1 diabetic patients, could reflect genetic influences; therefore, they were removed from in vivo environmental influences [24].

Our laboratories have established and largely exploited the model of primary cultures of human skin fibroblasts as a tool to study protein expression in type 1 diabetes [20,25,26]. In this study, we describe the application of a dynamic-SILAC approach to primary human fibroblasts derived from diabetic subjects with and without diabetic nephropathy with the purpose of extensively characterizing the dynamics and the abundance of the proteins and, whenever possible, deriving meaningful information regarding the energetic factors underlying the relationship between stability and protein abundance in this cellular model.

2. Results

2.1. Dynamic-SILAC Experiment and Determination of prOtein Half-Lives

To accurately estimate protein half-life in a model of diabetic nephropathy, we applied a dynamic-SILAC approach to fibroblasts obtained from 10 type 1 diabetic patients, 5 of whom were affected by diabetic nephropathy. Cells were cultured as described in Section 4, and the dynamic-SILAC experiment was performed in steady-state conditions, when cells were at confluence. To verify that no evident morphological changes occurred during the timeframe of the experiment, we regularly checked cell morphology and counted viable cells. Results for an exemplifying subject are reported in Table 1.

Table 1. Number of viable cells repeatedly counted during the experiment.

Time Point	Average Cell Number
1 h	5.26×10^6
2 h	5.68×10^6
4 h	5.66×10^6
7.5 h	5.56×10^6
24 h	5.98×10^6

The proteomic workflow adopted to estimate individual protein turnover rates is described in Figure 1. On average, 1287 proteins were confidently identified and quantified in each of the 10 subjects, with a standard deviation SD equal to 86 (more details about the number of identified proteins in each sample are reported in Supplementary Table S1). As described in the experimental procedures, to increase the robustness of the estimate of the turnover constant k, we decided to consider only those proteins that were quantified at 24 h and at least one out of the other two time points. After filtering, a total of 1661 different proteins from the 10 subjects were considered for the parameter k estimate.

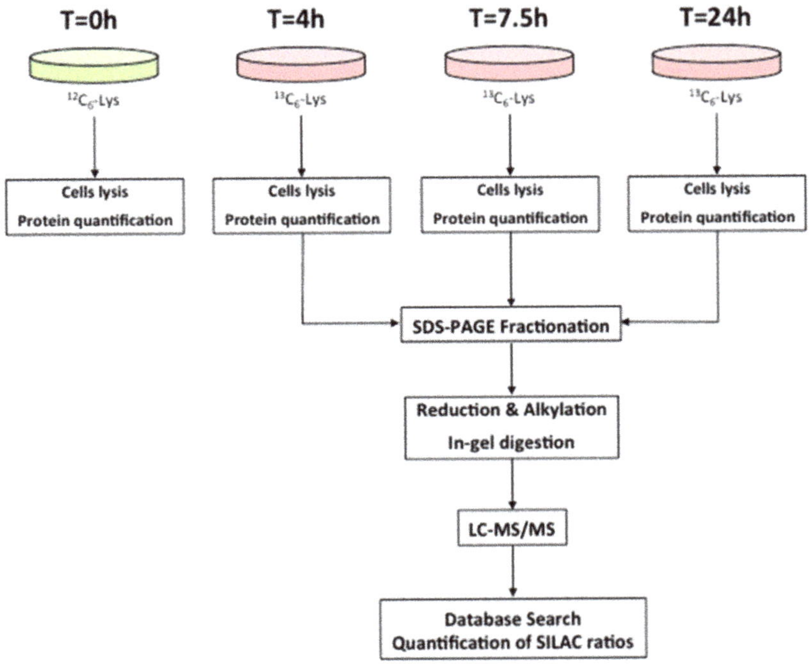

Figure 1. Proteomics workflow used to estimate turnover rates of individual proteins. Cells were grown in the SILAC heavy medium and collected at 4, 7.5, and 24 h. Following cell lysis and protein quantification, a fractionation step by SDS-PAGE was performed, and proteins were then digested and analyzed by LC-MS/MS as detailed in Section 4.

As specified in Section 4, protein quantification was obtained by the software as the median value of the quantification of all unique peptides belonging to each protein. For these proteins, a model of technical variability was derived from peptide measurements using these latter as replicated protein measurements (see Section 4). As shown in Figure 2, neither the standard deviation (SD, empty circles) nor the coefficient of variation (CV, solid circles) are constant; the former increases, whereas the latter decreases with the average of the heavy to light ratio (H/L).

Therefore, we excluded a constant SD and constant CV model for our data and adopted the following model describing the SD as a function of H/L ratio r:

$$SD = \sqrt{\alpha^2 + \beta^2 \cdot r^2} \quad (1)$$

where α and β are constants fixed at the values of 0.02 and 0.1, respectively. The fit of the model of technical variability is shown in Figure 2.

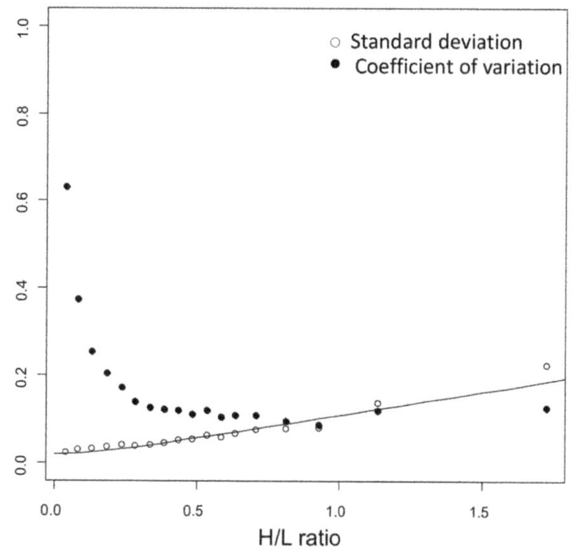

Figure 2. Model of technical variability. The standard deviation (SD, empty circles), and the coefficient of variation (CV, solid circles) are not constant with the average of the heavy to light ratio (H/L). Therefore, the model $SD = \sqrt{\alpha^2 + \beta^2 \cdot r^2}$ was used to fit the standard deviation. The model was derived from replicated measurements of peptides.

The rate constant parameter k of the 1661 proteins in our dataset was identified by fitting the H/L ratio r data across time (see details in Section 4), using the least square method weighted accordingly to the model of technical variability (Equation (1)). To assess the goodness of fit, we considered the CV of the parameter estimate, whose distribution is shown in Figure 3.

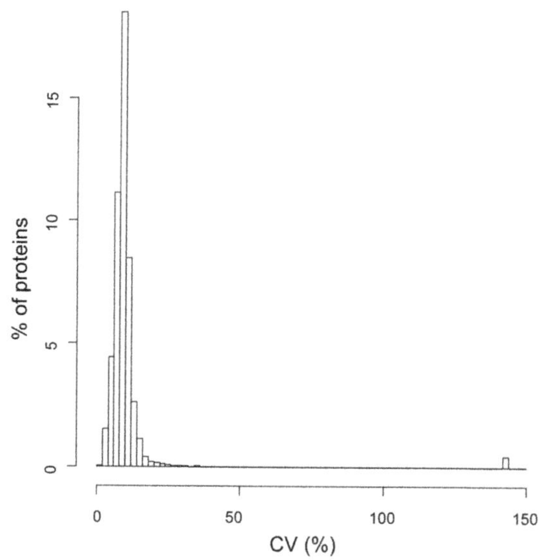

Figure 3. Coefficient of variation of the parameter estimate. Distribution of the CV of the parameter k estimate of the 1661 proteins in our dataset identified by fitting the H/L ratio across different time points.

From the distribution plot, it is clear that for a vast majority of proteins, the parameter k was estimated with very good precision, i.e., with a CV lower than 15%. A total of 1642 different proteins across all subjects (on average, 935 proteins for each subject, with a standard deviation SD equal to 119) showed a CV of the parameter estimate lower than 50% and were retained for further analysis (Table S2).

Figure 4 shows the average half-life distribution of the 1642 analyzed proteins, where the half-life $T_{1/2}$ is computed as $\log2/k$. Considering only the proteins for which a reliable estimate of k could be obtained in at least 2 out of 10 samples (in total 1338 proteins), the distribution of the mean values of $T_{1/2}$ ranges between 3 and 573 h, being on average equal to 59.9 h (Table S3). Note that the protein with the highest $T_{1/2}$ is keratin 5, probably a contaminant protein that could be filtered out of the dataset. In this case, the highest $T_{1/2}$ would be 477 h.

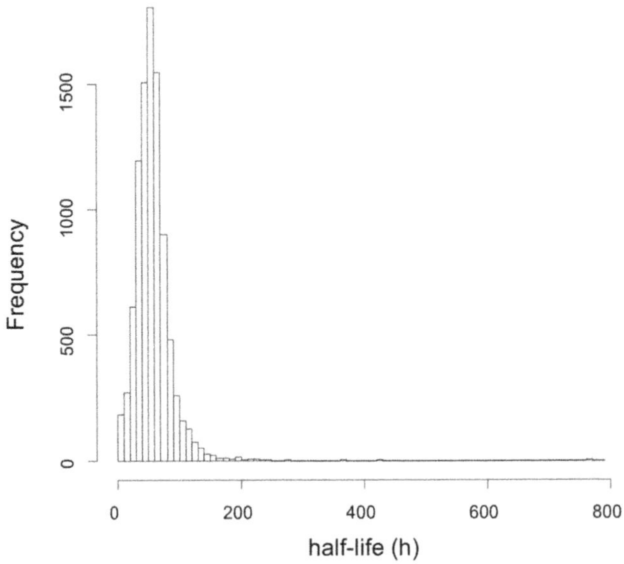

Figure 4. Distribution of the mean values of half-life $T_{1/2}$. The vast majority of proteins show a half-life between 3 and 200 h with an average of 59.9 h.

2.2. Protein Turnover in Diabetic-Controls vs. Diabetic-Nephropathic Subjects

After evaluating the normal distribution of data (Saphiro-Wilk test at 0.05 level), a t-test was performed on the $T_{1/2}$ value of 974 proteins, for which k was estimated in at least two subjects for each group, i.e., diabetic-controls vs. diabetic-nephropathic subjects. For 20 proteins a p-value ≤ 0.05 was found (Table 2), although it was not significant after correction for multiple testing. Probably the high variability of the data, mainly linked to the biological variability of the human subjects, together with the low numerosity of the dataset, does not allow for significant p-values after correction.

Gene set enrichment analysis was also performed to assess pathways associated with the two phenotypes. The 29 Reactome pathways [27,28] were found to be significantly associated with diabetic nephropathic subjects in terms of increased protein half-life with respect to diabetic controls. These are reported in Table S4 and discussed later in the text.

Table 2. Proteins with a significantly different half-life in Diabetic and Nephropathic subjects.

Gene Name	Protein Description	Average Half-Lives in h (SD) Diabetic	Nephropathic	*p*-Value	Valid Values Diabetic	Nephropathic
RAB13	RAB13, member RAS oncogene family	26.4 (1.5)	32.6 (0.1)	1.90×10^{-3}	5	2
RAB23	RAB23, member RAS oncogene family	33.0 (2.8)	42.9 (4.3)	1.10×10^{-2}	5	3
SQRDL	sulfide quinone reductase-like (yeast)	74.6 (12.7)	123.6 (16.5)	1.54×10^{-2}	4	2
PABPC1	poly(A) binding protein, cytoplasmic 1	41.2 (3.9)	49.1 (4.4)	1.85×10^{-2}	5	5
IPO7	importin 7	28.7 (1.8)	34.7 (3.9)	2.02×10^{-2}	4	5
NPM1	nucleophosmin (nucleolar phosphoprotein B23, numatrin)	48.2 (3.4)	55.6 (4.5)	2.10×10^{-2}	5	5
TCP1	t-complex 1	58.0 (6.0)	73.7 (12.1)	2.71×10^{-2}	5	5
EIF4A1	eukaryotic translation initiation factor 4A1	29.8 (1.8)	34.7 (4.0)	3.31×10^{-2}	5	5
CCT8	chaperonin containing TCP1, subunit 8 (theta)	57.8 (5.2)	68.9 (9.3)	3.34×10^{-2}	5	5
EIF4G1	eukaryotic translation initiation factor 4 gamma, 1	26.8 (1.3)	32.2 (3.8)	3.66×10^{-2}	4	4
TFRC	transferrin receptor	23.3 (2.9)	29.2 (3.2)	3.72×10^{-2}	4	5
TOP2B	topoisomerase (DNA) II beta 180kDa	33.0 (3.3)	25.9 (0.3)	3.93×10^{-2}	3	2
ACTB	actin, beta	65.3 (5.3)	89.8 (20.9)	4.12×10^{-2}	5	4
ACLY	ATP citrate lyase	35.2 (4.7)	43.5 (4.9)	4.25×10^{-2}	5	5
ST13P4	Suppression of tumorigenicity 13 pseudogene 4	21.8 (2.5)	37.6 (7.3)	4.36×10^{-2}	3	2
GSTM5	glutathione S-transferase mu 5	53.5 (5.9)	104.9 (18.5)	4.38×10^{-2}	2	2
HNRNPF	heterogeneous nuclear ribonucleoprotein F	29.8 (4.0)	38.1 (6.0)	4.80×10^{-2}	5	4
PA2G4	proliferation-associated 2G4, 38kDa	49.4 (3.7)	62.2 (12.5)	4.85×10^{-2}	5	5
CAPRIN1	cell cycle associated protein 1	14.7 (0.9)	16.2 (1.1)	4.89×10^{-2}	5	5
EIF4H	eukaryotic translation initiation factor 4H	28.9 (1.9)	37.6 (3.6)	4.90×10^{-2}	2	3

2.3. Protein Quantification in Diabetic-Controls vs. Diabetic-Nephropathic Subjects

Protein abundance relative quantification was performed on a total of 2226 different proteins across 10 subjects (Table S5). To note that protein half-life could be reliably assessed for 1664 proteins, since for the estimate of *k* we decided to use data associated with proteins identified at least at two temporal points, one of which was set to be 24 h. This limitation does not apply for the estimate of protein abundance, which therefore led to the quantification of a higher number of proteins. Abundance levels in the three time points indirectly confirmed the steady state hypothesis. Indeed, for each protein, the variance across different time points tends to be equal to or lower than the technical variation, measured as the variance observed across peptides matching the same protein in the same sample (Figure 5). Here it is clear that protein abundance does not significantly vary across the three time points, thus indicating that a steady state can be confidently assumed.

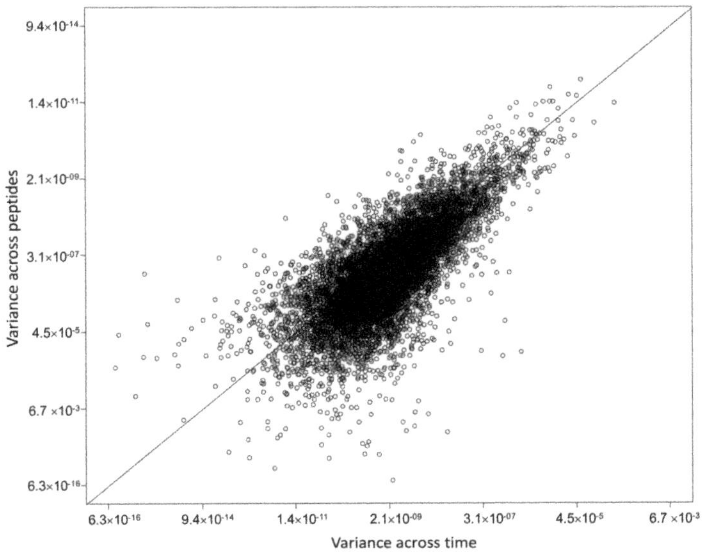

Figure 5. Variance across time vs. technical variance. The figure shows the variance across different time points (x axis) vs. the technical variation (y axis), measured as the variance observed across peptides matching the same protein in the same sample. The variance across time is equal to or lower than the technical variation, in line with the steady state assumption.

After confirming the normal distribution of data (Saphiro-Wilk test at 0.05 level), a t-test was performed on 1299 proteins, for which the relative abundance was measured on at least two subjects for each group, i.e., diabetic-controls vs. diabetic-nephropathic subjects (Table S5). For 40 proteins a p-value ≤ 0.05 was found (Table 3), although it was not significant after correction for multiple testing.

Table 3. Proteins with a significantly different abundance in Diabetic and Nephropathic subjects.

Gene Name	Protein Description	Fold Change (Nephropathic vs. Diabetic)	p-Value	Valid Values	
				Diabetic	Nephropathic
GLIPR2	GLI pathogenesis-related 2	−1.8	3.30×10^{-3}	4	4
RPS3A	ribosomal protein S3A	−1.6	3.90×10^{-3}	5	5
TRIM25	tripartite motif containing 25	−1.5	6.41×10^{-3}	3	4
SLC25A6	solute carrier family 25 (mitochondrial carrier; adenine nucleotide translocator), member 6	−1.7	6.95×10^{-3}	5	5
TUBB4A	tubulin, beta 4A class IVa	2.1	9.41×10^{-3}	5	5
NDUFB10	NADH dehydrogenase (ubiquinone) 1 beta subcomplex, 10, 22kDa	−1.7	1.21×10^{-2}	5	4
NPC2	Niemann-Pick disease, type C2	2.5	1.47×10^{-2}	2	4
SYNE1	spectrin repeat containing, nuclear envelope 1	−2.2	1.47×10^{-2}	5	5

Table 3. *Cont.*

Gene Name	Protein Description	Fold Change (Nephropathic vs. Diabetic)	p-Value	Valid Values	
				Diabetic	Nephropathic
SEC31A	SEC31 homolog A (S. cerevisiae)	1.6	1.55×10^{-2}	3	4
TUBB4B	tubulin, beta 4B class IVb	3.3	1.70×10^{-2}	2	4
SLC25A3	solute carrier family 25 (mitochondrial carrier; phosphate carrier), member 3	−1.7	2.40×10^{-2}	5	5
DNAJC8	DnaJ (Hsp40) homolog, subfamily C, member 8	−1.8	2.40×10^{-2}	5	4
COX4I1	cytochrome c oxidase subunit IV isoform 1	−1.7	2.42×10^{-2}	5	5
LGALS3	lectin, galactoside-binding, soluble, 3	−1.4	2.57×10^{-2}	5	5
NQO1	NAD(P)H dehydrogenase, quinone 1	−2.1	2.64×10^{-2}	5	5
SERPINB2	serpin peptidase inhibitor, clade B (ovalbumin), member 2	1.6	2.76×10^{-2}	2	4
XRCC6	X-ray repair complementing defective repair in Chinese hamster cells 6	−1.4	2.80×10^{-2}	5	5
SDHB	succinate dehydrogenase complex, subunit B, iron sulfur (Ip)	−1.6	2.85×10^{-2}	4	5
FHL2	four and a half LIM domains 2	−1.6	2.87×10^{-2}	4	4
RBMX	RNA binding motif protein, X-linked	−1.6	3.03×10^{-2}	2	3
PSMA7	proteasome (prosome, macropain) subunit, alpha type, 7	−2.1	3.18×10^{-2}	3	3
XRCC5	X-ray repair complementing defective repair in Chinese hamster cells 5 (double-strand-break rejoining)	−1.4	3.21×10^{-2}	5	5
SLC25A5	solute carrier family 25 (mitochondrial carrier; adenine nucleotide translocator), member 5	−1.9	3.34×10^{-2}	5	5
RNH1	ribonuclease/angiogenin inhibitor 1	1.4	3.43×10^{-2}	5	5
TUBA1C	tubulin, alpha 1c	1.7	3.57×10^{-2}	2	5
CAPNS1	calpain, small subunit 1	−1.8	3.71×10^{-2}	4	5
RRAS	related RAS viral (r-ras) oncogene homolog	−1.7	3.82×10^{-2}	5	5

Table 3. *Cont.*

Gene Name	Protein Description	Fold Change (Nephropathic vs. Diabetic)	*p*-Value	Valid Values	
				Diabetic	Nephropathic
CYB5A	cytochrome b5 type A (microsomal)	−1.8	3.90×10^{-2}	4	5
CTSD	cathepsin D	−1.4	3.95×10^{-2}	5	5
EEF1B2	eukaryotic translation elongation factor 1 beta 2	−1.6	4.08×10^{-2}	5	5
MAPK14	mitogen-activated protein kinase 14	−1.6	4.09×10^{-2}	2	3
PSMB2	proteasome (prosome, macropain) subunit, beta type, 2	−1.6	4.22×10^{-2}	2	3
MTPN	myotrophin	−1.6	4.30×10^{-2}	5	5
MYO1B	myosin IB	1.6	4.35×10^{-2}	5	5
APPL2	adaptor protein, phosphotyrosine interaction, PH domain and leucine zipper containing 2	−1.3	4.51×10^{-2}	5	5
RPS3	ribosomal protein S3	−1.7	4.55×10^{-2}	5	5
FKBP7	FK506 binding protein 7	−1.7	4.59×10^{-2}	5	5
QDPR	quinoid dihydropteridine reductase	−2.8	4.72×10^{-2}	4	3
HM13	histocompatibility (minor) 13	−2.0	4.74×10^{-2}	2	4
IMPA1	inositol(myo)-1(or 4)-monophosphatase 1	−2.6	4.99×10^{-2}	3	3

2.4. Characterization of Proteins with Similar Half-Life and Abundance

Given the large range of protein half-lives spanning between 3 and 573 h (Figure 4 and Table S3), downstream analysis was performed to acquire further information on the proteins characterized by a similar range of half-lives and abundance. Considering that only a few proteins showed a significant change in protein half-life between diabetic controls and nephropathic subjects (Table 2), we decided to use the average value of k calculated on all 10 subjects to cluster proteins that consistently show half-life within arbitrarily predefined intervals. These groups of proteins were analyzed for their interactions and functional enrichment using STRING v. 11.0, DAVID v. 6.7, and Revigo. Results were filtered to keep only significant enriched terms with a minimum of 4 counts and a p value < 0.001. The results in their full form are reported in Table S6 and graphically displayed in Figure S1. A concise summary of this analysis is also presented in Table 4, which reports the considered half-life intervals, the number of proteins falling in each interval, and their functional characterization.

According to this analysis, the most stable proteins (with $T_{1/2} > 80$ h) are those involved in energy metabolism, cellular respiration, chromatin organization, and DNA packaging, while the proteins characterized by a higher turnover (i.e., those with $T_{1/2} < 30$ h) are mainly secreted proteins such as collagens and proteins involved in extracellular matrix organization.

A similar analysis was also performed using the protein abundance calculated from the overall protein expression data. The range was divided into arbitrarily pre-defined intervals to cluster proteins with similar abundance. Since only a limited number of proteins show a significantly different abundance between diabetic controls and nephropathic subjects (Table 3), we decided to average the values of all subjects, keeping only the proteins for which an estimate could be obtained for at least 2 out of 10 patients. Figure 6 shows that our estimation of protein expression data for 1801 proteins spans four orders of magnitude

Table 4. Gene Ontology annotation of proteins grouped according to their average half-life. Proteins are grouped in 9 arbitrary ranges of half-life, and the most enriched and significant GO terms (Biological Process, Cellular Component, Molecular Function, and KEGG pathways) are listed for each group.

$T_{1/2}$ Range (h)	# of Proteins	Biological Processes	Cellular Components	Molecular Functions	KEGG Pathways
<20	64	Extracellular matrix organization; collagen metabolism; cell adhesion and motility	Extracellular matrix; vesicle; collagen trimer	Extracellular matrix structural constituent; Receptor binding; Protein binding	ECM-receptor interaction; Focal adhesion
20–30	82	Small GTPase mediated signal transduction; mRNA splicing; localization	Cytoplasm; Spliceosomal complex	RNA binding; GTPase activity	
30–40	157	Intracellular transport; Translation; RNA splicing; localization	Spliceosomal complex; Nuclear part; Cytoplasmic part; Ribonucleoprotein complex; EIF3 and EIF4F	Protein binding; RNA binding; Nucleotide binding; Nucleoside-triphosphatase activity	Spliceosome
40–50	193	Protein transport; Golgi vesicle transport; RNA splicing; Translation	Cytoplasm; Vesicle; Golgi; Cytoskeleton; EIF3; Ribonucleoprotein complex; Plasma membrane part	Protein binding; Nucleotide binding; RNA binding; actin binding; GTPase activity; Translation initiation factor activity	Endocytosis
50–60	227	Vesicle mediated transport; Cell cycle progress; Membrane and cytoskeleton organization; Translation elongation; Protein folding	Cytoplasm; Ribosome; Cytoskeleton; Endoplasmic reticulum;	Protein binding; Nucleotide binding; RNA binding; GTPase activity; Structural constituent of ribosome	Aminoacyl-tRNA biosynthesis; Ribosome
60–70	272	RNA processing; Translation; Protein metabolism; Ribosome biogenesis; Protein folding; Ras protein signal transduction	Cytosol; Mitochondria; Nuclear part; Ribosome; Cytoskeleton; Proteasome complex	Structural constituent of ribosome; actin and cytoskeletal protein binding; RNA binding; Protein binding; Threonine-type endopeptidase activity.	Ribosome; Proteasome
70–80	170	Carbohydrate and protein metabolism; Oxidation-reduction process; Cellular respiration; Protein folding; Translation elongation; Acetyl-CoA metabolism;	Cytoplasm; Mitochondrion; Cytoskeleton; Endoplasmic reticulum part; Proteasome complex; Melanosome	Peroxiredoxin activity; Oxidoreductase activity; endopeptidase activity; Isomerase activity; Cytoskeletal protein binding; Coenzyme binding;	Glycolysis/ Gluconeogenesis; Proteasome; Pyruvate metabolism; TCA cycle;
80–90	76	Carbohydrate metabolism; Generation of energy; Carboxylic acid, alcohol, and ketone metabolism;	Cytoplasm; Mitochondrion;	Catalytic activity; Monosaccharide binding; Oxidoreductase activity; Isomerase activity	Glycolysis/ Gluconeogenesis
>90	97	Generation of energy; Oxidative phosphorylation; Carboxylic acid, and ketone metabolism; Chromatin organization and DNA packaging	Cytoplasm; ATP synthase complex; Mitochondrion; Nucleosome; Nuclear membrane; Protein-DNA complex		

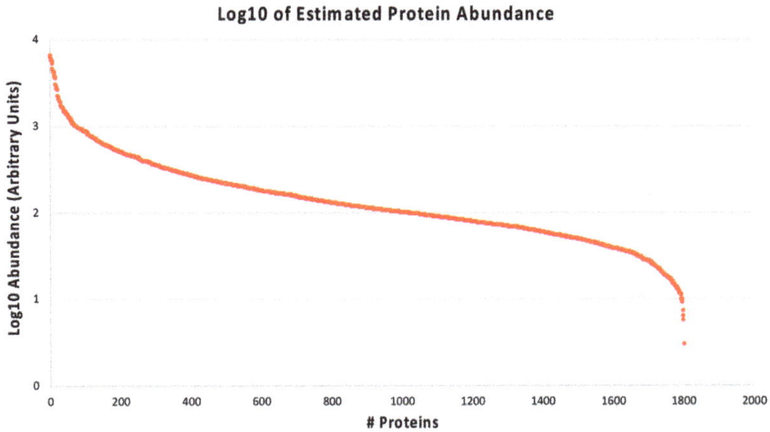

Figure 6. Distribution of protein abundances. The average protein abundance estimated for the entire protein set is shown in a logarithmic scale (Arbitrary Units) and spans 4 orders of magnitude.

This range was divided into 8 intervals, and proteins clustered in each interval were analyzed for their interactions and functional enrichment using STRING v. 11.0, DAVID v. 6.7, and Revigo. Results were filtered to keep only significant enriched terms with a minimum of 4 counts and a p value < 0.001. The complete results are reported in Table S7 and graphically displayed in Figure S2. A summary of this analysis is reported in Table 5, where Log2 abundance intervals, the number of proteins falling in each interval, and their functional characterization are listed.

Table 5. Gene Ontology annotation of proteins grouped according to their average abundance. Proteins are grouped in 8 arbitrary ranges of abundance (expressed as Log2 of arbitrary units), and the most enriched and significant GO terms (Biological Process, Cellular Component, Molecular Function, and KEGG pathways) are listed for each group.

Relative Abundance (Log2 A.U.)	# of Proteins	Biological Processes	Cellular Component	Molecular Function	KEGG Pathways
>9.0	202	Translation elongation; Protein folding; DNA packaging; Cytoskeleton organization; Cell redox homeostasis, Glycolytic process	Cytosol; Nucleus; Protein-DNA complex; Cytoskeleton; Vesicle; Large ribosomal subunit	RNA binding; Structural constituent of ribosome; GTP binding; Protein binding; Cytoskeletal and actin binding	Ribosome; Systemic lupus erythematosus; Pathogenic *Escherichia coli* infection
8.0–9.0	224	Translational elongation; RNA splicing; Cytoskeleton organization; Protein transport; Small GTPase mediated signal transduction; Generation of precursor metabolites and energy	Cytoplasm; Actin cytoskeleton; Small ribosomal subunit; Spliceosomal complex; Vesicle	Protein binding; GTP binding; RNA binding; Cytoskeletal protein and actin binding; Structural constituent of ribosome; Hydrogen ion transmembrane transporter activity	Ribosome; Spliceosome; Parkinson's disease

Table 5. *Cont.*

Relative Abundance (Log2 A.U.)	# of Proteins	Biological Processes	Cellular Component	Molecular Function	KEGG Pathways
7.5–8.0	180	Intracellular transport; Vesicle mediated transport; Localization; Vesicle and membrane organization; Small GTPase mediated signal transduction	Cytoplasm; Vesicle; Endoplasmic reticulum; Cytoskeleton; Arp2/3 protein complex; Proteasome core complex; Ribonucleoprotein complex	Protein binding; Actin binding; RNA binding; GTP binding; GTPase activity; Threonine-type peptidase activity	Proteasome; Pathogenic *Escherichia coli* infection
7.0–7.5	220	Intracellular transport; RNA processing and splicing; Translation; Membrane organization; Protein folding; Energy derivation by oxidation of organic compounds	Cytosol; Endoplasmic reticulum; Vesicles; Endomembrane system; Spliceosomal complex; EIF3 complex	Protein binding; RNA binding; Translation initiation factory activity;	
6.5–7.0	283	Nucleotide Metabolism; Heterocycle metabolism; Protein metabolism; Protein folding; RNA splicing; Redox processes; Cellular respiration; Response to oxidative stress	Cytoplasm; Intracellular organelle part; Macromolecular complex	RNA binding; Nucleotide binding; NADH dehydrogenase activity; Oxidoreductase activity; Protein binding; Cofactor binding	Aminoacyl-tRNA biosynthesis; Proteasome; Oxidative phosphorylation; Huntington's disease; Amino sugar and nucleotide sugar metabolism; Alzheimer's disease
6.0–6.5	261	Protein transport; Protein metabolism; Carboxylic acid metabolism; Amine metabolism; Mitotic cell cycle; regulation of ligase activity	Cytoplasm; Mitochondrion; Proteasome complex; Ribonucleoprotein complex; Organelle membrane	Protein binding; RNA binding; Nucleotide binding; Cytoskeletal proteins and actin binding; Translation initiation factor activity	Proteasome
5.0–6.0	298	Acetyl-CoA metabolism; Heterocycle metabolism; Nucleotide biosynthesis; carboxylic acid metabolism; Cellular respiration; Protein transport	Cytoplasm; Mitochondrion; Golgi apparatus part; Envelope	Catalytic activity; Hydrolase activity; Protein binding; Purine nucleotide binding; Pyrophosphatase activity	TCA cycle

Table 5. *Cont.*

Relative Abundance (Log2 A.U.)	# of Proteins	Biological Processes	Cellular Component	Molecular Function	KEGG Pathways
<5.0	135	mRNA transport; Protein transport; Cellular localization; Oxidation-reduction process	Cytoplasm; Nuclear part; Envelope	Purine nucleotide binding; Catalytic activity; Electron carrier activity	Valine, leucine, and isoleucine degradation

In general, when plotting average protein abundance vs. average protein turnover rate k (Figure 7, left upper and lower panels) or, alternatively, average relative protein abundance vs. average protein half-life (Figure 7, right upper and lower panels), several clusters of proteins can be easily identified. Based on pre-defined intervals, similar to those used in the preview analyses, it is possible to distinguish four main groups of proteins: (1) proteins with low turnover rate (long half-life) and high abundance (in red in Figure 7); (2) proteins with high turnover rate (short half-life) and high abundance (in blue); (3) proteins with high turnover rate (short half-life) and low abundance (in green); and finally (4) proteins with low turnover rate (long half-life) and low abundance (in pink).

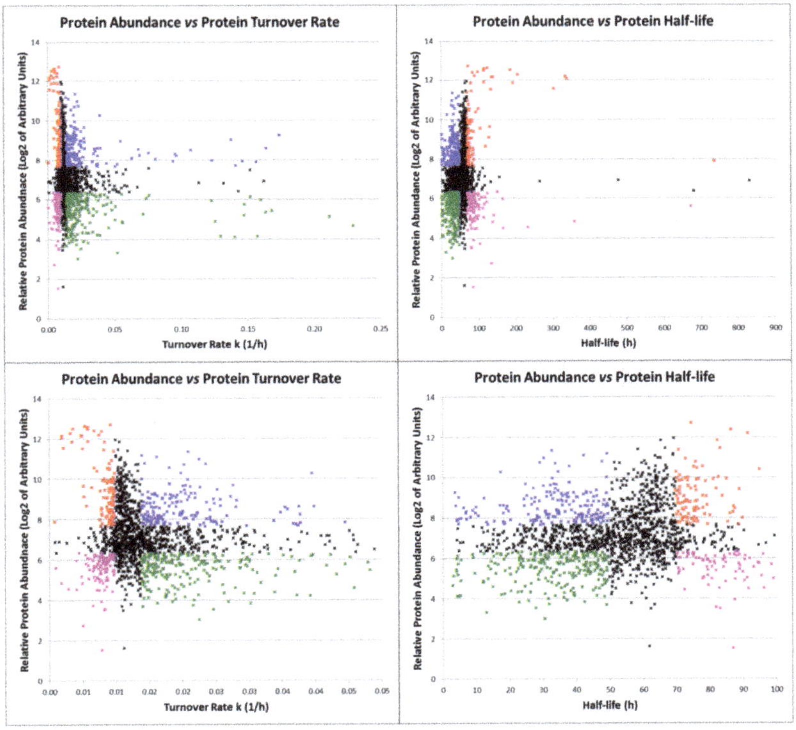

Figure 7. Distribution of turnover rates as a function of protein abundance. Left panels show the turnover rate constant k as a function of protein abundance, while right panels show the protein half-life as a function of protein abundance. Zoomed data are on display in the lower panels. Proteins characterized by a low turnover rate and high abundance are depicted in red, proteins with a high turnover rate and high abundance in blue, proteins with a high turnover rate and low abundance in green, and proteins with a low turnover rate and low abundance in pink.

The proteins belonging to each of these groups were analyzed with STRING to highlight possible interactions and functional enrichment. The results, graphically displayed in Figure S3 and detailed in Table S8, show that the first group is particularly enriched for cytoskeletal proteins, proteins involved in chromatin organization, and proteins related to energy production. The second group involves mainly proteins related to extracellular matrix organization, cellular response to stress, metabolism of mRNA, and vesicle mediated transport. The third group is enriched for secreted proteins and proteins involved in RNA splicing, membrane trafficking, and the cell cycle. Finally, to the fourth group belong mainly proteins involved in the metabolism of proteins, hydrocarbons, and lipids.

3. Discussion

In this work, we quantified protein turnover using in vitro cultured fibroblast cells harvested from ten diabetic subjects, five of whom had nephropathy.

We performed a dynamic SILAC experiment on the cultured fibroblasts that had reached confluence. At time T = 0, the regular culture medium was replaced with a medium containing the heavy amino acid $^{13}C_6$ Lysine, and cells were sampled at 4, 7.5, and 24 h (Figure 1). During the entire duration of the experiment, cells were regularly checked and no significant changes in their number or morphology could be detected (Table 1). Cells were lysed, proteins were subjected to SDS-PAGE to reduce sample complexity, and in-gel digestion was performed with the protease LysC. The choice of using only labeled lysine and not also labeled arginine, as it is usually done in SILAC experiments, was made to avoid any problem that could arise from the possible arginine-to-proline metabolic conversion [29–31] and that could therefore affect the estimate of turnover rate. The use of LysC as a digesting enzyme yields a lower number of protein identifications with respect to the classical digestion with trypsin, since less and larger peptides are generated. However, we preferred to acquire more robust data at the expense of a smaller dataset. In total, we performed 300 nLC-MS/MS analyses. SILAC H/L ratios at different times were used to estimate the turnover parameter k by using weighted least squares (as detailed in Section 4). The CV of the parameter estimate, used to assess the goodness of fit, was lower than 15% in the vast majority of cases, thus confirming that the adopted model is adequate to describe the data (Figure 3). With this strategy, we were able to confidently estimate the turnover parameter k (and therefore the half-life) for 1642 different proteins, generating a high-quality dataset of protein turnover rate from human fibroblasts isolated from type 1 diabetic patients (Table S2).

3.1. *Proteins with a Significantly Different Turnover in Nephropathic Subjects*

On the 974 proteins for which k was estimated in at least two subjects for each group, we could perform a t-test to highlight proteins with a significant ($p \leq 0.05$) different turnover rate between diabetic and nephropathic subjects. Only the 20 proteins listed in Table 2 turned out to have half-lives different in the two groups. By looking at the reported data, it is evident that all proteins except one (namely TOP2B) are characterized by longer half-lives in nephropathic subjects with respect to the diabetic controls. A screening of the literature reveals that almost all proteins listed in Table 2 have been reported to be implicated in nephropathy, often of diabetic origin.

3.1.1. Mesangial Proteins

The Rab family of small G proteins plays important roles in mediating vesicular membrane trafficking in eukaryotic cells [32,33], and more than 60 mammalian Rab proteins have been identified and characterized.

Rab13 and its effector protein, JRAB/MICAL-L2, are involved in the transport of the cell adhesion molecules occludin and claudins to the tight junctional area in epithelial cells [34,35]. Rab13 was identified in a gene expression profiling study to be altered in a population of macrophages from nephritic NZB/W mice [36].

Knocking-down or overexpressing Rab23 affected the expression of collagen in cultured mesangial cells, thus suggesting that Rab23 may be overexpressed in FSGS mice to suppress hedgehog signaling and/or influence collagen synthesis [37]. A proteomic study conducted on mesangial cells points to the possible involvement of Rab23 in a variety of cellular events, such as gene expression, signaling, protein synthesis, organ and tissue morphology, cellular movement, and contraction function [38].

3.1.2. Chaperone and Cytoskeleton Proteins

In this study, we identified TCP1 and CCT8 as proteins with an altered turnover rate in nephropathic subjects. The chaperonin-containing T-complex (TRiC/CCT complex) is a chaperone machinery that assists the folding of dozens of proteins, in particular those that appear to be slow-folding and aggregation-prone [39]. However, this complex has been known for a long time to fold actin and tubulin [40–42] and evidence suggests that disruptions of actin dynamics result in altered cytoskeletal organization [43]. Interestingly, TRiC/CCT was also identified in our Gene Set Enrichment Analysis as one of the major Reactome Pathways to be affected in our model of diabetic nephropathy (Table S4). It has been reported that the beta subunit of the complex may play a central role in mesangial cell hypo-contractility in diabetic nephropathy [44], while both CCT2 and CCT8 were found to be significantly altered in exosomes derived from primary human proximal tubular epithelial cells (PTEC) under diseased conditions [45].

Actin and tubulin are the major components of the cytoskeletal structure. A disassembly of the actin cytoskeleton and marked alterations of beta tubulin, a major component of microtubules, represent prominent features of DN [46,47]. Modifications of chaperone-like proteins have been previously detected in cultured fibroblasts from T1DM subjects with nephropathy, and they may be patho-physiologically related to the development of diabetic renal disease [20,48]. Changes in the cytoskeleton are key alterations in the pathophysiology of DN: substantial differences in cytoskeletal and cytoskeleton-related protein expression were found between normal subjects and T1DM patients with DN but not with T1DM patients without DN [20], suggesting that nephropathy, and not diabetes *per se*, was associated with the observed changes.

3.1.3. Proteins Associated to Hydrogen Sulfide (H_2S) Metabolism

Just like other gaseous compounds, such as nitric oxide (NO) and carbon monoxide (CO), H_2S is known to act as a signaling molecule [49,50] and modulate a vast array of biological functions [51]. The conversion of hydrogen sulfide to thiosulfate and sulfane is catalyzed by the mitochondrial protein sulfide quinone oxidoreductase (SQRDL or SQOR) with the help of a quinone, usually ubiquinone [52].

Ubiquinone (also known as Coenzyme Q10 or CoQ10) is involved in several processes (primarily the electron transport chain) and functions as a cofactor for several enzymes, among them the SQRDL protein. CoQ10 deficiency is the cause of several human diseases, and mutations in the *COQ8B* gene result mainly in the disruption of kidney function, causing a steroid-resistant nephrotic syndrome [53]. Interestingly, H_2S oxidation impairment causes CoQ10 associated nephrotic syndrome, a chronic kidney disease related to CoQ10 deficiency, and it has been shown that reduced SQOR levels lead to increased ROS production, thus contributing to oxidative stress in conditions of CoQ deficiency [54]. Our data show a strong increase in SQOR half-life in nephropathic subjects (123.6 h) with respect to the diabetic controls (74.6 h); however, the physiological significance of such a finding is difficult to grasp. On one hand, it may indicate a reduced enzyme efficiency and a higher oxidative stress. On the other hand, longer enzyme survival could instead determine a decrease in the H_2S levels and a reduced oxidative stress. Therefore, the increased half-life of this enzyme in the cultured fibroblasts from T1DM subjects with nephropathy could be interpreted as an attempt to activate a protective mechanism through a reduction of oxidative stress, inflammation, mesangial cell proliferation, and an inhibition of the renin-angiotensin system activity [55–57].

3.1.4. Proteins Involved in Translation and Kidney Hypertrophy

Kidney hypertrophy and matrix accumulation are associated with the development of long-term complications of diabetes [58], and translation has been reported to represent a potential biomarker for the prognosis of kidney disease [59].

We identified three subunits of the eIF4F complex (EIF4A1, EIF4G1, and EIF4H) and the PABPC1 protein, all of which had a turnover rate $\approx$20% greater in the nephropathic than in control subjects (Table 2). EIF4A is an ATP-dependent RNA helicase with low activity. However, the ATPase and helicase activities are strongly stimulated when EIF4A is in complex with eIF4G, eIF4E, eIF4B, and eIF4H [60].

Interestingly, accumulating evidence has highlighted a central role for translation in hypertrophy in models of diabetic nephropathy, both in vivo and in vitro [61,62]. Moreover, EIF4F has been reported to be a potential biomarker for membranous nephropathy prognosis [63], and PABPC1 is listed among the proteins associated with kidney diseases from the curated CTD Gene-Disease Associations dataset (http://ctdbase.org/detail.go?type=disease&acc=MESH:D007674 accessed on 1 July 2022).

3.1.5. Other Proteins with Altered Turnover Rate

Glutathione S-transferase Mu 5 protein (GSTM5) exhibits an almost doubled half-life in the nephropathic vs. the non-nephropathic T1DM subjects (Table 2). Little has been reported in the literature regarding the possible association between this protein and the development or progression of diabetic nephropathy, although other members of the same protein family have been reported as putative biomarkers of diabetic nephropathy [64,65].

The half-life of Heterogeneous Nuclear Ribonucleoprotein F (HNRNPF) in nephropathic subjects is about 30% greater than that in non-nephropathic subjects (Table 2). Very interestingly, this protein is known to exert a protective role against oxidative stress and to attenuate nephropathy progression in diabetic mice and possibly in human kidneys via stimulation of Sirtuin-1 expression [66]. Therefore, the reduced turnover rate might be explained as an attempt to mitigate and counteract the adverse effects of nephropathy. Moreover, HNRNPF has been suggested to be a potential target for the treatment of hypertension and kidney injury in diabetes [67,68].

The transferrin receptor (TFRC) also shows an increased half-life in nephropathic subjects. An altered expression of TFRC has been detected on mesangial cells in IgA nephropathy [69,70], and recently the *TFRC* gene was reported to be downregulated in tubules of samples derived from patients affected by chronic kidney diseases [71].

Caprin-1 is an ubiquitous protein highly expressed in dividing cells [72]. The *Caprin-1* gene has been found to be downregulated in B2R knockout (B2R$^{-/-}$) mice, a mouse model of diabetic nephropathy [73].

The Proliferation-associated protein 2G4 (PA2G4) and the topoisomerase DNA II beta (TOP2B) genes were found to be downregulated in the obstructive nephropathy of PAI-1–overexpressing mice [74]. PA2G4 was also evaluated as a potential biomarker in the serum of type 1 diabetes patients [75].

The ATP-Citrate Lyase (ACLY), the enzyme that converts citrate to acetyl-CoA, shows an increased half-life of about 25% in nephropathic subjects with respect to the diabetic controls. Interestingly, very recently, ACLY has been reported as a critical epigenetic regulator that promotes renal injury in obesity and type 2 diabetes [76,77], while other researchers have used two independent mouse models of kidney fibrosis to demonstrate that the AKT-dependent modulation of this enzyme is involved in kidney fibrogenesis and ECM deposition [78].

3.1.6. GSEA Reveals That Proteasomal Proteins Have Longer Half-Lives in Nephropathic Subjects

Although only 20 proteins show a significant difference in protein turnover rate between diabetic controls and nephropathic subjects, GSEA highlighted a number of Reactome Pathways that are significantly different (FDR q-value < 0.05) in the two groups of patients. In particular, the proteins belonging to the pathways, identified and reported in Table S4, show an increased half-life in in all nephropathic subjects. A scrutiny of the GSEA output highlights that a large part of the proteins that contribute to the significant pathways belong to the proteasomal complex.

The ubiquitin proteasome system (UPS) plays a central role in the pathogenesis and progression of various diseases, among which is diabetic nephropathy [79,80]. The UPS is predominantly involved in protein homeostasis through the ubiquitination and proteasomal degradation of proteins. However, ubiquitination is not only involved in proteasome degradation, but also regulates the participation of substrate proteins in a variety of cell signaling pathways [80]. Proteasome inhibition has been shown to attenuate diabetic nephropathy [81] and have a protective effect against renal dysfunction [82,83]. Moreover, the deletion of the proteasome activator genes, *PA28α* and *PA28β*, resulted in protection against renal injury and retinal microvascular injury in diabetic mouse models [84]. Other researchers reported that an increased level of ROS induced by hyperglycemia covalently modifies the 20S proteasome subunits, thus decreasing its activity in the diabetic kidney [85]. Moreover, proteasome inhibitors improve renal fibrosis in rats with obstructive nephropathy [86], reduce collagen production, proliferation, and inflammation in nasal fibroblasts [87], and seem to be effective for the treatment of nephropathy [88]. Therefore, our data showing an increased half-life for a high number of proteasomal proteins in nephropathic subjects corroborates data already reported in the literature and suggests that the UPS could be a potential target for treatment of diabetic nephropathy.

3.2. *Proteins with Different Abundance in Nephropathic Subjects*

Our analysis led to the identification of 40 proteins with a significantly different abundance in diabetic controls vs. nephropathic subjects (Table 3). Most of these proteins were already reported as related to DN and are listed among the proteins associated with kidney diseases in the CTD Gene-Disease Associations dataset. For example, the GLI pathogenesis-related 2 has a fold change of -1.8 in our dataset; curiously, it has been reported that GLIPR-2 is elevated in the kidneys of patients affected by DN [89] and that miR-30e targeting GLIPR-2 is downregulated in DN, while its overexpression inhibits GLIPR-2, thus protecting from renal fibrosis in DN [90]. The fact that we found the protein to have a lower abundance in DN with respect to the diabetic controls seems, therefore, to be in contrast with the previous observations. However, it is worth noting that GLIPR-2 was found to be elevated in nephropathic kidneys with respect to normal kidneys, while we observed a reduction with respect to diabetic subjects.

We identified three members of the SLC25A family (the phosphate carrier SCL25A3, and the adenine nucleotide translocators SLC25A5 and SLCA25A6), all of them with a reduced abundance in nephropathic subjects. SCL25A3 was found to be differentially expressed in sclerosis-prone ROP-Os/+ and sclerosis resistant C57-Os/+ mouse kidneys [91], and both *SLC25A5* and *SLCA25A6* genes are reported to be modulated in type 2 diabetic patients with end-stage renal disease [92]. Interestingly, other 3 proteins (NDUFB10, SDHB and COX4I1) known to be functionally related to the SLC25 complex were identified with a lower abundance in nephropathic subjects, and with a fold change very similar to that found in the SCL25 proteins. Moreover, other three proteins that function generally in the processes of electron transport (NQO1, CYB5A, and QDPR) were also found to have decreased abundance in DN patients. The quinoid dihydropteridine reductase (QDPR) has been suggested to be an important modulator of diabetic nephropathy through the regulation of the TGF-β1/Smad3 signaling pathway [93] and to play an important role as a protective factor against oxidative stress [94], while NQO1 polymorphism has been recently associated with the risk of diabetic nephropathy [95]. Altogether, these data suggest a possible impairment of general electron transfer activity—in particular in the mitochondria—in nephropathic subjects, in agreement with data reported in the literature [96,97].

Three members of the tubulin family (namely TUBB4A, TUBB4B, and TUBA1C) show an increased abundance, in agreement with previous data that demonstrated that tubulins are strongly modulated in nephropathic subjects [20]. Moreover, other cytoskeletal proteins and cytoskeletal-regulating proteins with different abundances were identified in our study, namely SYNE1, CAPNS1, MTPN, MYO1B, and RRAS. Microarray studies have highlighted CAPNS1 as being modulated in membranous nephropathy [63] and in immunoglobulin A nephropathy [98], while other studies have reported the importance of miR-375 (for which myotrophin MTPN is a target) for glucose homeostasis and as a potential biomarker in type 2 diabetes [99,100]. Finally, both SYNE1 and MYO1B have been described as related to diabetic nephropathy before [101–103].

Although, to the best of our knowledge, not much is known about a possible role of the DNA repair proteins, XRCC5 and XRCC6, in the context of diabetic nephropathy. Recently an association between other members of the same protein family (XRCC1 and XRCC3) and diabetic nephropathy has been suggested [104].

We also identified two proteasomal proteins (PSMA7 and PSMB2), with some other functional related proteins (RPS3, RPS3A, and EEF1B2), all of them with a lower abundance in nephropathic subjects. The role of proteasomal proteins in the context of nephropathy was discussed above (paragraph 3.1.6). Interestingly, RPS3 was described to be associated with diabetic nephropathy [92], while RPS3A was also reported to have an increased expression in membranous nephropathic kidneys [105].

Galectin-3 (LGALS3) is upregulated under diabetic conditions, providing protection toward tissue injury induced by advanced glycation end-products (AGEs) [106]. It has been considered as a possible therapeutic target for prevention and treatment of diabetes and its complications [107]. In our analysis, LGALS3 has a reduced abundance in nephropathic subjects, suggesting a possible lack of a protective effect in this group of patients. On the other hand, we found SerpinB2 to be more abundant in nephropathic subjects. Interestingly, reduced levels of SerpinB2 have been associated with the delayed development of diabetic nephropathy [108].

Other proteins found with an altered abundance in our study and known to be related to diabetic nephropathy are CTSD, FHL2, and Sec31A. This latter was associated with DN [109] and is involved in the inhibition of nerve regeneration in diabetic neuropathy [110]. CTSD expression was found to be altered in the renal tubular epithelium in patients with DN [111], and very recently, a urinary proteomic study conducted on a large cohort of type 1 diabetic subjects identified cathepsin D as a promising biomarker of rapid eGFR (estimated glomerular filtration rate) decline, which reflects kidney injury [112]. FLH2, a protein implicated in Wnt/β-catenin signaling, plays a crucial role in albuminuria and has been indicated as a potential therapeutic target against diabetic kidney damage and fibrotic kidney disease [113,114]. Finally, we found a member of the MAPK family, namely MAPK14, with a reduced abundance in nephropathic subjects with respect to diabetic controls. In agreement with our data, the implication of MAPK signaling in the development and progression of diabetic nephropathy has been amply documented [115,116].

3.3. *Proteins with Similar Half-Life and Abundance Are Functionally Related*

The data synthetically summarized in Table 4 and fully displayed in Table S6 and Figure S1 highlight the notion that proteins with similar half-lives are also very often functionally related. To perform this analysis, proteins were divided into 9 arbitrary groups based on their estimated half-lives, and the enrichment of gene ontology terms was assessed using the bioinformatic tools specified in Section 4. What emerges from this analysis did not come entirely as a surprise, since our data reveal that the most stable proteins (i.e., those with $T_{1/2} > 70$ h) are mainly mitochondrial, nuclear, and cytoskeletal proteins involved in energy metabolism, cellular respiration, structural functions, protein folding, translation, chromatin organization, and DNA packaging. All these basic and very important biological functions rely on proteins that are also characterized by a medium-to-high abundance (Tables 5 and S7 and Figure S2), and therefore their rapid turnover would require a very high energy consumption. In other words, the cell invests lots of energy in the synthesis of these proteins, and therefore their half-lives are conveniently long. On the other hand, if cells need to modulate the abundance of these proteins (either increasing or decreasing it) the adjustment to the new conditions cannot be obtained in a short time, therefore requiring a longer adaptation.

Proteins characterized by shorter half-lives (between 60 and 70 h) are mainly ribosomal and proteasomal, proteins related to RNA metabolism and to Ras signal transduction. Half-lives in the range 50–60 h are typical of proteins related to cell cycle, vesicle mediated transport, and actin cytoskeletal organization, while in the range 30–50 h fall predominantly proteins involved in mRNA processing, in small GTPase mediated signal transduction and Golgi vesicle transport. Finally, it was not completely surprising to identify mainly secreted proteins and proteins involved in extracellular matrix organization among those characterized by a higher turnover rate (i.e., those with $T_{1/2}$ < 30 h). Indeed, for the category of secreted proteins, our estimated half-life is given by the contribution of two different processes: the turnover rate and the rate of secretion. Since we only measured intracellular proteins, we cannot distinguish between these two processes, although the particularly short half-life of these proteins suggests that the rate of secretion is probably much faster than the intracellular turnover rate. Regarding this aspect, it is interesting to note that collagens have estimated half-lives that are generally shorter (although not statistically significant) in nephropathic subjects with respect to diabetic controls. This might reflect the notion that nephropathic conditions are characterized by increased matrix accumulation [58].

To further confirm, as already reported by others [13,117], that proteins involved in common biological processes tend to have similar turnover rates, we compared the half-lives of a number of proteins that are subunits of specific, well-characterized macromolecular complexes. Some examples of this analysis are reported in Figure 8, where the half-lives of proteasomal alpha subunits, subunits of the coatomer and of the chaperone protein TCP1 complex, and ribosomal subunits of the 40S complex are shown.

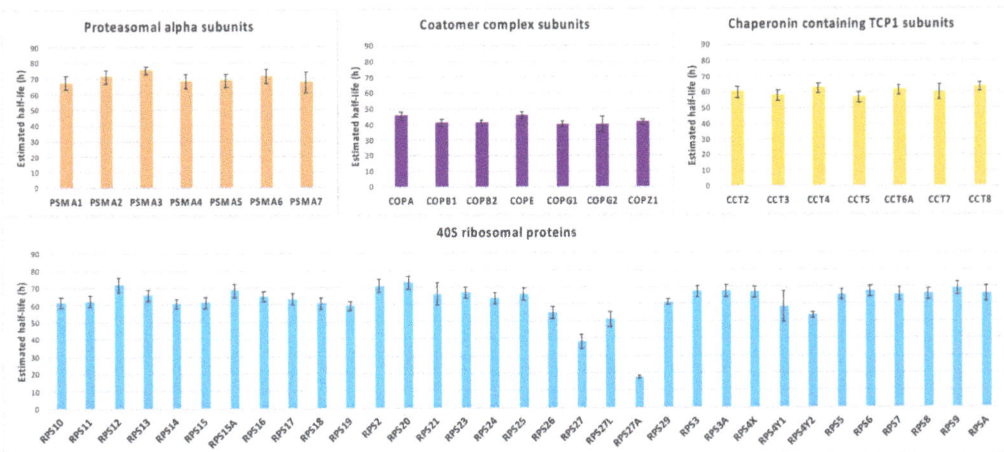

Figure 8. Half-lives of proteins belonging to selected macromolecular complexes. The graphs show the estimated half-lives for proteins that are part of specific, well-characterized macromolecular complexes. Error bars indicate standard errors.

It is evident that proteins belonging to the same functional complex have very similar estimated half-lives, with the remarkable exception of proteins RPS27A and RPS27 of the 40S ribosomal complex, which show a much faster turnover rate as compared to almost all the other subunits. Interestingly, the *RPS27A* gene codes for a single copy of ubiquitin fused to the ribosomal protein S27a; therefore, it is post-translationally regulated, and its turnover rate might therefore be strongly influenced by this process. RPS27 is a ribosomal protein with extra-ribosomal functions: it has been reported as involved in DNA repair, transcription, and signal transduction [118,119], and for these unique features, it is conceivable that its turnover rate is regulated independently of the other ribosomal subunits. To rule out the possibility that the constant trend visible in Figure 8 might be due to a fortunate coincidence, we compared the distribution of half-lives of the subunits of each complex with the distribution of 10 populations randomly generated starting from the same dataset and using the same number of proteins. The selection of the populations was performed automatically using the "Random" function of Excel. As shown in Figure 9, it is evident that in all cases the pattern relative to the randomly selected proteins is much more scattered compared to the behavior of the proteins belonging to some complex, indicating that the similar turnover rate estimated for subunits belonging to the same complex reflects a true cellular regulation.

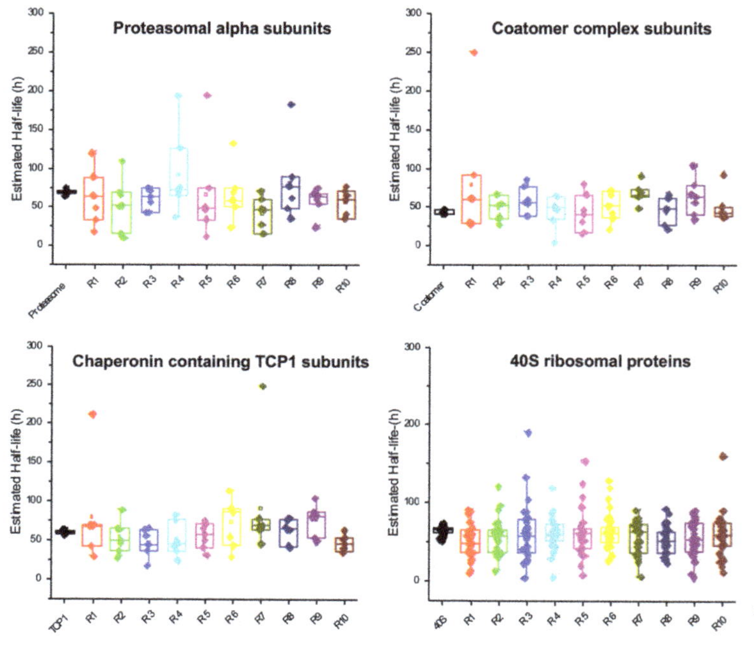

Figure 9. Box plots showing the distribution of half-lives for proteins belonging to the macromolecular complexes indicated in Figure 8, compared to the distribution of half-lives relative to 10 randomly selected populations of proteins generated automatically from the same dataset.

Similarly to what we did for the categorization of proteins with comparable turnover rate, we decided to group proteins based on their relative protein abundance. To this purpose, the abundance range was divided in arbitrarily pre-defined intervals, and GO-enriched terms were assessed for each cluster. Table 5 summarizes the results of this analysis (for more detailed results, see Table S6 and Figure S2), which show that the most abundant proteins are those related to DNA packaging, cytoskeletal organization, translation, RNA metabolism, and energy production. Protein folding and vesicle-mediated transport are mainly associated with proteins with an average medium/high abundance, while terms related to nucleotide metabolism, RNA splicing, cellular respiration, and the cell cycle are particularly enriched among proteins with a medium/low abundance. Finally, proteins characterized by low abundance are mainly related to the TCA cycle, mRNA and protein transport, nucleotide biosynthesis, and redox processes.

When protein half-lives are plotted as a function of abundance (Figure 7), some particularly interesting clusters of proteins emerge. Not surprisingly, proteins with both a long half-life and high abundance (in red in Figure 7) belong mainly to the category of structural proteins and proteins related to very basic and important cellular functions such as chromatin organization and energy production. Proteins with a short half-life and high abundance (in blue in Figure 7) require a particularly high energy consumption to be maintained and are mainly involved in extracellular matrix organization, cellular response to stress, metabolism of mRNA, and vesicle-mediated transport. Note that, as discussed above, extracellular matrix proteins are probably in this category (i.e., short half-life) only because we cannot distinguish between turnover rate and secretion rate. Beside proteins that are secreted, proteins with a short half-life and low abundance (in green in Figure 7) are mainly involved in RNA splicing, membrane trafficking, and the cell cycle. It is worth noting that the concentration of these proteins can be rapidly increased in case of need by simply limiting their degradation rate, leading to their fast accumulation. Finally, proteins with a long half-life and low abundance (in pink in Figure 7) are involved in the metabolism of proteins, hydrocarbons, and lipids. The cell does not require many copies of these proteins, but nevertheless they are involved in basic cellular functions and appear, therefore, particularly stable.

We must finally highlight that it is difficult to compare our results with similar data previously published by other research groups. For instance, when comparing the average protein half-lives obtained in this study across 10 subjects with those reported in the seminal work of Schwanhausser et al. on a single sample [9], it appears that the half-life that we estimate tends to be consistently lower, although the order of magnitude is pretty similar. These differences can be attributed to several reasons. First of all, we used patient-derived primary fibroblasts, whereas murine fibroblasts were used in [9], and moreover, we considered cells at confluence, whereas in Schwanhausser et al., the total protein abundance is assumed to double during the duration of one cell cycle. Another important difference is that the last time-point measured in our experiment is 24 h vs. 13.5 h in Schwanhausser et al. Given that the ratio r of proteins labeled with heavy and light amino acids increases slowly for high half-life proteins, having late time points in the experimental set-up should guarantee a better estimate of k for these kinds of proteins. Finally, while in Schwanhausser et al., a simple least square estimation is used, here a weighted least square estimate of parameter k was adopted.

A more reasonable comparison can be made with the data reported by Welle et al. [120], where protein turnover rate is estimated in immortalized human fibroblasts. Although both the analytical approach (classical dynamic-SILAC vs. a hyperplexing strategy) and the cellular model (patient-derived primary fibroblasts vs. immortalized fibroblasts) are not identical, the half-lives we estimated appear to be in good agreement with the data they have published (same average value of k and a correlation coefficient > 0.6), thus further supporting the reliability of our dataset.

4. Materials and Methods

4.1. Patients' Selection and Enrolment

We sought to quantify protein turnover in unperturbed fibroblast cells in a population of 10 diabetic subjects, five of whom had nephropathy. Five Caucasian T1DM patients with DN (i.e., with a urinary albumin excretion rate (AER) > 200 mg/min in sterile urine, not associated with other proteinuric diseases) and five T1DM patients without DN (AER < 20 mg/min) were recruited. The aims of the study were explained in detail, and each subject gave informed consent to the study. The protocol had been approved by the Ethical Committee of the Medical Faculty at the University of Padova and was performed according to the Helsinki Declaration (1983 revision).

The patients' characteristics are reported in detail elsewhere [20]. Age (means: 36–39 yrs), male/female ratio (2/3), body mass index (BMI) (means: 22–24 kg/m^2), diabetes duration (≈20 yrs), and glycated hemoglobin levels (means: 9–11%) were not different between the two groups of diabetic subjects. Their albumin excretion rate (AER) was determined on three timed overnight urine collections, by a turbidimetric method (Turbiquant Albumin Dade Behring, Marburg, Germany), and the median value was used for DN classification. The mean blood pressure was calculated as diastolic blood pressure plus one-third systolic (i.e., pulse) pressure. All drugs were suspended the day before the study.

4.2. Cell Collection and Culture

The fibroblasts were obtained by skin biopsies as described in detail elsewhere [25]. The skin explants were incubated at 37 °C after addition of HAM'S F-10 Nutrient Mixture medium (Sigma Aldrich, St. Louis, MO, USA) supplemented with 20% foetal bovine serum (FBS Sigma Aldrich), 1 mM glutamine, (Sigma Aldrich), 100 U/mL penicillin and 100 µg/mL streptomycin (Sigma Aldrich). The growth medium was changed every 3–4 days. Usually, the fibroblasts became visible after 4–5 days of culture, and they reached the confluence after about 3 weeks. Thereafter, the culture medium was aspirated, and cells were washed three times with PBS. The fibroblasts were recovered by adding 0.05% trypsin and 0.02% EDTA (Sigma-Aldrich), transferred into 75 cm^2 flasks and cultured with the culture medium containing 10% FBS. After the third passage, cells were frozen and kept in liquid nitrogen. Before each experiment, the fibroblasts were thawed and grown up to the 4th–5th passage.

4.3. Dynamic-SILAC Experiment, Sample Preparation and In-Gel Protein Digestion

For the dynamic SILAC experiment a custom-made Dulbecco's Modified Eagles Medium (DMEM) without L-arginine, L-lysine, and L-glutamine (Athena Enzyme systems, Baltimore, MD, USA) was used after adding L-arginine, L-glutamine, and L-lysine (Sigma) or $^{13}C_6$-Lysine (Cambridge Isotope Laboratories, Tewksbury, MA, USA), and 10% dialyzed foetal bovine serum (FBS, Invitrogen, Paisley, UK). To determine protein half-lives, a dynamic-SILAC approach was used. The fibroblasts were cultured in standard, light (L) DMEM medium until they reached confluence. Thereafter, the standard medium was removed, the cells were washed three times with 10 mL of phosphate-saline buffer (PBS, pH = 7.4, Sigma Aldrich), and the heavy medium (containing the$^{13}C_6$-Lysine) was added to the culture at time T = 0.

Cells were harvested at 4, 7.5, and 24 h, lysed by the addition of 70 µL of Tris-HCl 62.5 mM, pH 7.2, 1% SDS, and protease inhibitors (Protease Inhibitor Cocktails, Sigma Aldrich), and by repeated freeze-thaw cycles in liquid nitrogen. The samples were then centrifuged at 14,000 rpm for 15 min to remove cell debris, and the protein concentration in the supernatant was quantified by the Lowry method. Thereafter, 70 µg of total proteins for each time point and for each subject were loaded onto a 12% precast gel (NuPAGE, Invitrogen). The electrophoretic separation was performed by applying a constant voltage of 80 V for 30 min. Gel was then stained for 3 h with colloidal coomassie (SimplyBlue Safe Stain, Invitrogen) and then destained with water. Each gel lane was cut into five bands that were then subjected to in-gel enzymatic digestion with LysC protease (Promega, Madison, WI, USA) as described in [121].

4.4. LC-MS/MS Analysis

Each of the fractions obtained as specified above was analyzed by LC-MS/MS. Data were submitted for database search and quantification of SILAC H/L ratios. The analysis was conducted with a LTQ-Orbitrap XL mass spectrometer (Thermo Fisher Scientific) interfaced to a nano-HPLC Ultimate 3000 (Dionex—Thermo Fisher Scientific, Waltham, MA, USA). Samples were loaded into a 10 cm pico-frit capillary column (75 µm I.D., 15 µm tip, New Objective, Littleton, MA, USA) packed in-house with C18 material (Aeris peptide 3.6 um XB-C18, Phenomenex, Torrance, CA, USA), and peptides were separated by a linear gradient from 3% to 40% acetonitrile/0.1% formic acid in 40 min. The instrument operated in a Top4 data-dependent mode, with a full MS scan from 300 to 1700 Da acquired at high resolution (60,000) in the Orbitrap, followed by 4 MS/MS spectra of the most intense ions acquired in the linear trap.

To increase the number of identifications and robustness of quantification, each sample was analyzed twice. After the first round of analysis, all data files were searched against the human section of the Uniprot database (as specified below). All peptides that were identified with high confidence were used to create a static exclusion list that was then inserted into the instrument method. All samples were analyzed again under identical chromatographic and instrumental conditions, but with the application of the static exclusion list. The second round of analysis allows for an increase in the number of protein groups and unique peptides confidently identified, thus increasing the robustness of quantification. A series of representative Venn diagrams showing the performance of this analytical strategy are reported in Figure S4. In total, 300 LC-MS/MS analyses were performed (2 analyses for each of the five gel bands, for the three time points, and for the 10 patients).

4.5. Protein Identification and Quantification

All raw files generated in the study were analyzed with the software Proteome Discoverer (version 1.2, Thermo Fisher Scientific) interfaced to a Mascot server (version 2.2.4, Matrix Science, Chicago, IL, USA). The search was performed against the human section of the Uniprot Database (www.uniprot.org, accessed on 1 April 2013) using the MudPIT protocol. LysC was selected as the digesting enzyme, with up to one missed cleavage allowed. Precursor and fragment tolerances were set at 10 ppm and 0.6 Da, respectively. Carbamidomethyl cysteine was selected as a static modification, while methionine oxidation and $^{13}C_6$-lysine were set as variable modifications. Data were filtered based on the search against a corresponding randomized database, and the false discovery rate (FDR) was calculated by the software. Only proteins identified with at least 2 unique peptides with high confidence (>99%) were considered positive hits. SILAC ratios were calculated by the software for each identified peptide, and peptides were grouped into protein families according to the principle of maximum parsimony. Protein quantification was calculated as the median value of the quantification of all peptides belonging to the same protein family. For each cell line, all the data obtained from the five gel bands, both with and without the application of the excluding list, were merged into a single msf output file. Msf files relative to the three time points for each cell line were finally merged into a single multi-report file.

4.6. Kinetic Analyses

As derived in the following equations, for constant incorporation rates, the logarithm of the SILAC H/L ratios increases linearly with time; therefore, protein half-lives can be obtained by properly fitting H/L rations measured at different time points. Five biological replicates were available for each group to assess statistical significance.

Proteins labeled with light amino acids (P_L) are assumed to decay exponentially with the degradation rate constant k (Equation (2)).

$$P_L = P_{TOT} \cdot e^{-kt} \quad (2)$$

Our experiment is conducted with cells at confluence and at constant volume. Under the hypothesis of steady state, i.e., P_{TOT}, k and protein synthesis constant in time, no amino acid recycling, and assuming a mono-compartmental model, the protein labelled with heavy amino acids (P_H) can be expressed as the difference between the total number of a specific protein (P_{TOT}) and P_L as in Equation (3).

$$P_H = P_{TOT} - P_L = P_{TOT} \cdot \left(1 - e^{-kt}\right) \quad (3)$$

The rate constant of the protein decay k can then be obtained by fitting the model of the ratio r of protein labeled with heavy and light amino acids at different time points (Equation (4)).

$$\mathrm{r} = \frac{P_H}{P_L} = \frac{1 - e^{-kt}}{e^{-kt}} \tag{4}$$

By taking the natural logarithm (ln), Equation (4) can be transformed into:

$$\ln(\mathrm{r} + 1) = kt \tag{5}$$

Proteins whose SILAC ratio was not available for the time of 24 h, and for at least one of the other time points (4 or 7.5 h), were filtered out. For the remaining proteins, the parameter k was identified by fitting the H/L ratio r data according to Equation (5), using the weighted least square method. Weights were calculated as the inverse of the variance of ln(r + 1) data, starting from Equation (1) and using the error propagation rules.

Equation (1) is a model of the technical variability of r derived from peptide measurements, using these latter as replicate protein measurements. In more detail, the standard deviation (SD) and the coefficient of variation (CV) of the H/L ratio r were calculated for each protein, for each subject, and for each time point. Measurements with CV% higher than 50% were excluded from downstream analysis. The range of r values were then divided into intervals of the same bin size (0.05) or containing at least 10 protein measurements, and, for each interval, the median of the SDs and the CVs was considered to fit a model of technical variability.

The goodness of fit was evaluated using the precision of parameter k estimates; parameters with coefficients of variation higher than 50% were considered unreliable.

Once k is determined, the protein half-life, i.e., the time required for the amount of the protein to fall to half its initial value if the synthesis is zero, can be calculated as:

$$T_{1/2} = \frac{\ln(2)}{k} \tag{6}$$

4.7. Protein Abundance

In parallel, we quantified the relative protein abundance using the sum of peak intensities of all peptides matching a specific protein divided by the number of observed peptides for that protein and by the total sum of peaks in each LC-MS/MS run [9]. Under the hypothesis of steady state, protein abundance levels were averaged across the three time points.

4.8. Bioinformatic and Statistical Analysis

A number of bioinformatic tools were used to assess whether proteins characterized by similar half-lives or abundance tend to share interacting partners and be associated with similar Gene Ontology (GO) terms. For this purpose, our datasets were analyzed with STRING v. 11.0 [122] to highlight physical/functional interactions among proteins and with David Bioinformatic Resources v. 6.7 [123,124] and Revigo [125] to underline and graphically visualize enriched GO terms associated with the different classes of proteins.

Gene set enrichment analysis (GSEA) was also performed on our data using MSigDB Canonical pathways gene set collection [126]. The GSEA was used to determine whether the members of a given gene set were associated with a group. If a gene set had a positive enrichment score, a significant number of its gene members had higher expression in one of the predefined groups, and the gene set was termed "enriched". A 1000 random sample permutations were carried out, and the significance threshold was set at FDR < 0.05. All comparisons between groups were performed using the two-tailed Student t-test for unpaired data.

5. Conclusions

In this manuscript, we describe the application of a dynamic-SILAC approach to study the turnover rate and the relative abundance of proteins using a cellular model of diabetic nephropathy. Under the hypothesis of steady state, no amino acid recycling, and assuming a mono-compartmental model, we adopted a model describing the SD as a function of heavy to light ratio and estimated the parameter k using the least square method weighted accordingly to the model of technical variability. We could reliably estimate the turnover rate for more than 1600 proteins and the relative abundance for more than 2200 individual proteins. Several of these turned out to be significantly different in either half-life or abundance between nephropathic subjects and diabetic controls. Many of these proteins were already known to be related to diabetic complications and therefore represent possible biomarkers or therapeutic targets. However, beside the aspects strictly related to the pathological condition, the data collected in this study represent a reliable compendium of protein half-lives in human fibroblasts and a rich source of important information related to basic cell biology.

Supplementary Materials: The following supporting information can be downloaded at: https://www.mdpi.com/article/10.3390/ijms24032811/s1.

Author Contributions: Conceptualization, G.A. and P.T.; methodology, G.A., B.D.C. and G.M.T.; formal analysis, B.D.C.; investigation, L.P. and E.I.; resources, G.A., P.T. and G.M.T.; writing—original draft preparation, G.A.; writing—review and editing, G.A., B.D.C., P.T. and G.M.T.; visualization, G.A. and B.D.C.; supervision, G.A., P.T. and G.M.T.; project administration, P.T.; funding acquisition, G.A. and P.T. All authors have read and agreed to the published version of the manuscript.

Funding: This research was funded by the University of Padova, grant number CPDA101844.

Institutional Review Board Statement: The study was conducted in accordance with the Declaration of Helsinki, and approved by the Ethical Committee of Padova's City and University Hospital (protocol code 1469P; date of approval: 25 October 2007).

Informed Consent Statement: Informed consent was obtained from all subjects involved in the study.

Data Availability Statement: The mass spectrometry proteomics data have been deposited to the ProteomeXchange Consortium via the PRIDE partner repository with the dataset identifier PXD038541.

Acknowledgments: The authors wish to thank Roberto Trevisan (Azienda Socio Sanitaria Territoriale Papa Giovanni XXIII, Bergamo, Italy) and Monica Vedovato (Department of Medicine, University of Padova, Italy) for their support in the collection of patient-derived biopsies.

Conflicts of Interest: The authors declare no conflict of interest.

References

1. Rozanova, S.; Barkovits, K.; Nikolov, M.; Schmidt, C.; Urlaub, H.; Marcus, K. Quantitative Mass Spectrometry-Based Proteomics: An Overview. *Methods Mol. Biol.* **2021**, *2228*, 85–116. [CrossRef] [PubMed]
2. Ross, A.B.; Langer, J.D.; Jovanovic, M. Proteome Turnover in the Spotlight: Approaches, Applications, and Perspectives. *Mol. Cell Proteom.* **2021**, *20*, 100016. [CrossRef] [PubMed]
3. Labbadia, J.; Morimoto, R.I. The biology of proteostasis in aging and disease. *Annu. Rev. Biochem.* **2015**, *84*, 435–464. [CrossRef] [PubMed]
4. Zheng, Q.; Huang, T.; Zhang, L.; Zhou, Y.; Luo, H.; Xu, H.; Wang, X. Dysregulation of Ubiquitin-Proteasome System in Neurodegenerative Diseases. *Front. Aging Neurosci.* **2016**, *8*, 303. [CrossRef]
5. Abu Ahmad, Y.; Oknin-Vaisman, A.; Bitman-Lotan, E.; Orian, A. From the Evasion of Degradation to Ubiquitin-Dependent Protein Stabilization. *Cells.* **2021**, *10*, 2374. [CrossRef]
6. Kaur, N.; Raja, R.; Ruiz-Velasco, A.; Liu, W. Cellular Protein Quality Control in Diabetic Cardiomyopathy: From Bench to Bedside. *Front. Cardiovasc. Med.* **2020**, *7*, 585309. [CrossRef]
7. Waterlow, J.C. Whole-body protein turnover in humans—Past, present, and future. *Annu. Rev. Nutr.* **1995**, *15*, 57–92. [CrossRef]
8. Doherty, M.K.; Hammond, D.E.; Clague, M.J.; Gaskell, S.J.; Beynon, R.J. Turnover of the human proteome: Determination of protein intracellular stability by dynamic SILAC. *J. Proteome. Res.* **2009**, *8*, 104–112. [CrossRef]
9. Schwanhäusser, B.; Busse, D.; Li, N.; Dittmar, G.; Schuchhardt, J.; Wolf, J.; Chen, W.; Selbach, M. Global quantification of mammalian gene expression control. *Nature* **2011**, *473*, 337–342, Erratum in *Nature* **2013**, *495*, 126–127. [CrossRef]

10. Fornasiero, E.F.; Mandad, S.; Wildhagen, H.; Alevra, M.; Rammner, B.; Keihani, S.; Opazo, F.; Urban, I.; Ischebeck, T. Sakib, M.S.; et al. Precisely measured protein lifetimes in the mouse brain reveal differences across tissues and subcellula fractions. *Nat. Commun.* **2018**, *9*, 4230. [CrossRef]
11. McClatchy, D.B.; Martínez-Bartolomé, S.; Gao, Y.; Lavallée-Adam, M.; Yates, J.R., III. Quantitative analysis of global protein stability rates in tissues. *Sci. Rep.* **2020**, *10*, 15983. [CrossRef]
12. Ong, S.E.; Blagoev, B.; Kratchmarova, I.; Kristensen, D.B.; Steen, H.; Pandey, A.; Mann, M. Stable isotope labeling by amino acids in cell culture, SILAC, as a simple and accurate approach to expression proteomics. *Mol. Cell Proteomics.* **2002**, *1*, 376–386 [CrossRef]
13. Mathieson, T.; Franken, H.; Kosinski, J.; Kurzawa, N.; Zinn, N.; Sweetman, G.; Poeckel, D.; Ratnu, V.S.; Schramm, M. Becher, I.; et al. Systematic analysis of protein turnover in primary cells. *Nat. Commun.* **2018**, *9*, 689. [CrossRef]
14. Lerman, O.Z.; Galiano, R.D.; Armour, M.; Levine, J.P.; Gurtner, G.C. Cellular dysfunction in the diabetic fibroblast: Impairment in migration, vascular endothelial growth factor production, and response to hypoxia. *Am. J. Pathol.* **2003**, *162*, 303–315. [CrossRef]
15. Brem, H.; Golinko, M.S.; Stojadinovic, O.; Kodra, A.; Diegelmann, R.F.; Vukelic, S.; Entero, H.; Coppock, D.L.; Tomic-Canic, M. Primary cultured fibroblasts derived from patients with chronic wounds: A methodology to produce human cell lines and test putative growth factor therapy such as GMCSF. *J. Transl. Med.* **2008**, *6*, 75. [CrossRef]
16. Jones, A.R.I.V.; Coleman, E.L.; Husni, N.R.; Deeney, J.T.; Raval, F.; Steenkamp, D.; Dooms, H.; Nikolajczyk, B.S.; Corkey, B.E. Type 1 diabetes alters lipid handling and metabolism in human fibroblasts and peripheral blood mononuclear cells. *PLoS ONE* **2017** *12*, e0188474. [CrossRef]
17. Al-Rikabi, A.H.A.; Tobin, D.J.; Riches-Suman, K.; Thornton, M.J. Dermal fibroblasts cultured from donors with type 2 diabetes mellitus retain an epigenetic memory associated with poor wound healing responses. *Sci. Rep.* **2021**, *11*, 1474. [CrossRef]
18. LaPointe, M.S.; Batlle, D. Cultured skin fibroblasts as an in vitro model to assess phenotypic features in subjects with diabetic nephropathy. *Am. J. Kidney Dis.* **2001**, *38*, 646–648. [CrossRef]
19. Caramori, M.L.; Kim, Y.; Fioretto, P.; Huang, C.; Rich, S.S.; Miller, M.E.; Russell, G.B.; Mauer, M. Cellular basis of diabetic nephropathy: IV. Antioxidant enzyme mRNA expression levels in skin fibroblasts of type 1 diabetic sibling pairs. *Nephrol. Dial. Transplant.* **2006**, *21*, 3122–3126. [CrossRef]
20. Millioni, R.; Iori, E.; Puricelli, L.; Arrigoni, G.; Vedovato, M.; Trevisan, R.; James, P.; Tiengo, A.; Tessari, P. Abnormal cytoskeletal protein expression in cultured skin fibroblasts from type 1 diabetes mellitus patients with nephropathy: A proteomic approach *Proteomics. Clin. Appl.* **2008**, *2*, 492–503. [CrossRef]
21. Selby, N.M.; Taal, M.W. An updated overview of diabetic nephropathy: Diagnosis, prognosis, treatment goals and latest guidelines *Diabetes Obes. Metab.* **2020**, *22* (Suppl. S1), 3–15. [CrossRef] [PubMed]
22. Esposito, S.; Toni, G.; Tascini, G.; Santi, E.; Berioli, M.G.; Principi, N. Environmental Factors Associated With Type 1 Diabetes. *Front. Endocrinol.* **2019**, *10*, 592. [CrossRef] [PubMed]
23. Freedman, B.I.; Bostrom, M.; Daeihagh, P.; Bowden, D.W. Genetic factors in diabetic nephropathy. *Clin. J. Am. Soc. Nephrol.* **2007**, *2*, 1306–1316. [CrossRef] [PubMed]
24. Millioni, R.; Puricelli, L.; Iori, E.; Trevisan, R.; Tessari, P. Skin fibroblasts as a tool for identifying the risk of nephropathy in the type 1 diabetic population. *Diabetes Metab. Res. Rev.* **2012**, *28*, 62–70. [CrossRef] [PubMed]
25. Puricelli, L.; Iori, E.; Millioni, R.; Arrigoni, G.; James, P.; Vedovato, M.; Tessari, P. Proteome analysis of cultured fibroblasts from type 1 diabetic patients and normal subjects. *J. Clin. Endocrinol. Metab.* **2006**, *91*, 3507–3514. [CrossRef] [PubMed]
26. Iori, E.; Millioni, R.; Puricelli, L.; Arrigoni, G.; Lenzini, L.; Trevisan, R.; James, P.; Rossi, G.P.; Pinna, L.A.; Tessari, P. Glycolytic enzyme expression and pyruvate kinase activity in cultured fibroblasts from type 1 diabetic patients with and without nephropathy. *Biochim. Biophys. Acta* **2008**, *1782*, 627–633. [CrossRef]
27. Fabregat, A.; Jupe, S.; Matthews, L.; Sidiropoulos, K.; Gillespie, M.; Garapati, P.; Haw, R.; Jassal, B.; Korninger, F.; May, B.; et al. The Reactome Pathway Knowledgebase. *Nucleic. Acids. Res.* **2018**, *46*, D649–D655. [CrossRef]
28. Gillespie, M.; Jassal, B.; Stephan, R.; Milacic, M.; Rothfels, K.; Senff-Ribeiro, A.; Griss, J.; Sevilla, C.; Matthews, L.; Gong, C.; et al. The reactome pathway knowledgebase 2022. *Nucleic. Acids. Res.* **2022**, *50*, D687–D692. [CrossRef]
29. Hwang, S.I.; Lundgren, D.H.; Mayya, V.; Rezaul, K.; Cowan, A.E.; Eng, J.K.; Han, D.K. Systematic characterization of nuclear proteome during apoptosis: A quantitative proteomic study by differential extraction and stable isotope labeling. *Mol. Cell Proteom.* **2006**, *5*, 1131–1145. [CrossRef]
30. Blagoev, B.; Mann, M. Quantitative proteomics to study mitogen-activated protein kinases. *Methods* **2006**, *40*, 243–250. [CrossRef]
31. Ong, S.E.; Kratchmarova, I.; Mann, M. Properties of 13C-substituted arginine in stable isotope labeling by amino acids in cell culture (SILAC). *J. Proteome. Res.* **2003**, *2*, 173–181. [CrossRef]
32. Takai, Y.; Sasaki, T.; Matozaki, T. Small GTP-binding proteins. *Physiol. Rev.* **2001**, *81*, 153–208. [CrossRef]
33. Zerial, M.; McBride, H. Rab proteins as membrane organizers. *Nat. Rev. Mol. Cell Biol.* **2001**, *2*, 107–117, Erratum in *Nat. Rev. Mol. Cell Biol.* **2001**, *2*, 216. [CrossRef]
34. Morimoto, S.; Nishimura, N.; Terai, T.; Manabe, S.; Yamamoto, Y.; Shinahara, W.; Miyake, H.; Tashiro, S.; Shimada, M.; Sasaki, T. Rab13 mediates the continuous endocytic recycling of occludin to the cell surface. *J. Biol. Chem.* **2005**, *280*, 2220–2228. [CrossRef]
35. Yamamura, R.; Nishimura, N.; Nakatsuji, H.; Arase, S.; Sasaki, T. The interaction of JRAB/MICAL-L2 with Rab8 and Rab13 coordinates the assembly of tight junctions and adherens junctions. *Mol. Biol. Cell* **2008**, *19*, 971–983. [CrossRef]

36. Bethunaickan, R.; Berthier, C.C.; Ramanujam, M.; Sahu, R.; Zhang, W.; Sun, Y.; Bottinger, E.P.; Ivashkiv, L.; Kretzler, M.; Davidson, A. A unique hybrid renal mononuclear phagocyte activation phenotype in murine systemic lupus erythematosus nephritis. *J. Immunol.* **2011**, *186*, 4994–5003. [CrossRef]
37. Huang, T.H.; Shui, H.A.; Ka, S.M.; Tang, B.L.; Chao, T.K.; Chen, J.S.; Lin, Y.F.; Chen, A. Rab 23 is expressed in the glomerulus and plays a role in the development of focal segmental glomerulosclerosis. *Nephrol. Dial. Transplant.* **2009**, *24*, 743–754. [CrossRef]
38. Huang, T.H.; Ka, S.M.; Hsu, Y.J.; Shui, H.A.; Tang, B.L.; Hu, K.Y.; Chang, J.L.; Chen, A. Rab23 plays a role in the pathophysiology of mesangial cells–a proteomic analysis. *Proteomics* **2011**, *11*, 380–394. [CrossRef]
39. Yam, A.Y.; Xia, Y.; Lin, H.T.; Burlingame, A.; Gerstein, M.; Frydman, J. Defining the TRiC/CCT interactome links chaperonin function to stabilization of newly made proteins with complex topologies. *Nat. Struct. Mol. Biol.* **2008**, *15*, 1255–1262. [CrossRef]
40. Frydman, J.; Nimmesgern, E.; Erdjument-Bromage, H.; Wall, J.S.; Tempst, P.; Hartl, F.U. Function in protein folding of TRiC, a cytosolic ring complex containing TCP-1 and structurally related subunits. *EMBO J.* **1992**, *11*, 4767–4778. [CrossRef]
41. Gao, Y.; Thomas, J.O.; Chow, R.L.; Lee, G.H.; Cowan, N.J. A cytoplasmic chaperonin that catalyzes beta-actin folding. *Cell* **1992**, *69*, 1043–1050. [CrossRef] [PubMed]
42. Yaffe, M.B.; Farr, G.W.; Miklos, D.; Horwich, A.L.; Sternlicht, M.L.; Sternlicht, H. TCP1 complex is a molecular chaperone in tubulin biogenesis. *Nature* **1992**, *358*, 245–248. [CrossRef] [PubMed]
43. Brackley, K.I.; Grantham, J. Activities of the chaperonin containing TCP-1 (CCT): Implications for cell cycle progression and cytoskeletal organisation. *Cell Stress Chaperones.* **2009**, *14*, 23–31. [CrossRef] [PubMed]
44. Wu, C.Z.; Chang, L.C.; Lin, Y.F.; Hung, Y.J.; Pei, D.; Chen, J.S. Chaperonin-containing t-complex protein-1 subunit β as a possible biomarker for the phase of glomerular hyperfiltration of diabetic nephropathy. *Dis. Markers.* **2015**, *2015*, 548101. [CrossRef]
45. Wang, X.; Wilkinson, R.; Kildey, K.; Potriquet, J.; Mulvenna, J.; Lobb, R.J.; Möller, A.; Cloonan, N.; Mukhopadhyay, P.; Kassianos, A.J.; et al. Unique molecular profile of exosomes derived from primary human proximal tubular epithelial cells under diseased conditions. *J. Extracell. Vesicles.* **2017**, *6*, 1314073. [CrossRef]
46. Sanai, T.; Sobka, T.; Johnson, T.; el-Essawy, M.; Muchaneta-Kubara, E.C.; Ben Gharbia, O.; el Oldroyd, S.; Nahas, A.M. Expression of cytoskeletal proteins during the course of experimental diabetic nephropathy. *Diabetologia* **2000**, *43*, 91–100. [CrossRef]
47. Shimoni, Y.; Rattner, J.B. Type 1 diabetes leads to cytoskeleton changes that are reflected in insulin action on rat cardiac K(+) currents. *Am. J. Physiol. Endocrinol. Metab.* **2001**, *281*, E575–E585. [CrossRef]
48. Tessari, P.; Puricelli, L.; Iori, E.; Arrigoni, G.; Vedovato, M.; James, P.; Coracina, A.; Millioni, R. Altered chaperone and protein turnover regulators expression in cultured skin fibroblasts from type 1 diabetes mellitus with nephropathy. *J. Proteome. Res.* **2007**, *6*, 976–986. [CrossRef]
49. Li, L.; Rose, P.; Moore, P.K. Hydrogen sulfide and cell signaling. *Annu. Rev. Pharmacol. Toxicol.* **2011**, *51*, 169–187. [CrossRef]
50. Kabil, O.; Vitvitsky, V.; Banerjee, R. Sulfur as a signaling nutrient through hydrogen sulfide. *Annu. Rev. Nutr.* **2014**, *34*, 171–205. [CrossRef]
51. Rose, P.; Moore, P.K.; Zhu, Y.Z. H2S biosynthesis and catabolism: New insights from molecular studies. *Cell Mol. Life Sci.* **2017**, *74*, 1391–1412. [CrossRef]
52. Olson, K.R. Mitochondrial adaptations to utilize hydrogen sulfide for energy and signaling. *J. Comp. Physiol. B* **2012**, *182*, 881–897. [CrossRef]
53. Stefely, J.A.; Pagliarini, D.J. Biochemistry of Mitochondrial Coenzyme Q Biosynthesis. *Trends Biochem. Sci.* **2017**, *42*, 824–843. [CrossRef]
54. Kleiner, G.; Barca, E.; Ziosi, M.; Emmanuele, V.; Xu, Y.; Hidalgo-Gutierrez, A.; Qiao, C.; Tadesse, S.; Area-Gomez, E.; Lopez, L.C.; et al. CoQ10 supplementation rescues nephrotic syndrome through normalization of H2S oxidation pathway. *Biochim. Biophys. Acta Mol. Basis. Dis.* **2018**, *1864*, 3708–3722. [CrossRef]
55. Zhou, X.; Feng, Y.; Zhan, Z.; Chen, J. Hydrogen sulfide alleviates diabetic nephropathy in a streptozotocin-induced diabetic rat model. *J. Biol. Chem.* **2014**, *289*, 28827–28834. [CrossRef]
56. Yu, Y.T.; Yang, J.W. Protective Effect of Hydrogen Sulfide in the Secondary Renal Disease and Renal Transplantation. *Sheng Li Ke Xue Jin Zhan.* **2016**, *47*, 93–96.
57. Dugbartey, G.J. Diabetic nephropathy: A potential savior with 'rotten-egg' smell. *Pharmacol. Rep.* **2017**, *69*, 331–339. [CrossRef]
58. Hostetter, T.H. Progression of renal disease and renal hypertrophy. *Annu. Rev. Physiol.* **1995**, *57*, 263–278. [CrossRef]
59. Kasinath, B.S.; Feliers, D.; Sataranatarajan, K.; Ghosh Choudhury, G.; Lee, M.J.; Mariappan, M.M. Regulation of mRNA translation in renal physiology and disease. *Am. J. Physiol. Renal. Physiol.* **2009**, *297*, F1153–F1165. [CrossRef]
60. Marintchev, A.; Edmonds, K.A.; Marintcheva, B.; Hendrickson, E.; Oberer, M.; Suzuki, C.; Herdy, B.; Sonenberg, N.; Wagner, G. Topology and regulation of the human eIF4A/4G/4H helicase complex in translation initiation. *Cell* **2009**, *136*, 447–460. [CrossRef]
61. Kasinath, B.S.; Mariappan, M.M.; Sataranatarajan, K.; Lee, M.J.; Ghosh Choudhury, G.; Feliers, D. Novel mechanisms of protein synthesis in diabetic nephropathy–role of mRNA translation. *Rev. Endocr. Metab. Disord.* **2008**, *9*, 255–266. [CrossRef] [PubMed]
62. Grice, B. What I didn't see before. *Nursing* **1991**, *21*, 93–94. [CrossRef] [PubMed]
63. Zhou, G.; Xin, G.; Zhang, W.; Zhang, X. Biomarker prediction for membranous nephropathy prognosis by microarray analysis. *Nephrology* **2019**, *24*, 526–533. [CrossRef] [PubMed]
64. Togashi, Y.; Miyamoto, Y. Urinary cystatin C as a biomarker for diabetic nephropathy and its immunohistochemical localization in kidney in Zucker diabetic fatty (ZDF) rats. *Exp. Toxicol. Pathol.* **2013**, *65*, 615–622. [CrossRef]

65. Brennan, E.P.; Ehrich, M.; O'Donovan, H.; Brazil, D.P.; Crean, J.K.; Murphy, M.; Sadlier, D.M.; Martin, F.; Godson, C.; van den Boom, D.; et al. DNA methylation profiling in cell models of diabetic nephropathy. *Epigenetics* **2010**, *5*, 396–401. [CrossRef]
66. Lo, C.S.; Shi, Y.; Chenier, I.; Ghosh, A.; Wu, C.H.; Cailhier, J.F.; Ethier, J.; Lattouf, J.B.; Filep, J.G.; Ingelfinger, J.R.; et al. Heterogeneous Nuclear Ribonucleoprotein F Stimulates Sirtuin-1 Gene Expression and Attenuates Nephropathy Progression in Diabetic Mice. *Diabetes* **2017**, *66*, 1964–1978. [CrossRef]
67. Ghosh, A.; Abdo, S.; Zhao, S.; Wu, C.H.; Shi, Y.; Lo, C.S.; Chenier, I.; Alquier, T.; Filep, J.G.; Ingelfinger, J.R.; et al. Insulin Inhibits Nrf2 Gene Expression via Heterogeneous Nuclear Ribonucleoprotein F/K in Diabetic Mice. *Endocrinology* **2017**, *158*, 903–919. [CrossRef]
68. Lo, C.S.; Chang, S.Y.; Chenier, I.; Filep, J.G.; Ingelfinger, J.R.; Zhang, S.L.; Chan, J.S. Heterogeneous nuclear ribonucleoprotein F suppresses angiotensinogen gene expression and attenuates hypertension and kidney injury in diabetic mice. *Diabetes* **2012**, *61*, 2597–2608. [CrossRef]
69. Moura, I.C.; Centelles, M.N.; Arcos-Fajardo, M.; Malheiros, D.M.; Collawn, J.F.; Cooper, M.D.; Monteiro, R.C. Identification of the transferrin receptor as a novel immunoglobulin (Ig)A1 receptor and its enhanced expression on mesangial cells in IgA nephropathy. *J. Exp. Med.* **2001**, *194*, 417–425. [CrossRef]
70. Berthelot, L.; Papista, C.; Maciel, T.T.; Biarnes-Pelicot, M.; Tissandie, E.; Wang, P.H.; Tamouza, H.; Jamin, A.; Bex-Coudrat, J.; Gestin, A.; et al. Transglutaminase is essential for IgA nephropathy development acting through IgA receptors. *J. Exp. Med.* **2012**, *209*, 793–806. [CrossRef]
71. Zhou, L.T.; Qiu, S.; Lv, L.L.; Li, Z.L.; Liu, H.; Tang, R.N.; Ma, K.L.; Liu, B.C. Integrative Bioinformatics Analysis Provides Insight into the Molecular Mechanisms of Chronic Kidney Disease. *Kidney Blood Press Res.* **2018**, *43*, 568–581. [CrossRef]
72. Grill, B.; Wilson, G.M.; Zhang, K.X.; Wang, B.; Doyonnas, R.; Quadroni, M.; Schrader, J.W. Activation/division of lymphocytes results in increased levels of cytoplasmic activation/proliferation-associated protein-1: Prototype of a new family of proteins. *J. Immunol.* **2004**, *172*, 2389–2400. [CrossRef]
73. Jaffa, M.A.; Kobeissy, F.; Al Hariri, M.; Chalhoub, H.; Eid, A.; Ziyadeh, F.N.; Jaffa, A.A. Global renal gene expression profiling analysis in B2-kinin receptor null mice: Impact of diabetes. *PLoS ONE* **2012**, *7*, e44714. [CrossRef]
74. Matsuo, S.; López-Guisa, J.M.; Cai, X.; Okamura, D.M.; Alpers, C.E.; Bumgarner, R.E.; Peters, M.A.; Zhang, G.; Eddy, A.A. Multifunctionality of PAI-1 in fibrogenesis: Evidence from obstructive nephropathy in PAI-1-overexpressing mice. *Kidney Int.* **2005**, *67*, 2221–2238, Erratum in *Kidney Int.* **2005**, *68*, 910. [CrossRef]
75. Zhi, W.; Sharma, A.; Purohit, S.; Miller, E.; Bode, B.; Anderson, S.W.; Reed, J.C.; Steed, R.D.; Steed, L.; Hopkins, D.; et al. Discovery and validation of serum protein changes in type 1 diabetes patients using high throughput two dimensional liquid chromatography-mass spectrometry and immunoassays. *Mol. Cell Proteom.* **2011**, *10*, M111.012203. [CrossRef]
76. Deb, D.K.; Chen, Y.; Sun, J.; Wang, Y.; Li, Y.C. ATP-citrate lyase is essential for high glucose-induced histone hyperacetylation and fibrogenic gene upregulation in mesangial cells. *Am. J. Physiol. Renal. Physiol.* **2017**, *313*, F423–F429. [CrossRef]
77. Chen, Y.; Deb, D.K.; Fu, X.; Yi, B.; Liang, Y.; Du, J.; He, L.; Li, Y.C. ATP-citrate lyase is an epigenetic regulator to promote obesity-related kidney injury. *FASEB J.* **2019**, *33*, 9602–9615. [CrossRef]
78. Imamura, M.; Moon, J.S.; Chung, K.P.; Nakahira, K.; Muthukumar, T.; Shingarev, R.; Ryter, S.W.; Choi, A.M.; Choi, M.E. RIPK3 promotes kidney fibrosis via AKT-dependent ATP citrate lyase. *JCI Insight.* **2018**, *3*, e94979. [CrossRef]
79. Gao, C.; Huang, W.; Kanasaki, K.; Xu, Y. The role of ubiquitination and sumoylation in diabetic nephropathy. *Biomed. Res. Int.* **2014**, *2014*, 160692. [CrossRef]
80. Goru, S.K.; Kadakol, A.; Gaikwad, A.B. Hidden targets of ubiquitin proteasome system: To prevent diabetic nephropathy. *Pharmacol. Res.* **2017**, *120*, 170–179. [CrossRef]
81. Huang, W.; Yang, C.; Nan, Q.; Gao, C.; Feng, H.; Gou, F.; Chen, G.; Zhang, Z.; Yan, P.; Peng, J.; et al. The proteasome inhibitor, MG132, attenuates diabetic nephropathy by inhibiting SnoN degradation in vivo and in vitro. *Biomed. Res. Int.* **2014**, *2014*, 684765. [CrossRef] [PubMed]
82. Zeng, W.; Qi, W.; Mu, J.; Wei, Y.; Yang, L.L.; Zhang, Q.; Wu, Q.; Tang, J.Y.; Feng, B. MG132 protects against renal dysfunction by regulating Akt-mediated inflammation in diabetic nephropathy. *Sci. Rep.* **2019**, *9*, 2049. [CrossRef] [PubMed]
83. Luo, Z.F.; Qi, W.; Feng, B.; Mu, J.; Zeng, W.; Guo, Y.H.; Pang, Q.; Ye, Z.L.; Liu, L.; Yuan, F.H. Prevention of diabetic nephropathy in rats through enhanced renal antioxidative capacity by inhibition of the proteasome. *Life Sci.* **2011**, *88*, 512–520. [CrossRef] [PubMed]
84. Yadranji Aghdam, S.; Mahmoudpour, A. Proteasome Activators, PA28α and PA28β, Govern Development of Microvascular Injury in Diabetic Nephropathy and Retinopathy. *Int. J. Nephrol.* **2016**, *2016*, 3846573. [CrossRef]
85. Queisser, M.A.; Yao, D.; Geisler, S.; Hammes, H.P.; Lochnit, G.; Schleicher, E.D.; Brownlee, M.; Preissner, K.T. Hyperglycemia impairs proteasome function by methylglyoxal. *Diabetes* **2010**, *59*, 670–678. [CrossRef]
86. Tashiro, K.; Tamada, S.; Kuwabara, N.; Komiya, T.; Takekida, K.; Asai, T.; Iwao, H.; Sugimura, K.; Matsumura, Y.; Takaoka, M.; et al. Attenuation of renal fibrosis by proteasome inhibition in rat obstructive nephropathy: Possible role of nuclear factor kappaB. *Int. J. Mol. Med.* **2003**, *12*, 587–592. [CrossRef]
87. Pujols, L.; Fernández-Bertolín, L.; Fuentes-Prado, M.; Alobid, I.; Roca-Ferrer, J.; Agell, N.; Mullol, J.; Picado, C. Proteasome inhibition reduces proliferation, collagen expression, and inflammatory cytokine production in nasal mucosa and polyp fibroblasts. *J. Pharmacol. Exp. Ther.* **2012**, *343*, 184–197. [CrossRef]

88. Neubert, K.; Meister, S.; Moser, K.; Weisel, F.; Maseda, D.; Amann, K.; Wiethe, C.; Winkler, T.H.; Kalden, J.R.; Manz, R.A.; et al. The proteasome inhibitor bortezomib depletes plasma cells and protects mice with lupus-like disease from nephritis. *Nat. Med.* **2008**, *14*, 748–755. [CrossRef]
89. Huang, S.; Liu, F.; Niu, Q.; Li, Y.; Liu, C.; Zhang, L.; Ni, D.; Pu, X. GLIPR-2 overexpression in HK-2 cells promotes cell EMT and migration through ERK1/2 activation. *PLoS ONE.* **2013**, *8*, e58574. [CrossRef]
90. Zhao, D.; Jia, J.; Shao, H. miR-30e targets GLIPR-2 to modulate diabetic nephropathy: In vitro and in vivo experiments. *J. Mol. Endocrinol.* **2017**, *59*, 181–190. [CrossRef]
91. El-Meanawy, A.; Schelling, J.R.; Iyengar, S.K.; Hayden, P.; Barathan, S.; Goddard, K.; Pozuelo, F.; Elashi, E.; Nair, V.; Kretzler, M.; et al. Identification of nephropathy candidate genes by comparing sclerosis-prone and sclerosis-resistant mouse strain kidney transcriptomes. *BMC Nephrol.* **2012**, *13*, 61. [CrossRef]
92. Yang, H.; Zhang, X.; Xin, G. Investigation of mechanisms of mesenchymal stem cells for treatment of diabetic nephropathy via construction of a miRNA-TF-mRNA network. *Ren. Fail.* **2018**, *40*, 136–145. [CrossRef]
93. Gu, Y.; Gong, Y.; Zhang, H.; Dong, X.; Zhao, T.; Burczynski, F.J.; Wang, G.; Sun, S.; Zhu, B.; Han, W.; et al. Regulation of transforming growth factor beta 1 gene expression by dihydropteridine reductase in kidney 293T cells. *Biochem. Cell Biol.* **2013**, *91*, 187–193. [CrossRef]
94. Si, Q.; Sun, S.; Gu, Y. A278C mutation of dihydropteridine reductase decreases autophagy via mTOR signaling. *Acta Biochim. Biophys. Sin.* **2017**, *49*, 706–712. [CrossRef]
95. Sharma, M.; Mehndiratta, M.; Gupta, S.; Kalra, O.P.; Shukla, R.; Gambhir, J.K. Genetic association of NAD(P)H quinone oxidoreductase (NQO1*2) polymorphism with NQO1 levels and risk of diabetic nephropathy. *Biol. Chem.* **2016**, *397*, 725–730. [CrossRef]
96. Finsterer, J.; Scorza, F.A. Renal manifestations of primary mitochondrial disorders. *Biomed. Rep.* **2017**, *6*, 487–494. [CrossRef]
97. Emma, F.; Montini, G.; Parikh, S.M.; Salviati, L. Mitochondrial dysfunction in inherited renal disease and acute kidney injury. *Nat. Rev. Nephrol.* **2016**, *12*, 267–280. [CrossRef]
98. Cox, S.N.; Serino, G.; Sallustio, F.; Blasi, A.; Rossini, M.; Pesce, F.; Schena, F.P. Altered monocyte expression and expansion of non-classical monocyte subset in IgA nephropathy patients. *Nephrol. Dial. Transplant.* **2015**, *30*, 1122–1232. [CrossRef]
99. Mafi, A.; Aghadavod, E.; Mirhosseini, N.; Mobini, M.; Asemi, Z. The effects of expression of different microRNAs on insulin secretion and diabetic nephropathy progression. *J. Cell Physiol.* **2018**, *234*, 42–50. [CrossRef]
100. Higuchi, C.; Nakatsuka, A.; Eguchi, J.; Teshigawara, S.; Kanzaki, M.; Katayama, A.; Yamaguchi, S.; Takahashi, N.; Murakami, K.; Ogawa, D.; et al. Identification of circulating miR-101, miR-375 and miR-802 as biomarkers for type 2 diabetes. *Metabolism* **2015**, *64*, 489–497. [CrossRef]
101. Scott, L.J.; Mohlke, K.L.; Bonnycastle, L.L.; Willer, C.J.; Li, Y.; Duren, W.L.; Erdos, M.R.; Stringham, H.M.; Chines, P.S.; Jackson, A.U.; et al. A genome-wide association study of type 2 diabetes in Finns detects multiple susceptibility variants. *Science* **2007**, *316*, 1341–1345. [CrossRef] [PubMed]
102. Liljedahl, L.; Pedersen, M.H.; Norlin, J.; McGuire, J.N.; James, P. N-glycosylation proteome enrichment analysis in kidney reveals differences between diabetic mouse models. *Clin. Proteomics.* **2016**, *13*, 22. [CrossRef] [PubMed]
103. Fu, F.; Wei, X.; Liu, J.; Mi, N. Bioinformatic analysis of specific genes in diabetic nephropathy. *Ren. Fail.* **2015**, *37*, 1219–1224. [CrossRef] [PubMed]
104. Yesil-Devecioglu, T.; Dayan, A.; Demirtunc, R.; Sardas, S. Role of DNA repair genes XRCC3 and XRCC1 in predisposition to type 2 diabetes mellitus and diabetic nephropathy. *Endocrinol. Diabetes Nutr.* **2019**, *66*, 90–98, (In English, Spanish). [CrossRef] [PubMed]
105. Wu, C.C.; Chen, J.S.; Huang, C.F.; Chen, C.C.; Lu, K.C.; Chu, P.; Sytwu, H.K.; Lin, Y.F. Approaching biomarkers of membranous nephropathy from a murine model to human disease. *J. Biomed. Biotechnol.* **2011**, *2011*, 581928. [CrossRef]
106. Iacobini, C.; Amadio, L.; Oddi, G.; Ricci, C.; Barsotti, P.; Missori, S.; Sorcini, M.; Di Mario, U.; Pricci, F.; Pugliese, G. Role of galectin-3 in diabetic nephropathy. *J. Am. Soc. Nephrol.* **2003**, *14* (Suppl. S3), S264–S270. [CrossRef]
107. Li, Y.; Li, T.; Zhou, Z.; Xiao, Y. Emerging roles of Galectin-3 in diabetes and diabetes complications: A snapshot. *Rev. Endocr. Metab. Disord.* **2022**, *23*, 569–577. [CrossRef]
108. Nicholas, S.B.; Aguiniga, E.; Ren, Y.; Kim, J.; Wong, J.; Govindarajan, N.; Noda, M.; Wang, W.; Kawano, Y.; Collins, A.; et al. Plasminogen activator inhibitor-1 deficiency retards diabetic nephropathy. *Kidney Int.* **2005**, *67*, 1297–1307. [CrossRef]
109. Guo, J.; Rackham, O.J.L.; Sandholm, N.; He, B.; Österholm, A.M.; Valo, E.; Harjutsalo, V.; Forsblom, C.; Toppila, I.; Parkkonen, M.; et al. Whole-Genome Sequencing of Finnish Type 1 Diabetic Siblings Discordant for Kidney Disease Reveals DNA Variants associated with Diabetic Nephropathy. *J. Am. Soc. Nephrol.* **2020**, *31*, 309–323. [CrossRef]
110. Jacobson, S.L.; Piper, H.M. Cell cultures of adult cardiomyocytes as models of the myocardium. *J. Mol. Cell Cardiol.* **1986**, *18*, 661–678. [CrossRef]
111. Du, F.; Wang, T.; Li, S.; Meng, X.; Zhang, H.Y.; Li, D.T.; Du, Z.X.; Wang, H.Q. Cathepsin D protects renal tubular cells from damage induced by high glucose independent of its enzymatic activity. *Am. J. Transl. Res.* **2017**, *9*, 5528–5537.
112. Limonte, C.P.; Valo, E.; Drel, V.; Natarajan, L.; Darshi, M.; Forsblom, C.; Henderson, C.M.; Hoofnagle, A.N.; Ju, W.; Kretzler, M.; et al. Urinary Proteomics Identifies Cathepsin D as a Biomarker of Rapid eGFR Decline in Type 1 Diabetes. *Diabetes Care* **2022**, *45*, 1416–1427. [CrossRef]

113. Li, S.Y.; Huang, P.H.; Tarng, D.C.; Lin, T.P.; Yang, W.C.; Chang, Y.H.; Yang, A.H.; Lin, C.C.; Yang, M.H.; Chen, J.W.; et al. Four-and-a-Half LIM Domains Protein 2 Is a Coactivator of Wnt Signaling in Diabetic Kidney Disease. *J. Am. Soc. Nephrol.* **2015**, *26*, 3072–3084. [CrossRef]
114. Cai, T.; Sun, D.; Duan, Y.; Qiu, Y.; Dai, C.; Yang, J.; He, W. FHL2 promotes tubular epithelial-to-mesenchymal transition through modulating β-catenin signalling. *J. Cell Mol. Med.* **2018**, *22*, 1684–1695. [CrossRef]
115. Adhikary, L.; Chow, F.; Nikolic-Paterson, D.J.; Stambe, C.; Dowling, J.; Atkins, R.C.; Tesch, G.H. Abnormal p38 mitogen-activated protein kinase signalling in human and experimental diabetic nephropathy. *Diabetologia* **2004**, *47*, 1210–1222. [CrossRef]
116. Toyoda, M.; Suzuki, D.; Honma, M.; Uehara, G.; Sakai, T.; Umezono, T.; Sakai, H. High expression of PKC-MAPK pathway mRNAs correlates with glomerular lesions in human diabetic nephropathy. *Kidney Int.* **2004**, *66*, 1107–1114. [CrossRef]
117. Price, J.C.; Guan, S.; Burlingame, A.; Prusiner, S.B.; Ghaemmaghami, S. Analysis of proteome dynamics in the mouse brain. *Proc. Natl. Acad. Sci. USA* **2010**, *107*, 14508–14513, Erratum in *Proc. Natl. Acad. Sci. USA* **2014**, *111*, 3645. [CrossRef]
118. Fernandez-Pol, J.A.; Hamilton, P.D.; Klos, D.J. Essential viral and cellular zinc and iron containing metalloproteins as targets for novel antiviral and anticancer agents: Implications for prevention and therapy of viral diseases and cancer. *Anticancer Res.* **2001**, *21*, 931–957.
119. Fernandez-Pol, J.A.; Klos, D.J.; Hamilton, P.D. Metallopanstimulin gene product produced in a baculovirus expression system is a nuclear phosphoprotein that binds to DNA. *Cell Growth Differ.* **1994**, *5*, 811–825.
120. Welle, K.A.; Zhang, T.; Hryhorenko, J.R.; Shen, S.; Qu, J.; Ghaemmaghami, S. Time-resolved Analysis of Proteome Dynamics by Tandem Mass Tags and Stable Isotope Labeling in Cell Culture (TMT-SILAC) Hyperplexing. *Mol. Cell Proteom.* **2016**, *15*, 3551–3563. [CrossRef]
121. Resmini, G.; Rizzo, S.; Franchin, C.; Zanin, R.; Penzo, C.; Pegoraro, S.; Ciani, Y.; Piazza, S.; Arrigoni, G.; Sgarra, R.; et al. HMGA1 regulates the Plasminogen activation system in the secretome of breast cancer cells. *Sci. Rep.* **2017**, *7*, 11768. [CrossRef] [PubMed]
122. Szklarczyk, D.; Gable, A.L.; Lyon, D.; Junge, A.; Wyder, S.; Huerta-Cepas, J.; Simonovic, M.; Doncheva, N.T.; Morris, J.H.; Bork, P.; et al. STRING v11: Protein-protein association networks with increased coverage, supporting functional discovery in genome-wide experimental datasets. *Nucleic. Acids Res.* **2019**, *47*, D607–D613. [CrossRef] [PubMed]
123. Huang, W.; Sherman, B.T.; Lempicki, R.A. Bioinformatics enrichment tools: Paths toward the comprehensive functional analysis of large gene lists. *Nucleic. Acids Res.* **2009**, *37*, 1–13. [CrossRef] [PubMed]
124. Huang, W.; Sherman, B.T.; Lempicki, R.A. Systematic and integrative analysis of large gene lists using DAVID bioinformatics resources. *Nat. Protoc.* **2009**, *4*, 44–57. [CrossRef]
125. Supek, F.; Bošnjak, M.; Škunca, N.; Šmuc, T. REVIGO summarizes and visualizes long lists of gene ontology terms. *PLoS ONE.* **2011**, *6*, e21800. [CrossRef]
126. Subramanian, A.; Tamayo, P.; Mootha, V.K.; Mukherjee, S.; Ebert, B.L.; Gillette, M.A.; Paulovich, A.; Pomeroy, S.L.; Golub, T.R.; Lander, E.S.; et al. Gene set enrichment analysis: A knowledge-based approach for interpreting genome-wide expression profiles. *Proc. Natl. Acad. Sci. USA* **2005**, *102*, 15545–15550. [CrossRef]

International Journal of
Molecular Sciences

MDPI

Article

Estimating Total Quantitative Protein Content in *Escherichia coli, Saccharomyces cerevisiae*, and HeLa Cells

Georgii V. Dolgalev [1], Taras A. Safonov [2], Viktoriia A. Arzumanian [1], Olga I. Kiseleva [1] and Ekaterina V. Poverennaya [1,2,*]

1 Institute of Biomedical Chemistry, Moscow 119281, Russia
2 X-BIO Institute, University of Tyumen, 6 Volodarskogo St., Tyumen 625003, Russia
* Correspondence: k.poverennaya@ibmc.msk.ru

Abstract: The continuous improvement of proteomic techniques, most notably mass spectrometry, has generated quantified proteomes of many organisms with unprecedented depth and accuracy. However, there is still a significant discrepancy in the reported numbers of total protein molecules per specific cell type. In this article, we explore the results of proteomic studies of *Escherichia coli, Saccharomyces cerevisiae*, and HeLa cells in terms of total protein copy numbers per cell. We observe up to a ten-fold difference between reported values. Investigating possible reasons for this discrepancy, we conclude that neither an unmeasured fraction of the proteome nor biases in the quantification of individual proteins can explain the observed discrepancy. We normalize protein copy numbers in each study using a total protein amount per cell as reported in the literature and create integrated proteome maps of the selected model organisms. Our results indicate that cells contain from one to three million protein molecules per μm^3 and that protein copy density decreases with increasing organism complexity.

Keywords: quantitative proteomics; absolute protein abundance; total protein; *E. coli*; *S. cerevisiae*; HeLa

Citation: Dolgalev, G.V.; Safonov, T.A.; Arzumanian, V.A.; Kiseleva, O.I.; Poverennaya, E.V. Estimating Total Quantitative Protein Content in *Escherichia coli, Saccharomyces cerevisiae*, and HeLa Cells. *Int. J. Mol. Sci.* **2023**, *24*, 2081. https://doi.org/10.3390/ijms24032081

Academic Editor: Paolo Iadarola

Received: 6 December 2022
Revised: 16 January 2023
Accepted: 18 January 2023
Published: 20 January 2023

1. Introduction

The difference in observed phenotypes between cells of different organisms and between individual cells of multicellular organisms can be largely explained by differences in their proteome composition [1]. However, the proteome remains much more difficult to quantify than the genome or the transcriptome since it is significantly more complex—the number of different proteoforms in a typical human cell is estimated to be well into the hundreds of thousands or even millions, and individual proteins can exhibit extremely dynamic behavior [2,3]. Additionally, in contrast to genomics, proteomics currently lacks methods for amplifying the signal from proteins present in low numbers, which makes their quantification difficult [4].

Despite all the challenges, the field of proteomics has demonstrated great progress in the quantification of proteomes of different organisms. This progress is, for the most part, attributed to advancements in mass spectrometry that now has sufficient power to quantify tens of thousands of individual proteins in a single experiment [5]. This has resulted in the publication of hundreds of high-quality proteomes of various cell types [6] or even whole tissues [7]. Several alternative approaches to mass spectrometry have also been developed for the purpose of proteome quantification, for instance, fluorescence-based quantification [8], immunoblotting [9] and ribosomal profiling [10]. However, in contrast to mass spectrometry, these methods are less universal. One of the most interesting questions of quantitative proteomics is how many protein molecules in total are present per particular cell type. This question can be approached in two different ways. The first way is to measure the total protein mass per cell and average protein mass, which allows one to calculate an approximate number of proteins per cell by simple division of these two values. Since this calculation requires relatively simple biochemical measurements, such an approach was used well before the development of robust strategies for whole proteome quantification [11]. Another way to approach this question is to quantify levels of individual proteins, and then sum them up to obtain a total protein copy number per cell, which is where mass spectrometry is the method of choice.

A rigorous attempt to combine and contrast these two approaches was conducted by R. Milo in 2013 [12]. Surprisingly, estimates of the total number of protein copies calculated using protein density per cell and average protein mass contrasted with estimates of total protein copy number from whole proteome quantification experiments. This difference could be as high as ten fold for some cell types. Since the publication of the article in 2013, a plethora of new whole proteome quantification datasets, as well as updated measurements of physiological parameters of various cell types, have been published. This opens an opportunity to explore the proteomes of common model organisms more rigorously and to arrive at consensus values of total protein copies per cell, as well as copy numbers of individual proteins.

2. Results

2.1. Overview of Selected Proteomic Studies of *E. coli*, *S. cerevisiae*, and HeLa Cells

To achieve our goal of understanding how many protein molecules were present per cell, we chose three well studied model organisms for research: *E. coli* as an example of a relatively simple prokaryotic cell, *S. cerevisiae* as a unicellular eukaryote organism, and HeLa cell line as a cell from a complex multicellular organism such as human.

For the chosen cell types, we searched for published proteomic studies that reported copies per cell for these model organisms. We only selected studies that quantified a significant portion of the proteome and performed original calculations of protein copy numbers. In addition, we also required that the studies use similar growth conditions for particular cell types. In total, we found 21 such studies, with 7 studies for each model organism (Table 1). These studies primarily utilized shotgun mass spectrometry. Additionally, one study by Lawless et al. performed targeted protein quantification using the single reaction monitoring (SRM) technique for an impressive number of proteins (>1000) [13]. While mass spectrometry is the method of choice for proteome quantification, three of the selected studies utilized alternative methods such as immunoblotting (Ghaemmaghami et al., 2003 [9]), single molecule fluorescence-based protein abundance quantification (Taniguchi et al., 2010 [8]), and ribosomal profiling (Li et al., 2014 [10]).

Table 1. Overview of selected proteomic studies of *E. coli*, *S. cerevisiae*, and HeLa.

Cell Type	Study	Code	Method	Strategy for Calculation of Protein Copies	Proteins Quantified	Total Protein Copies per Cell	Ref.
E. coli	Taniguchi et al., 2010	TA10	Fluorescence	Single-molecule fluorescence calibration	1018	94,571	[8]
	Valgepea et al., 2013	VA13	Shotgun MS	iBAQ, calibration with standards (UPS2) * and total protein per cell	1179	4,293,284	[14]
	Li et al., 2014	LI14	Ribosomal profiling	Relative protein synthesis rates multiplied by total protein per cell	3883	5,627,623	[10]
	Wisniewski et al., 2014	WI14	Shotgun MS	TPA and total protein per cell	2261	1,321,542	[15]
	Soufi et al., 2015	SO15	Shotgun MS	iBAQ, calibration with standards (UPS2) and cell count	1913	11,214,979	[16]

Table 1. *Cont.*

Cell Type	Study	Code	Method	Strategy for Calculation of Protein Copies	Proteins Quantified	Total Protein Copies per Cell	Ref.
	Schmidt et al., 2016	SC16	Shotgun MS	LFQ, calibration with standards and cell count	2355	5,070,410	[6]
	Radzikowski et al., 2016	RA16	Shotgun MS	LFQ and total protein per cell	1959	2,220,410	[17]
S. cerevisiae	Ghaemmaghami et al., 2003	GH03	Western blotting	Calibration with standards and cell count	3868	46,664,471	[9]
	Kulak et al., 2014	KU14	Shotgun MS	TPA and total protein per cell	4570	48,114,163	[18]
	Lawless et al., 2016	LA16	SRM	Standards (QconCAT) and cell count	1167	56,322,039	[13]
	Lahtvee et al., 2017	LT17	Shotgun MS	iBAQ, calibration with standards (UPS2) and cell count	1788	96,407,334	[19]
	Martin-Perez et al., 2017	MP17	Shotgun MS	"Proteomic ruler"	3318	250,751,159	[20]
	Wang et al., 2019	WA19	Shotgun MS	"Proteomic ruler"	2582	71,802,810	[21]
	Xia et al., 2022	XI22	Shotgun MS	iBAQ, calibration with standards (UPS2) and total protein per cell	2526	73,823,343	[22]
HeLa	Nagaraj et al., 2011	NA11	Shotgun MS	iBAQ and total protein per cell	8078	2,007,666,667	[23]
	Wisniewski et al., 2012	WI12	Shotgun MS	TPA and total protein per cell (estimate)	8094	8,236,921,797	[24]
	Kulak et al., 2014	KU14	Shotgun MS	TPA and total protein per cell (estimate)	9677	2,982,812,197	[18]
	Hein et al., 2015	HE15	Shotgun MS	LFQ and total protein per cell	8804	2,916,903,614	[25]
	Itzhak et al., 2016	IT16	Shotgun MS	"Proteomic ruler"	8710	7,837,554,944	[26]
	Bekker-Jensen et al., 2017	BJ17	Shotgun MS	iBAQ and calibration with standards	14178	4,077,816,932	[5]
	Morgenstern et al., 2021	MO21	Shotgun MS	TPA and total protein per cell	8436	4,883,462,397	[27]

* Data in Valgepea et al., 2013 are based on rescaling of data from Arike et al., 2012 [28], who performed absolute quantification of *E. coli* cells using the iBAQ approach. Abbreviations: iBAQ—intensity-based absolute quantification; UPS2—universal protein standard 2; LFQ—label-free quantification; TPA—total protein approach; SRM—single reaction monitoring; QconCAT—quantification conCATamer

To obtain copies per cell, the first step in proteomic studies is to derive absolute abundance values for each quantified protein, i.e., values that are directly proportional to protein concentration in the cell. This is the most important step for studies that use mass spectrometry. As peptides have different efficiencies of ionization, their individual abundance alone cannot be directly converted to protein abundance [4]. A classic way to circumvent this problem is to spike-in labeled peptides or proteins with known concentration, which allows us to infer the absolute abundance of their unlabeled counterparts by comparison [29]. However, this is an extremely expensive and laborious approach if more than a few proteins have to be quantified, and only one study in our selection performed such a procedure for all proteins, which was by Lawless et al., 2016 [13]. However, most of the other studies opted for a label-free quantification approach, which typically consisted of inferring absolute protein abundance based on some form of integration of peptide-level data. Such approaches, for example, total protein approach (TPA) [15] or intensity-based absolute quantification (iBAQ) [30] have important technical differences but are nonetheless considered to be similar for our goals. Regarding the studies that did not utilize mass spectrometry, Western blotting and fluorescence only require calibration with standards of known quantity, although certain care must be taken when converting chemiluminescence/fluorescence intensity to protein abundance [31]. One remaining study by Li et al. utilized ribosomal profiling to estimate absolute protein synthesis rates that were directly proportional to protein abundance if the influence of post-translational control on protein levels was negligible, as was true for *E. coli* cells [10].

The second step to obtaining protein copy numbers consists of converting absolute abundance values to protein copies per cell, which is required for mass spectrometry and other approaches alike. Again, at this step, there is significant difference in possible approaches. The first approach is to introduce a limited set of labeled proteins into the sample and build a calibration curve, thus, inferring protein abundances in copies/moles in the sample, which was done, for instance, by Schmidt et al. [6]. After this step, however, to obtain protein copy numbers per cell, cell counts that were used as input must be measured either by flow cytometry or plate counting. Alternatively, if the total protein content in grams per cell is known, absolute abundance values can be converted to protein copies by normalization of the sum of intensities to total protein per cell. One other approach, termed "proteomic ruler" relies on the fact that the number of histones associated with DNA is constant for a specific type of eukaryotic cell, and thus, protein copies can be calculated by normalizing the summed intensity of histones to the protein/DNA ratio per cell [32]. To summarize, selected proteomic studies of *E. coli*, *S. cerevisiae*, and HeLa cells utilize various multi-step approaches to estimate the number of protein copies per cell, and discrepancies at each step can lead to a compound effect on the difference in copy numbers of individual proteins and, thus potentially, total protein copy number per cell.

2.2. Comparison of Reported Total Protein Copy Numbers for Selected Model Organisms

In terms of quantified proteins, studies have quantified variable numbers of them, ranging from 1000 to almost 14,000 proteins (Figure 1). For each of the selected studies, we extracted unadjusted values of protein copies per cell and summed them to obtain the total number of protein copies per cell for each study. The results for each model organism demonstrate significant discrepancies in terms of reported total protein copy numbers (Figure 1).

As expected, there is no correlation observed between the number of quantified proteins in a particular study and the calculated number of total protein copies per cell for each model organism (Pearson's r = 0.28 for *E. coli*, 0.05 for *S. cerevisiae*, and −0.26 for HeLa). Consequently, some sort of normalization procedure is required before the results of these studies can be integrated into a global proteomic map of the selected model organisms, thus, obtaining a consensus value of total protein copies per cell for a particular model organism and condition.

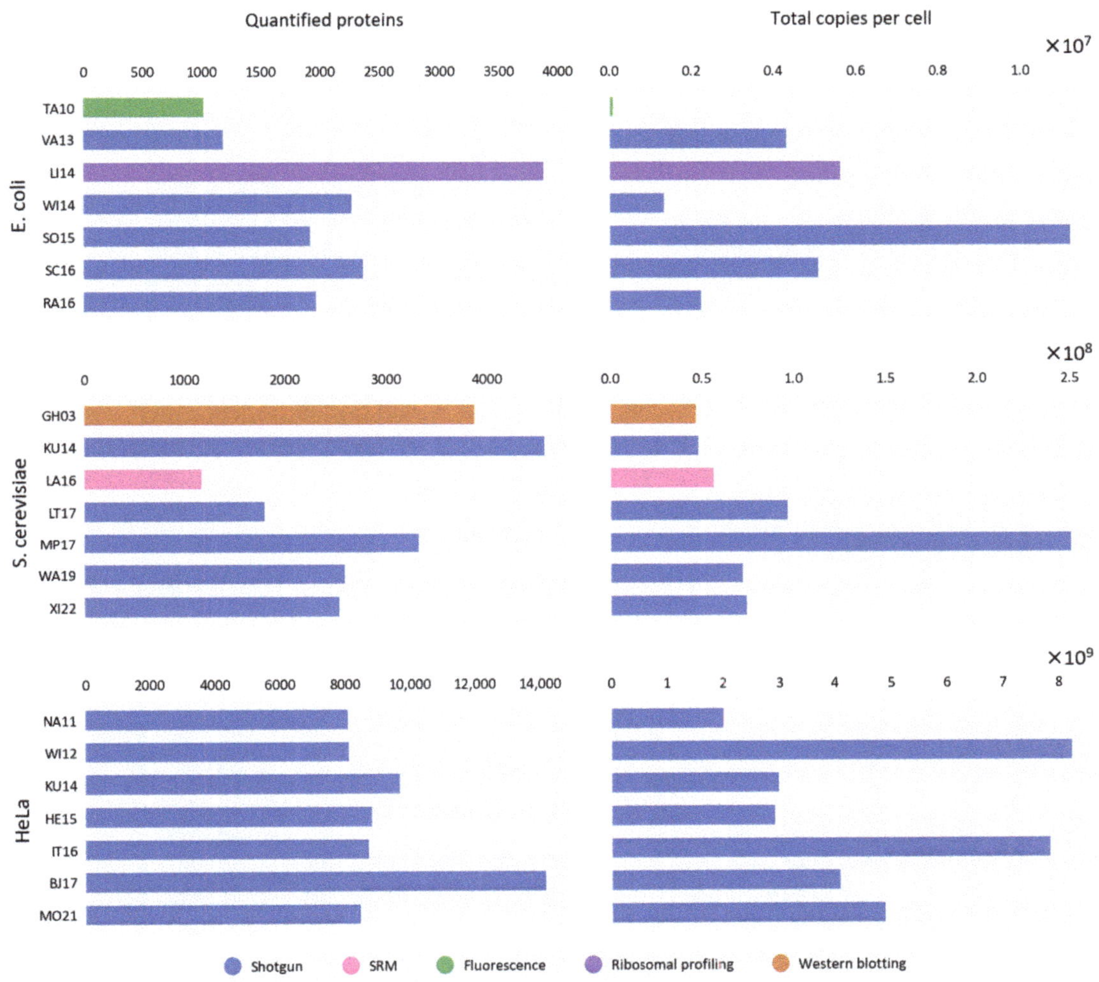

Figure 1. Results of proteomic studies in terms of the number of quantified proteins and total protein copies per cell for *E. coli*, *S. cerevisiae*, and HeLa.

2.3. Exploration of Possible Reasons for the Discrepancy in Reported Total Protein Copy Numbers

To enable cross-study comparisons, first, we mapped all identifiers in the study to the universal UniProt IDs from the most recent release of the UniProt database. As a result, some entries were lost due to possible nomenclature updates or because it was a protein group, rather than a single protein that was quantified. However, despite these losses, our procedure led to the assignment of >95% of all identifiers in most of the studies and led to no more than 10% loss of protein copies reported, which was negligible for our needs (Table S1).

To understand the source of the observed differences, we considered whether proteomic studies can omit a certain part of the proteome from identification due to method-specific biases. Since shotgun mass spectrometry and ribosomal profiling represent untargeted approaches to protein identification, biases in protein detection can leave a certain part of the proteome undetected, depending on the specific method employed. In contrast, targeted approaches such as Western blotting and fluorescence imaging can omit certain proteins, which may have high copy numbers, from the initial selection of targets altogether.

To analyze the extent of this "hidden proteome", we quantified how proteins identified in all untargeted studies (core proteins) contributed to the reported total protein copy numbers. We additionally excluded a study by Lahtvee et al. for *S. cerevisiae* from the calculation of the core because this study quantified significantly fewer proteins than in other untargeted studies of *S. cerevisiae* proteome, and exclusion of this dataset improved the contribution of core proteins to the total copy numbers, which was not true when excluding one of any other untargeted studies. We identified 934 proteins present in all untargeted studies of *E. coli* proteome (22% of all predicted protein-coding genes [33]), 1945 core proteins for *S. cerevisiae* cells (32% of all predicted protein-coding genes [34]), and 5051 proteins (24% of predicted number of human protein-coding genes [35]) for HeLa cells. Despite representing from a quarter to a third of all quantified proteins in each study, core proteins account for more than 70–80% of total protein copies per cell for all model organisms (Figure 2). If we relax our definition of core proteins to include proteins present in at least n−1 study (core 1), the contribution of such proteins increases to almost more than 90% of total protein copies per cell for the majority of studies (Figure 2). Thus, we reason that the most frequently detected proteins contribute to the majority of overall copies irrespective of the study. Additionally, we observe that untargeted studies quantify not all of the core proteins (Table S1), which may be the reason for the lower estimates of total protein copy numbers in these datasets.

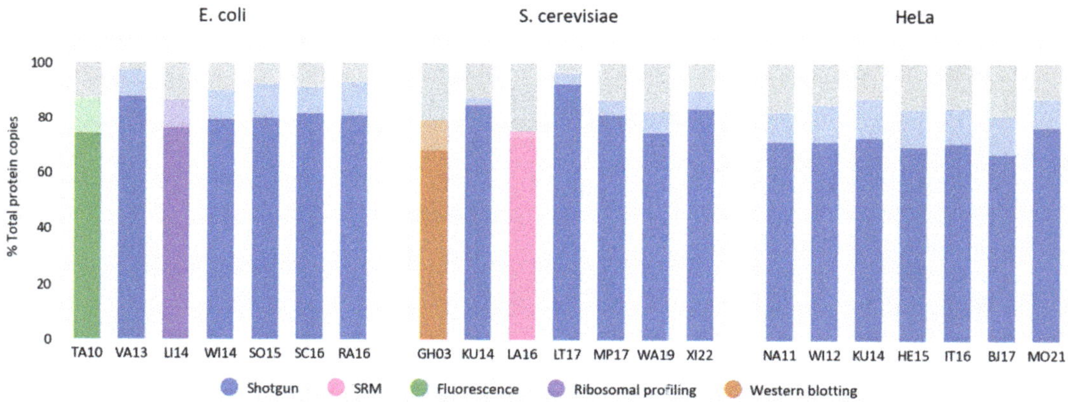

Figure 2. Contribution of proteins quantified in all untargeted datasets for a particular model organism (core proteins) to the reported total protein copy numbers as a percent of total copies. Less saturated portions of the bars represent contributions of proteins which are present in at least n − 1 untargeted dataset for a particular model organism (core 1) to total protein copies.

Nevertheless, it is unlikely but possible that some systematic bias prevents all studies from detecting a certain part of the proteome. To estimate the contribution of this hypothetical part of the proteome to the overall numbers, we analyzed how the frequency of detection of proteins (as being present in a certain number of datasets) correlates with their averaged normalized expression (Figure 3). The results indicate that there is a clear trend of underrepresentation of proteins with low biological abundance. Therefore, the fraction of unseen proteins likely represents proteins with very low expression values that cannot have such a profound effect on the total protein copy numbers.

To explore the possible discrepancies in quantification of individual proteins, we calculated pairwise Pearson's correlation for selected datasets (Figure 4). The overall levels of correlation are moderate (mean Pearson's r = 0.63 for *E. coli*, 0.57 for *S. cerevisiae*, and 0.70 for Hela cells). Moderate levels of pairwise correlations are typical for mass spectrometry-based experiments, although, of course, a better level is expected since identical cell types in almost identical conditions are considered. Interestingly, a targeted immunoblot-based study of *S. cerevisiae* by Ghaemmaghami et al. demonstrated lower-than-average levels of correlation with mass spectrometry studies. Additionally, a fluorescence-based study of the *E. coli* proteome by Taniguchi et al. showed a moderate level of correlation with other datasets despite quantifying a significantly smaller number of proteins in total. It is also surprising that we do not observe an increased level of correlation when taking into consideration only core proteins (lower left corner of heatmaps in Figure 4).

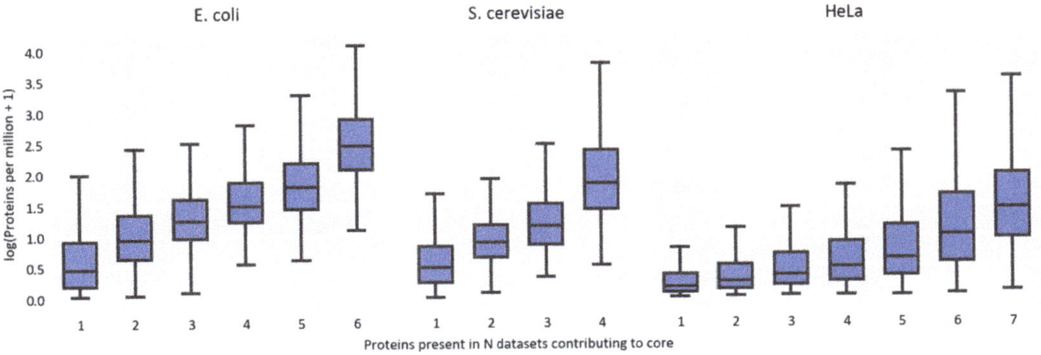

Figure 3. Normalized expression of most and least commonly detected proteins. Proteins per million denote protein copy number divided by the sum of all protein copies in the dataset and multiplied by a million.

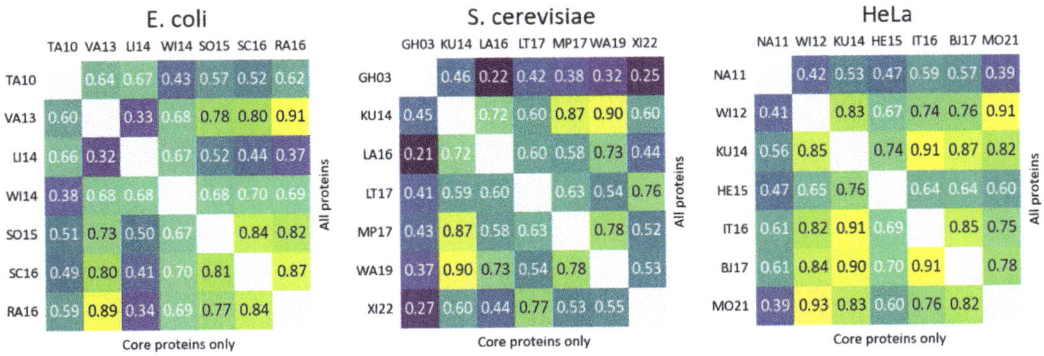

Figure 4. Pairwise Pearson's correlation of protein levels between individual datasets for all shared proteins (**top right half**) or core proteins (**lower left half**).

In conclusion, we reason that although abundances of specific proteins can be the reason for some of the discrepancies observed between total protein copy numbers, this factor alone cannot explain a ten-fold difference in some cases. Therefore, we believe that neither the unidentified portion of the proteome nor differences in quantification of particular proteins can produce the observed effect. In turn, it is likely that some sort of a study-specific coefficient can be applied to normalize quantitative information between different studies of a particular model organism.

2.4. Estimation of Total Protein Copy Numbers per Cell from Total Protein Mass per Cell

To produce consensus values of protein abundances per cell, we opted for a simple, but powerful approach of estimating the total number of protein copies per cell based on the total protein mass per cell and average protein mass.

As our first step, we calculated average protein molecular masses as reported in the datasets (Figure 5). Our results aligned with the well established observation that prokaryotes have a lower average protein mass (around 30 kDa) than eukaryotes (around 40 kDa) [12], although we also noticed a small difference in the median protein mass between *S. cerevisiae* and HeLa cells, which may have reflected the relative extent of cell complexity between yeast and humans.

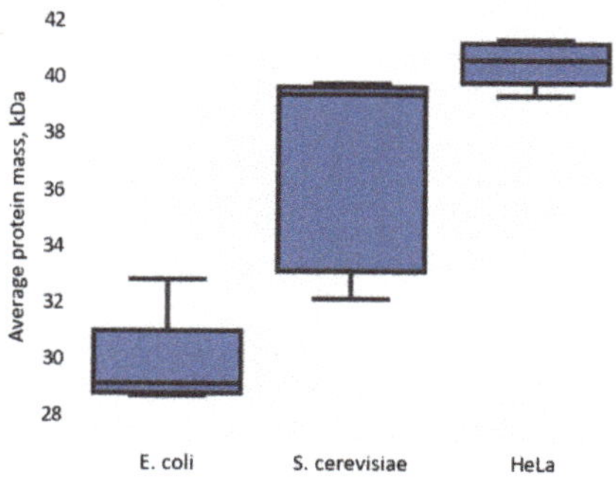

Figure 5. Distribution of average protein masses in the selected proteomic studies of *E. coli*, *S. cerevisiae*, and HeLa.

Next, we analyzed the reports of measured total protein weight per cell of *E. coli*, *S. cerevisiae*, and HeLa cells. We selected close to average measurements that were preferentially obtained in recent years. For *E. coli* cells, we took the value of 280 ng per cell, obtained by Schmidt et al. for our conditions of interest [6]. For *S. cerevisiae*, values from 4.9 to 6.4 pg were reported and we selected the average, i.e., 5.65 pg [36]. For HeLa cells, values from 200 to 300 pg were reported [18,37]. We took the average of these values for our calculations. Our selected values differed slightly from those taken by Milo [12]. Next, we estimated the total number of protein copies per cell by simply dividing the total protein mass per cell by the median average protein mass obtained for the selected cell types. We obtained the following numbers of total protein copies per cell: 5,774,718 for *E. coli*, 86,196,924 for *S. cerevisiae*, and 3,698,713,400 for HeLa cells. Comparing the obtained results with reported protein copy numbers from the selected studies, we noticed that our estimates were close to the average of reported values (Figure 6). Therefore, we reasoned that it was correct to normalize reported protein copies per cell to our calculated values for the chosen model organisms.

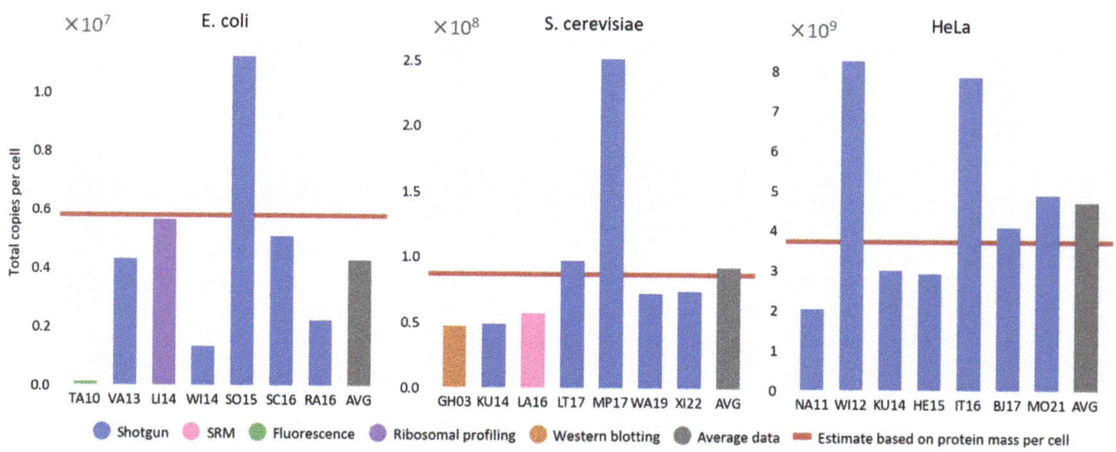

Figure 6. Comparisons of reported numbers of total protein copies per cell and the average values with estimates from total protein mass per cell.

To enable direct comparisons of the cellular proteome numbers between such different organisms, we converted the values of total protein copies per cell to total protein copies per unit of volume, μm^3, by dividing our estimates by a typical cellular volume in conditions typical for the selected studies. While the measurements of cell volume for *S. cerevisiae* (42 μm^3 [38,39]) and for HeLa cells (2425 μm^3 [40]) are fairly consistent in the available literature, reported values for *E. coli* cells display significant discrepancy, ranging from 1 μm^3 to almost 3 μm^3 in similar growth conditions [17,41–43]. *E. coli* and bacteria, in general, have a well documented phenomenon of demonstrating significant changes in cell volume depending on the growth rate. However, this is unlikely to explain the observed discrepancy as the aforementioned estimates were calculated for cells growing in similar conditions, and thus, with a similar growth rate. As a consensus value, we selected 2.15 μm^3, reported by Radzikowski et al. [17] (see also supplementary note in Schmidt et al. [6]). For the selected cell volumes, we obtained values of 2,685,915 protein copies per μm^3 for *E. coli*, 2,034,146 protein copies per μm^3 *S. cerevisiae*, and 1,525,243 protein copies per μm^3 for HeLa. As can be seen, there is a clear trend of decreasing protein copies per unit of cell volume with increasing organism complexity, although it is not as profound as estimated previously [12].

2.5. Normalization and Integration of Protein Copy Numbers in the Selected Studies

To enable the normalization of protein copy numbers according to a calculated total number of copies per cell, we developed and applied a normalization procedure to all datasets. Since individual studies quantified different numbers of proteins in total, simply normalizing the sum of all protein copies to an estimated number of total copies per cell would yield relatively increased values for datasets that quantified lower numbers of proteins. Accordingly, we decided to base our normalization procedure on the contribution of core proteins to the number of total copies. For each cell type, we calculated a minimal contribution of core proteins to total copy numbers across untargeted studies. Then, we normalized the sum of all core protein copies in each dataset to our estimated total number of protein copies per cell multiplied by the minimal observed contribution of core proteins to the total copy number. For untargeted studies, we also multiplied an estimated sum of core protein copy numbers by a proportion of core proteins that were detected in the untargeted study. Finally, we calculated the average copy number for each detected protein, thus, obtaining an integrated proteome for each cell type (Table S3). Our data include estimated copy number values for 3852 proteins in *E. coli* (91% of all predicted protein-coding genes [33]), 4680 proteins for *S. cerevisiae* (77% of all predicted protein-coding genes [34]), and 12,653 proteins for HeLa (60% of all predicted protein-coding genes [35]).

The sums of our averaged results are, as expected, close to the calculated number of proteins per cell type, differing no more than 10% from initial estimates (5,852,319 for E. coli, 81,627,580 for S. cerevisiae, and 3,360,824,528 for HeLa cells). Small differences can be explained by discrepancies in quantification of individual proteins, plus the fact that not all predicted proteins of the model organisms are present in the resulting proteomes. Additionally, our results correlate well with a previously published integrated analysis of *S. cerevisiae* proteome partly based on the datasets also selected for our study (Pearson's r = 0.82) [44]; however, our results are almost twice as high in terms of total protein copies per cell despite more proteins being included in the published dataset. In terms of median copy numbers for selected cell types, we calculated 145 median protein copies for *E. coli*, 1888 median protein copies for *S. cerevisiae*, and 15,654 median protein copies for HeLa cells.

To provide an additional measure of the integrity of the resulting consensus proteomes, we explored how copy numbers of ribosomal proteins correlate with the published numbers of ribosome complexes per cell. Ribosomal proteins are produced in equimolar concentrations in the cell and their numbers are tightly regulated [45]. For *E. coli*, we assume an approximate number of 31,739 ribosomes per cell in selected conditions (see Section 4). For *S. cerevisiae*, a value of 220,000 ribosomes per cell has been reported [46]. Finally, for HeLa cells, differing values have been reported: 3.3 million [47], 4 million [37], and 9.5 million [48]. We selected a median of these values. Generally, median copy numbers of ribosomal proteins (adjusted to the ribosomal composition, see Section 4) in the integrated proteomes are close to the selected numbers of ribosomal complexes except for *E. coli*, which is a bit lower (Figure 7). We take this as an indication that the scale of our transformation is in agreement with supplementary quantitative data for these cell types.

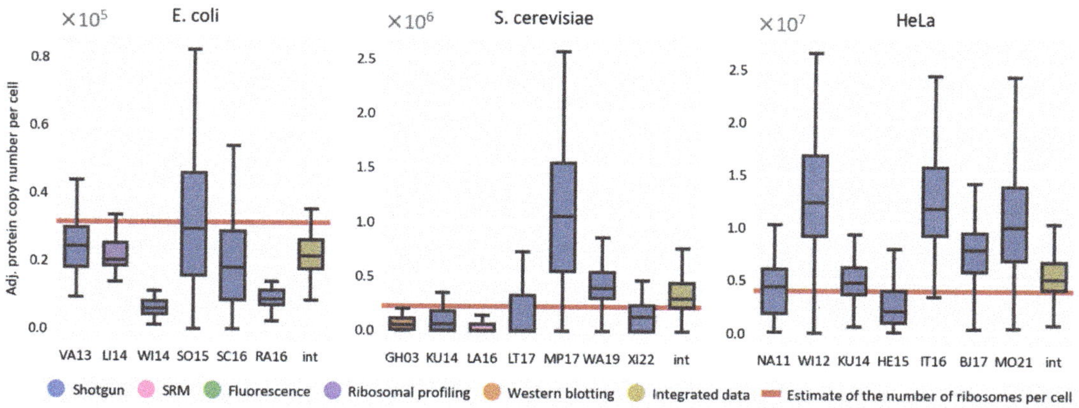

Figure 7. Distribution of adjusted copy numbers of ribosomal proteins in selected studies as well as in consensus proteomes. Adjusted protein copy number per cell denotes copy number of the ribosomal protein adjusted to the protein's stoichiometry in the ribosome (see Section 4).

However, it must be noted that the results of experiments which utilize mass spectrometry display considerable differences between copy numbers of ribosomal proteins. Lower copy numbers than expected can be attributed to some loss of material during sample preparation. However, considering the aforementioned phenomenon of tight balancing of the levels of ribosomal proteins [45], it is surprising to see ribosomal proteins with significantly higher copy numbers than expected. Nevertheless, it must also be taken into consideration that estimates of the number of ribosomes can noticeably vary in the literature (as is the case with HeLa). We expect that these results might prove to be stimulating to re-examine the levels of ribosomes and ribosomal protein copies in future studies.

3. Discussion

The present study explores the discrepancy in total protein copy numbers reported in whole proteome quantification studies. This discrepancy has been noticed before [12,49]. However, since then, more datasets have been published for common model organisms, such as the ones selected in our study, i.e., *E. coli*, *S. cerevisiae*, and HeLa cells. Exploring the results of 21 proteomic studies for these cell types, it seems likely that a variation in reported numbers is associated with subtle miscalculation performed at different steps of proteome quantification such as cell counting. First, selected datasets share proteins with the highest copy numbers per cell, which suggests that proteins not quantified in any study are present in low copies and would not have a large influence on the final result. Additionally, while we observe moderate levels of correlation for levels of individual proteins between the datasets, this factor alone cannot explain the observed difference in total protein copies per cell.

While there are more datasets than the ones used in this study, we prioritized datasets that performed original calculations of protein copy numbers. In fact, several published proteome quantification studies that have reported protein abundance in copies per cell used a presumed total number of protein copies for normalization [50,51]. While this is a reasonable approach, these studies used a lower total protein copy number than the one that can be calculated from the total protein mass per cell, as done in the present study. This is explained by the fact that previous publications that calculated the total number of protein copies per cell used different values of parameters, such as average protein mass [11], or performed a targeted quantification that did not include some of the highly expressed proteins in initial target selection [9]. Altogether, these observations highlight the importance of re-analysis and normalization of protein copy numbers. Our method of normalization of protein copies to a calculated total number of total protein copies per cell is based on total protein mass per cell and average protein mass. In the end, we obtained consensus proteome maps of *E. coli*, *S. cerevisiae*, and HeLa cells.

Additional attention should be paid to the HeLa proteome. Despite the fact that the HeLa proteome is supposed to be much more complex than proteomes of simpler organisms such as *E. coli* and *S. cerevisiae*, we observe similar, if not higher levels of correlation between individual datasets for HeLa. We believe that the reason for this is our data processing strategy that focused only on canonical proteins (master proteins [2]). Despite the fact that alternative splicing is known to diversify proteomes in higher eukaryotes, current mass spectrometry methods capture predominantly canonical proteins, which usually have the highest expression among all protein isoforms produced from one gene [52]. We reasoned that our canonical proteome map is a good enough consensus proteome that would be useful for most applications, but it is likely that, in the near future, isoform-resolved proteome maps of HeLa and other cells of higher eukaryotes will be published.

4. Materials and Methods

4.1. Data Processing

For all data processing steps, Python v.3.11 was used. Canonical proteome data for *E. coli*, *S. cerevisiae*, and HeLa were downloaded from UniProt (accessed on 17 August 2022). Data for absolute protein copy numbers per cell were extracted from a supplementary dataset for each of the selected whole proteome quantification studies. The exact supplementary datasets used are listed in Supplementary Table S1. For ID assignment, first, we considered whether a dataset contained UniProt IDs. If yes, we tested whether all UniProt IDs were present in downloaded proteomes from UniProt. In the case of protein groups, we contracted the protein group only to review canonical proteins. If the entry remained a protein group, it was excluded from further cross-study comparisons. Since there had been a change in nomenclature of several proteins since the publication of many of the selected studies, some UniProt IDs could not be assigned. To try and infer the UniProt ID, in these cases, we used B numbers for *E. coli* and ORF IDs for *S. cerevisiae*, if available, to infer UniProt IDs. If, in this case, the ID could not be assigned, the entry was excluded from cross-study comparisons and data integration. The results of ID assignment in terms of the proportion of correctly assigned IDs for each study are found in Supplementary Table S1.

4.2. Core Protein Assignment and Calculation of Pairwise Correlations

Proteins which were quantified in all of the selected datasets for a particular model organism were defined as the core proteins. Core 1 proteins were defined as proteins which were quantified in at least $n - 1$ dataset for a particular model organism. To perform calculations of the core, as well as pairwise Pearson's correlation, only entries with successfully assigned UniProt IDs were considered. Pairwise Pearson's correlations were calculated using a pearsonr function from the scipy.stats module. Heatmaps for correlations were created using a heatmap function from the seaborn module.

4.3. Normalization of Protein Copy Numbers in Individual Datasets

To calculate the total number of protein copies per cell from total protein mass per cell, we divided the literature-derived estimate of total protein mass per cell for each of the selected model organisms by the median average protein mass in the selected whole proteome quantification studies. Estimates of total protein mass per cell used were 280 ng for *E. coli* [6], 5.65 pg for *S. cerevisiae* [36], and 250 pg for HeLa cells, obtained as the average of values from several sources [18,37].

Next, the following formula was used for the calculation of the normalization factor for individual datasets:

$$f = (T \times A \times I \times C)/S,$$

where T is the total number of protein copies per cell type as estimated from total protein mass per cell, A is the proportion of proteins in the dataset with successfully assigned UniProt IDs, I is the number of core proteins present in the dataset divided by total number of calculated core proteins for cell type, C is the minimal proportion of contribution of core proteins to total protein copy number across all studies which contributed to calculation of core for cell type, S is the sum of copies of proteins which belong to core in the dataset.

Normalized protein copy numbers in datasets were obtained by multiplication of original copy numbers by the dataset-specific normalization factor f.

4.4. Estimation of the Number of Ribosomes in *E. coli* in Selected Conditions

To estimate the number of ribosomes in cells of *E. coli* for the selected conditions, we used the results of Bakshi et al., who counted 27,000 ribosomes per μm^3 on average in *E. coli* K-12 cells growing in rich medium (growth rate, $\mu = 1.11$) [53]. For *E. coli*, it has been estimated that the mass fraction of ribosomal proteins increased almost linearly with growth rates of $\mu > 0.3$ [54]. Accordingly, we calculated the number of ribosomes per μm^3 for *E. coli* growing in minimal glucose medium (as is in most of the selected studies) by multiplying 27,000 by the relation of growth rates between our datasets and the Bakshi et al. study. For *E. coli* strain MG1655 cells growing in glucose minimal medium, a growth rate of 0.6 has been reported [41], therefore, we obtained a value of 14,595 ribosomes per μm^3. Considering that *E. coli* strain MG1655 has been reported to have a volume of around 2.15 μm^3 in corresponding conditions [17], we arrived at a final value of 31,739 ribosomes per cell.

4.5. Calculating Adjusted Copy Numbers of Ribosomal Proteins

To provide estimates of the number of ribosomes based on the data for copy numbers of individual proteins, it was required to account for organism-specific differences in the organization of ribosomes. First, in *E. coli* ribosomes, 50S ribosomal protein L7/L12 (UniProt ID P0A7K2) is present in 4 copies per ribosome [55]. Accordingly, we divided the copy numbers of this protein by 4 to calculate the distribution of ribosomal proteins for *E. coli* as a proxy for the number of ribosomes.

In *S. cerevisiae*, multiple ribosomal proteins are encoded by pairs of genes and differ slightly in the sequences of the resulting isoforms [56]. Accordingly, to provide estimates for the number of ribosomes based on the data for copy numbers of individual proteins, it was required to summarize copy numbers of such pairs of proteins to arrive at copy numbers of "ribosomal parts" rather than individual proteins. We performed summation according to our custom data based on the information from the UniProt database (Table S2).

In the case of HeLa cells, similar summations were performed only for RPS4X and RPS4Y proteins, which are the only alternative ribosomal proteins in human cytoplasmic ribosomes [57].

5. Conclusions

In conclusion, our study integrates multiple published protein abundance data for *E. coli, S. cerevisiae,* and HeLa cells. Despite significant differences in reported total protein copy numbers, we conclude that these datasets quantify dominant parts of the proteome, and thus, can be normalized to total protein mass per cell. Our calculations indicate that a typical *E. coli* cell contains around 6 million protein molecules, a *S. cerevisiae* cell contains approximately 80 million proteins, and a HeLa cell contains around 3.4 billion protein copies. In terms of protein copy density, it decreases from *E. coli* (which contains 2.5 million protein molecules per μm^3) to *S. cerevisiae* (which has 2 million protein molecules per μm^3) to HeLa cells (which have 1.5 million protein molecules per μm^3). Our results generally agreed with some of the previously published estimates and, in the case of *S. cerevisiae*, improved the estimate of the total number of proteins, which was twice as low in some of the previous work. In general, we believe our integrated proteome datasets will be a useful resource for the scientific community.

Supplementary Materials: The following supporting information can be downloaded at: https://www.mdpi.com/article/10.3390/ijms24032081/s1.

Author Contributions: Conceptualization, G.V.D., T.A.S., and E.V.P.; Data curation, V.A.A.; Formal analysis, G.V.D.; Funding acquisition, E.V.P.; Investigation, G.V.D. and V.A.A.; Methodology, G.V.D. and T.A.S.; Project administration, E.V.P.; Supervision, O.I.K. and E.V.P.; Visualization, G.V.D.; Writing—original draft, G.V.D.; Writing—review and editing, O.I.K. and E.V.P. All authors have read and agreed to the published version of the manuscript.

Funding: This work was funded by the Ministry of Education and Science of the Russian Federation, grant number 075-15-2021-933 ("Avogadro" large-scale research facilities, unique project ID: RF00121X0004).

Data Availability Statement: The data presented in this study are available as Supplementary Materials.

Conflicts of Interest: The authors declare no conflict of interest.

References

1. Bludau, I.; Aebersold, R. Proteomic and Interactomic Insights into the Molecular Basis of Cell Functional Diversity. *Nat. Rev. Mol. Cell Biol.* **2020**, *21*, 327–340. [CrossRef] [PubMed]
2. Ponomarenko, E.A.; Poverennaya, E.V.; Ilgisonis, E.V.; Pyatnitskiy, M.A.; Kopylov, A.T.; Zgoda, V.G.; Lisitsa, A.V.; Archakov, A.I. The Size of the Human Proteome: The Width and Depth. *Int. J. Anal. Chem.* **2016**, *2016*, 7436849. [CrossRef]
3. Aebersold, R.; Agar, J.N.; Amster, I.J.; Baker, M.S.; Bertozzi, C.R.; Boja, E.S.; Costello, C.E.; Cravatt, B.F.; Fenselau, C.; Garcia, B.A.; et al. How Many Human Proteoforms Are There? *Nat. Chem. Biol.* **2018**, *14*, 206–214. [CrossRef] [PubMed]
4. Aebersold, R.; Mann, M. Mass-Spectrometric Exploration of Proteome Structure and Function. *Nature* **2016**, *537*, 347–355. [CrossRef] [PubMed]
5. Bekker-Jensen, D.B.; Kelstrup, C.D.; Batth, T.S.; Larsen, S.C.; Haldrup, C.; Bramsen, J.B.; Sørensen, K.D.; Høyer, S.; Ørntoft, T.F.; Andersen, C.L.; et al. An Optimized Shotgun Strategy for the Rapid Generation of Comprehensive Human Proteomes. *Cell Syst.* **2017**, *4*, 587–599.e4. [CrossRef] [PubMed]
6. Schmidt, A.; Kochanowski, K.; Vedelaar, S.; Ahrné, E.; Volkmer, B.; Callipo, L.; Knoops, K.; Bauer, M.; Aebersold, R.; Heinemann, M. The Quantitative and Condition-Dependent *Escherichia coli* Proteome. *Nat. Biotechnol.* **2016**, *34*, 104–110. [CrossRef]
7. Niu, L.; Geyer, P.E.; Gupta, R.; Santos, A.; Meier, F.; Doll, S.; Wewer Albrechtsen, N.J.; Klein, S.; Ortiz, C.; Uschner, F.E.; et al. Dynamic Human Liver Proteome Atlas Reveals Functional Insights into Disease Pathways. *Mol. Syst. Biol.* **2022**, *18*, e10947. [CrossRef]
8. Taniguchi, Y.; Choi, P.J.; Li, G.-W.; Chen, H.; Babu, M.; Hearn, J.; Emili, A.; Xie, X.S. Quantifying *E. coli* Proteome and Transcriptome with Single-Molecule Sensitivity in Single Cells. *Science* **2010**, *329*, 533–538. [CrossRef] [PubMed]
9. Ghaemmaghami, S.; Huh, W.-K.; Bower, K.; Howson, R.W.; Belle, A.; Dephoure, N.; O'Shea, E.K.; Weissman, J.S. Global Analysis of Protein Expression in Yeast. *Nature* **2003**, *425*, 737–741. [CrossRef]
10. Li, G.-W.; Burkhardt, D.; Gross, C.; Weissman, J.S. Quantifying Absolute Protein Synthesis Rates Reveals Principles Underlying Allocation of Cellular Resources. *Cell* **2014**, *157*, 624–635. [CrossRef]
11. Futcher, B.; Latter, G.I.; Monardo, P.; McLaughlin, C.S.; Garrels, J.I. A Sampling of the Yeast Proteome. *Mol. Cell. Biol.* **1999**, *19*, 7357–7368. [CrossRef] [PubMed]
12. Milo, R. What Is the Total Number of Protein Molecules per Cell Volume? A Call to Rethink Some Published Values. *Bioessays* **2013**, *35*, 1050–1055. [CrossRef] [PubMed]
13. Lawless, C.; Holman, S.W.; Brownridge, P.; Lanthaler, K.; Harman, V.M.; Watkins, R.; Hammond, D.E.; Miller, R.L.; Sims, P.F.G.; Grant, C.M.; et al. Direct and Absolute Quantification of over 1800 Yeast Proteins via Selected Reaction Monitoring. *Mol. Cell. Proteom. MCP* **2016**, *15*, 1309–1322. [CrossRef] [PubMed]
14. Valgepea, K.; Adamberg, K.; Seiman, A.; Vilu, R. *Escherichia coli* Achieves Faster Growth by Increasing Catalytic and Translation Rates of Proteins. *Mol. Biosyst.* **2013**, *9*, 2344–2358. [CrossRef] [PubMed]
15. Wiśniewski, J.R.; Rakus, D. Multi-Enzyme Digestion FASP and the 'Total Protein Approach'-Based Absolute Quantification of the *Escherichia coli* Proteome. *J. Proteom.* **2014**, *109*, 322–331. [CrossRef] [PubMed]
16. Soufi, B.; Krug, K.; Harst, A.; Macek, B. Characterization of the *E. coli* Proteome and Its Modifications during Growth and Ethanol Stress. *Front. Microbiol.* **2015**, *6*, 103. [CrossRef]
17. Radzikowski, J.L.; Vedelaar, S.; Siegel, D.; Ortega, Á.D.; Schmidt, A.; Heinemann, M. Bacterial Persistence Is an Active σS Stress Response to Metabolic Flux Limitation. *Mol. Syst. Biol.* **2016**, *12*, 882. [CrossRef]
18. Kulak, N.A.; Pichler, G.; Paron, I.; Nagaraj, N.; Mann, M. Minimal, Encapsulated Proteomic-Sample Processing Applied to Copy-Number Estimation in Eukaryotic Cells. *Nat. Methods* **2014**, *11*, 319–324. [CrossRef]
19. Lahtvee, P.-J.; Sánchez, B.J.; Smialowska, A.; Kasvandik, S.; Elsemman, I.E.; Gatto, F.; Nielsen, J. Absolute Quantification of Protein and MRNA Abundances Demonstrate Variability in Gene-Specific Translation Efficiency in Yeast. *Cell Syst.* **2017**, *4*, 495–504.e5. [CrossRef]
20. Martin-Perez, M.; Villén, J. Determinants and Regulation of Protein Turnover in Yeast. *Cell Syst.* **2017**, *5*, 283–294.e5. [CrossRef]

21. Wang, Y.; Weisenhorn, E.; MacDiarmid, C.; Andreini, C.; Bucci, M.; Taggart, J.; Banci, L.; Russell, J.; Coon, J.J.; Eide, D.J. The Cellular Economy of the *Saccharomyces cerevisiae* Zinc Proteome. *Metallomics* **2018**, *10*, 1755–1776. [CrossRef] [PubMed]
22. Xia, J.; Sánchez, B.J.; Chen, Y.; Campbell, K.; Kasvandik, S.; Nielsen, J. Proteome Allocations Change Linearly with the Specific Growth Rate of *Saccharomyces cerevisiae* under Glucose Limitation. *Nat. Commun.* **2022**, *13*, 2819. [CrossRef] [PubMed]
23. Nagaraj, N.; Wisniewski, J.R.; Geiger, T.; Cox, J.; Kircher, M.; Kelso, J.; Pääbo, S.; Mann, M. Deep Proteome and Transcriptome Mapping of a Human Cancer Cell Line. *Mol. Syst. Biol.* **2011**, *7*, 548. [CrossRef] [PubMed]
24. Wiśniewski, J.R.; Ostasiewicz, P.; Duś, K.; Zielińska, D.F.; Gnad, F.; Mann, M. Extensive Quantitative Remodeling of the Proteome between Normal Colon Tissue and Adenocarcinoma. *Mol. Syst. Biol.* **2012**, *8*, 611. [CrossRef] [PubMed]
25. Hein, M.Y.; Hubner, N.C.; Poser, I.; Cox, J.; Nagaraj, N.; Toyoda, Y.; Gak, I.A.; Weisswange, I.; Mansfeld, J.; Buchholz, F.; et al. A Human Interactome in Three Quantitative Dimensions Organized by Stoichiometries and Abundances. *Cell* **2015**, *163*, 712–723. [CrossRef]
26. Itzhak, D.N.; Tyanova, S.; Cox, J.; Borner, G.H. Global, Quantitative and Dynamic Mapping of Protein Subcellular Localization. *eLife* **2016**, *5*, e16950. [CrossRef]
27. Morgenstern, M.; Peikert, C.D.; Lübbert, P.; Suppanz, I.; Klemm, C.; Alka, O.; Steiert, C.; Naumenko, N.; Schendzielorz, A.; Melchionda, L.; et al. Quantitative High-Confidence Human Mitochondrial Proteome and Its Dynamics in Cellular Context. *Cell Metab.* **2021**, *33*, 2464–2483.e18. [CrossRef]
28. Arike, L.; Valgepea, K.; Peil, L.; Nahku, R.; Adamberg, K.; Vilu, R. Comparison and Applications of Label-Free Absolute Proteome Quantification Methods on *Escherichia coli*. *J. Proteom.* **2012**, *75*, 5437–5448. [CrossRef]
29. Kopylov, A.T.; Ponomarenko, E.A.; Ilgisonis, E.V.; Pyatnitskiy, M.A.; Lisitsa, A.V.; Poverennaya, E.V.; Kiseleva, O.I.; Farafonova, T.E.; Tikhonova, O.V.; Zavialova, M.G.; et al. 200+ Protein Concentrations in Healthy Human Blood Plasma: Targeted Quantitative SRM SIS Screening of Chromosomes 18, 13, Y, and the Mitochondrial Chromosome Encoded Proteome. *J. Proteome Res.* **2019**, *18*, 120–129. [CrossRef]
30. Schwanhäusser, B.; Busse, D.; Li, N.; Dittmar, G.; Schuchhardt, J.; Wolf, J.; Chen, W.; Selbach, M. Global Quantification of Mammalian Gene Expression Control. *Nature* **2011**, *473*, 337–342. [CrossRef]
31. Taylor, S.C.; Berkelman, T.; Yadav, G.; Hammond, M. A Defined Methodology for Reliable Quantification of Western Blot Data. *Mol. Biotechnol.* **2013**, *55*, 217–226. [CrossRef]
32. Wiśniewski, J.R.; Hein, M.Y.; Cox, J.; Mann, M. A "Proteomic Ruler" for Protein Copy Number and Concentration Estimation without Spike-in Standards. *Mol. Cell. Proteom. MCP* **2014**, *13*, 3497–3506. [CrossRef] [PubMed]
33. Serres, M.H.; Gopal, S.; Nahum, L.A.; Liang, P.; Gaasterland, T.; Riley, M. A Functional Update of the *Escherichia coli* K-12 Genome. *Genome Biol.* **2001**, *2*, research0035.1. [CrossRef] [PubMed]
34. Lin, D.; Yin, X.; Wang, X.; Zhou, P.; Guo, F.-B. Re-Annotation of Protein-Coding Genes in the Genome of Saccharomyces Cerevisiae Based on Support Vector Machines. *PLoS ONE* **2013**, *8*, e64477. [CrossRef]
35. Salzberg, S.L. Open Questions: How Many Genes Do We Have? *BMC Biol.* **2018**, *16*, 94. [CrossRef] [PubMed]
36. Siwiak, M.; Zielenkiewicz, P. A Comprehensive, Quantitative, and Genome-Wide Model of Translation. *PLoS Comput. Biol.* **2010**, *6*, e1000865. [CrossRef]
37. Macromolecular Components of, E. coli and HeLa Cells—KZ. Available online: https://www.thermofisher.com/ru/en/home/references/ambion-tech-support/rna-tools-and-calculators/macromolecular-components-of-e.html (accessed on 21 November 2022).
38. Jorgensen, P.; Nishikawa, J.L.; Breitkreutz, B.-J.; Tyers, M. Systematic Identification of Pathways That Couple Cell Growth and Division in Yeast. *Science* **2002**, *297*, 395–400. [CrossRef] [PubMed]
39. Di Bartolomeo, F.; Malina, C.; Campbell, K.; Mormino, M.; Fuchs, J.; Vorontsov, E.; Gustafsson, C.M.; Nielsen, J. Absolute Yeast Mitochondrial Proteome Quantification Reveals Trade-off between Biosynthesis and Energy Generation during Diauxic Shift. *Proc. Natl. Acad. Sci. USA* **2020**, *117*, 7524–7535. [CrossRef] [PubMed]
40. Zhao, L.; Kroenke, C.D.; Song, J.; Piwnica-Worms, D.; Ackerman, J.J.H.; Neil, J.J. Intracellular Water Specific MR of Microbead-Adherent Cells: The HeLa Cell Intracellular Water Exchange Lifetime. *NMR Biomed.* **2008**, *21*, 159–164. [CrossRef]
41. Volkmer, B.; Heinemann, M. Condition-Dependent Cell Volume and Concentration of Escherichia Coli to Facilitate Data Conversion for Systems Biology Modeling. *PLoS ONE* **2011**, *6*, e23126. [CrossRef]
42. Taheri-Araghi, S.; Bradde, S.; Sauls, J.T.; Hill, N.S.; Levin, P.A.; Paulsson, J.; Vergassola, M.; Jun, S. Cell-Size Control and Homeostasis in Bacteria. *Curr. Biol.* **2015**, *25*, 385–391. [CrossRef] [PubMed]
43. Basan, M.; Zhu, M.; Dai, X.; Warren, M.; Sévin, D.; Wang, Y.-P.; Hwa, T. Inflating Bacterial Cells by Increased Protein Synthesis. *Mol. Syst. Biol.* **2015**, *11*, 836. [CrossRef] [PubMed]
44. Ho, B.; Baryshnikova, A.; Brown, G.W. Unification of Protein Abundance Datasets Yields a Quantitative *Saccharomyces cerevisiae* Proteome. *Cell Syst.* **2018**, *6*, 192–205.e3. [CrossRef] [PubMed]
45. Perry, R.P. Balanced production of ribosomal proteins. *Gene* **2007**, *401*, 1–3. [CrossRef]
46. Yamaguchi, M.; Namiki, Y.; Okada, H.; Mori, Y.; Furukawa, H.; Wang, J.; Ohkusu, M.; Kawamoto, S. Structome of *Saccharomyces cerevisiae* Determined by Freeze-Substitution and Serial Ultrathin-Sectioning Electron Microscopy. *J. Electron. Microsc.* **2011**, *60*, 321–335. [CrossRef] [PubMed]
47. Duncan, R.; Hershey, J.W. Identification and Quantitation of Levels of Protein Synthesis Initiation Factors in Crude HeLa Cell Lysates by Two-Dimensional Polyacrylamide Gel Electrophoresis. *J. Biol. Chem.* **1983**, *258*, 7228–7235. [CrossRef]

48. Wolf, S.F.; Schlessinger, D. Nuclear Metabolism of Ribosomal RNA in Growing, Methionine-Limited, and Ethionine-Treated HeLa Cells. *Biochemistry* **1977**, *16*, 2783–2791. [CrossRef]
49. von der Haar, T. A Quantitative Estimation of the Global Translational Activity in Logarithmically Growing Yeast Cells. *BMC Syst. Biol.* **2008**, *2*, 87. [CrossRef]
50. Lu, P.; Vogel, C.; Wang, R.; Yao, X.; Marcotte, E.M. Absolute Protein Expression Profiling Estimates the Relative Contributions of Transcriptional and Translational Regulation. *Nat. Biotechnol.* **2007**, *25*, 117–124. [CrossRef]
51. Peng, M.; Taouatas, N.; Cappadona, S.; van Breukelen, B.; Mohammed, S.; Scholten, A.; Heck, A.J.R. Protease Bias in Absolute Protein Quantitation. *Nat. Methods* **2012**, *9*, 524–525. [CrossRef]
52. Ezkurdia, I.; Rodriguez, J.M.; Carrillo-de Santa Pau, E.; Vázquez, J.; Valencia, A.; Tress, M.L. Most Highly Expressed Protein-Coding Genes Have a Single Dominant Isoform. *J. Proteome Res.* **2015**, *14*, 1880–1887. [CrossRef] [PubMed]
53. Bakshi, S.; Siryaporn, A.; Goulian, M.; Weisshaar, J.C. Superresolution Imaging of Ribosomes and RNA Polymerase in Live Escherichia Coli Cells. *Mol. Microbiol.* **2012**, *85*, 21–38. [CrossRef] [PubMed]
54. Hu, X.-P.; Dourado, H.; Schubert, P.; Lercher, M.J. The Protein Translation Machinery Is Expressed for Maximal Efficiency in Escherichia Coli. *Nat. Commun.* **2020**, *11*, 5260. [CrossRef] [PubMed]
55. Diaconu, M.; Kothe, U.; Schlünzen, F.; Fischer, N.; Harms, J.M.; Tonevitsky, A.G.; Stark, H.; Rodnina, M.V.; Wahl, M.C. Structural Basis for the Function of the Ribosomal L7/12 Stalk in Factor Binding and GTPase Activation. *Cell* **2005**, *121*, 991–1004. [CrossRef]
56. Woolford, J.L.; Baserga, S.J. Ribosome Biogenesis in the Yeast Saccharomyces Cerevisiae. *Genetics* **2013**, *195*, 643–681. [CrossRef]
57. Nakao, A.; Yoshihama, M.; Kenmochi, N. RPG: The Ribosomal Protein Gene Database. *Nucleic Acids Res.* **2004**, *32*, D168–D170. [CrossRef]

International Journal of
Molecular Sciences

MDPI

Article

Software Tool for Visualization and Validation of Protein Turnover Rates Using Heavy Water Metabolic Labeling and LC-MS

Henock M. Deberneh and Rovshan G. Sadygov *

Department of Biochemistry and Molecular Biology, The University of Texas Medical Branch, 301 University Blvd, Galveston, TX 77555, USA
* Correspondence: rgsadygo@utmb.edu; Tel.: +1-409-772-3287

Abstract: Metabolic stable isotope labeling followed by liquid chromatography coupled with mass spectrometry (LC-MS) is a powerful tool for in vivo protein turnover studies of individual proteins on a large scale and with high throughput. Turnover rates of thousands of proteins from dozens of time course experiments are determined by data processing tools, which are essential components of the workflows for automated extraction of turnover rates. The development of sophisticated algorithms for estimating protein turnover has been emphasized. However, the visualization and annotation of the time series data are no less important. The visualization tools help to validate the quality of the model fits, their goodness-of-fit characteristics, mass spectral features of peptides, and consistency of peptide identifications, among others. Here, we describe a graphical user interface (GUI) to visualize the results from the protein turnover analysis tool, d2ome, which determines protein turnover rates from metabolic D_2O labeling followed by LC-MS. We emphasize the specific features of the time series data and their visualization in the GUI. The time series data visualized by the GUI can be saved in JPEG format for storage and further dissemination.

Keywords: in vivo protein turnover; heavy water metabolic labeling; isotope distribution; time series of isotope labeling; graphical user interface for mass spectral data

Citation: Deberneh, H.M.; Sadygov, R.G. Software Tool for Visualization and Validation of Protein Turnover Rates Using Heavy Water Metabolic Labeling and LC-MS. *Int. J. Mol. Sci.* **2022**, *23*, 14620. https://doi.org/10.3390/ijms232314620

Academic Editor: Paolo Iadarola

Received: 19 October 2022
Accepted: 18 November 2022
Published: 23 November 2022

1. Introduction

Cellular proteins are in a dynamic equilibrium. Protein concentrations are maintained while they are continuously synthesized and degraded. The equilibria are tissue-specific, and they shift during organismal development, aging, and diseases. Metabolic stable isotope labeling followed by liquid chromatography and mass spectrometry (LC-MS) has been a powerful tool to study in vivo protein turnover on a large scale and high throughput [1,2]. As a labeling agent, heavy water (drinking water enriched in D_2O) is easy to use, cost-efficient, and does not require adaptation period [3]. Low (<8%) concentrations of D_2O enrichments are normally used in drinking water [4]. It results in the composite spectra of unlabeled and labeled forms of a peptide in MS1. Statistical [5] and analytical [6] approaches to de-convolve the spectra have been described. A recent study [7] revealed that the precursor enrichment in D_2O labeling was nearly instantaneous, and a single exponential curve was sufficient for the modeling. In contrast, the precursor enrichment in heavy amino acid labeling [8] was delayed and tissue specific [7]. Therefore, the modeling of label incorporation in amino acid labeling was more complex and required more parameters.

Since the data are generated for thousands of proteins from tens of thousands of peptides at every time point of labeling, manual data processing is impractical. Several publicly available software tools [7,9,10] have been developed to process the mass spectral data and database search results to automate protein turnover rate estimations. The turnover rates are obtained from the exponential decay modeling of the monoisotopic relative isotope abundance (RIA). The tools generate protein turnover rates and goodness-of-fit (GOF) measures of the model, such as coefficient of determination (R^2), Pearson correlation, and standard deviation, among others. The results are normally reported in tables and saved in output files formatted in csv format. Though the turnover estimation tools automate data processing, the visualization and annotation of the results are as important. The csv formatted files can be read and processed using scripts in R [11] or Phyton environments. However, it requires familiarity with these environments. Therefore, a graphical user interface (GUI) to enter input data, easily access results for each protein/peptide, and obtain information about the statistical GOF measures is important. A protein turnover estimation software tool [12] for metabolic labeling with a heavy amino acid ($^{13}C_6$-Lys) contained a GUI, ApplE Turnover, to facilitate the data analysis. Another tool for protein turnover estimation from [5,5,5-$^2H_3$] Lue labeled samples, TurnoveR [13], used the functions of Skyline [14], an MS data analysis platform. Here, we report on our implementation of a GUI for a software tool, d2ome [10], to estimate protein turnover rates from D_2O labeling. The GUI facilitates several manually laborious steps in the data input, the selection of data processing parameters, and, importantly, it plots experimental time points and theoretical fit, shows the GOF measures, and spectral features (mass-to-charge ratio, m/z, the monoisotopic abundance, charge state, and the number of exchangeable hydrogens) of the peptide and its amino acid sequence. Every protein can be located by an easy search or from a drop-down list of alphabetically sorted protein names. mzML [15] (mass spectral) and mzid [16] (database search results) files are automatically matched in the input. Considering that the data for protein turnover is highly voluminous, the GUI will facilitate the data analysis, visualization, and validation of the results.

2. Results and Discussions

The time series data used in protein turnover studies is more complex than the static proteomics data. Thus, in static proteomics, the proteome is normally characterized by peptide sequence and its post-translational modifications, abundance, chromatographic retention time, m/z, and charge state. In contrast, protein turnover data, in addition to the listed information, requires the number of exchangeable hydrogens, body water enrichment in deuterium, the number of experiments in which the peptide was quantified, GOF measures (R^2, RMSE, SD) between the experimental data and the theoretical fit, the monoisotopic abundance, and the accuracy of the isotope distribution between the estimated and LC-MS data for the unlabeled sample. A typical workflow of experimental and data processing steps is shown in Figure 1.

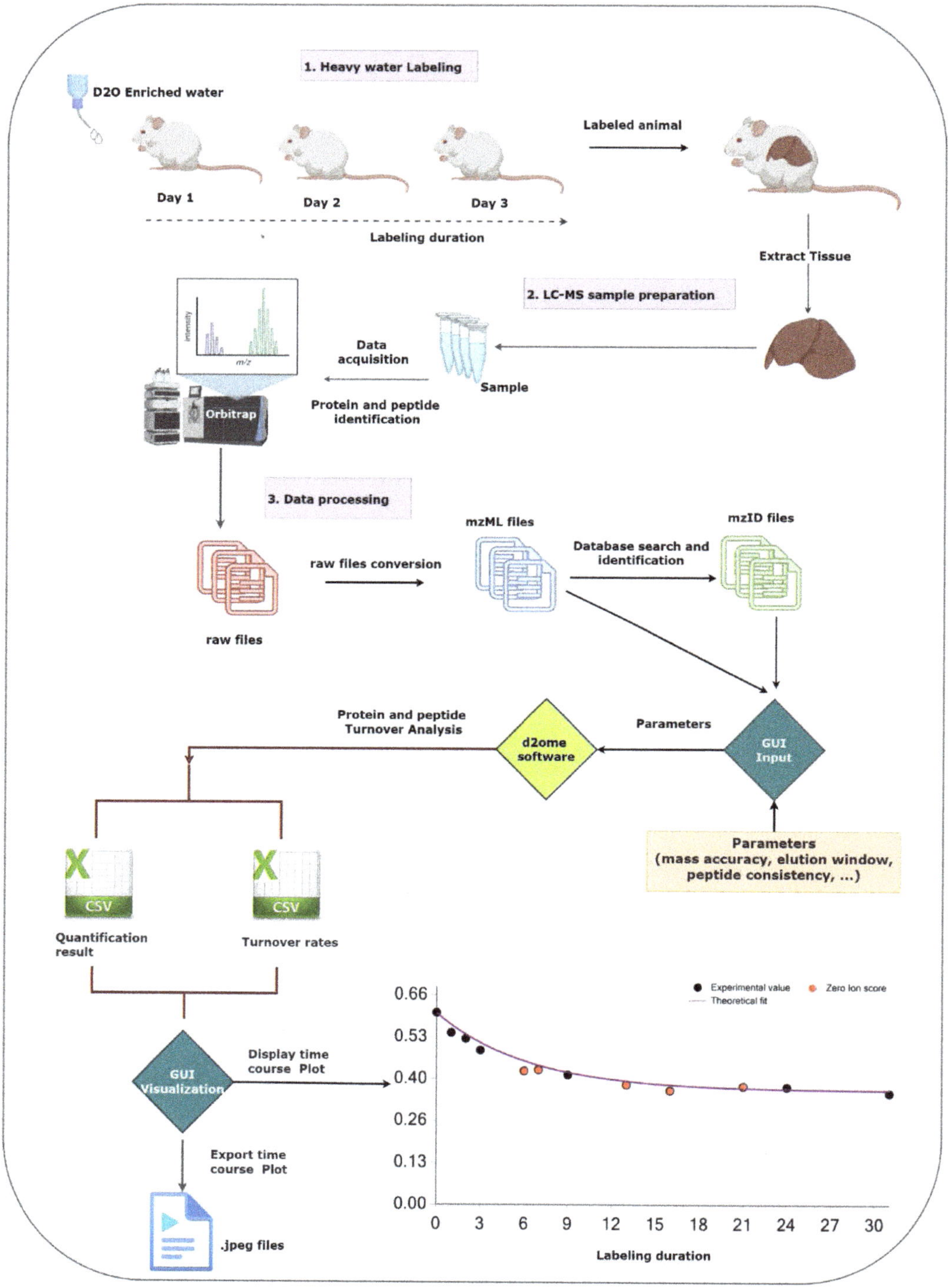

Figure 1. The figure shows the workflow of the main experimental and data processing of protein turnover studies and the role of the graphical user interface (GUI). The GUI aids in setting up the protein turnover rate estimations (creating input data from mzML and mzid files) and visualizing the results to facilitate the validations.

2.1. Data Input and Data Processing Parameters

Figure 2 presents a sample screenshot of the GUI for data and parameter input. The GUI enables users to input sets of database search results and corresponding mass spectral data, body water enrichment in deuterium, peptide and protein consistencies (the minimum number of experiments in which a peptide and a protein are identified in MS/MS), and the corresponding labeling duration in a tabular format. The software has a feature to automatically populate pairs of input files from the source folder by matching the file names. It allows users to reload all configurations from the previous runs for re-runs with different parameters. The GUI allows users to optimize the quantification results by customizing the input parameters. The parameters include spectral mass accuracy, retention time window for peak detection and integration, the threshold of peptide score (Mascot [17] Ion score), and expectation.

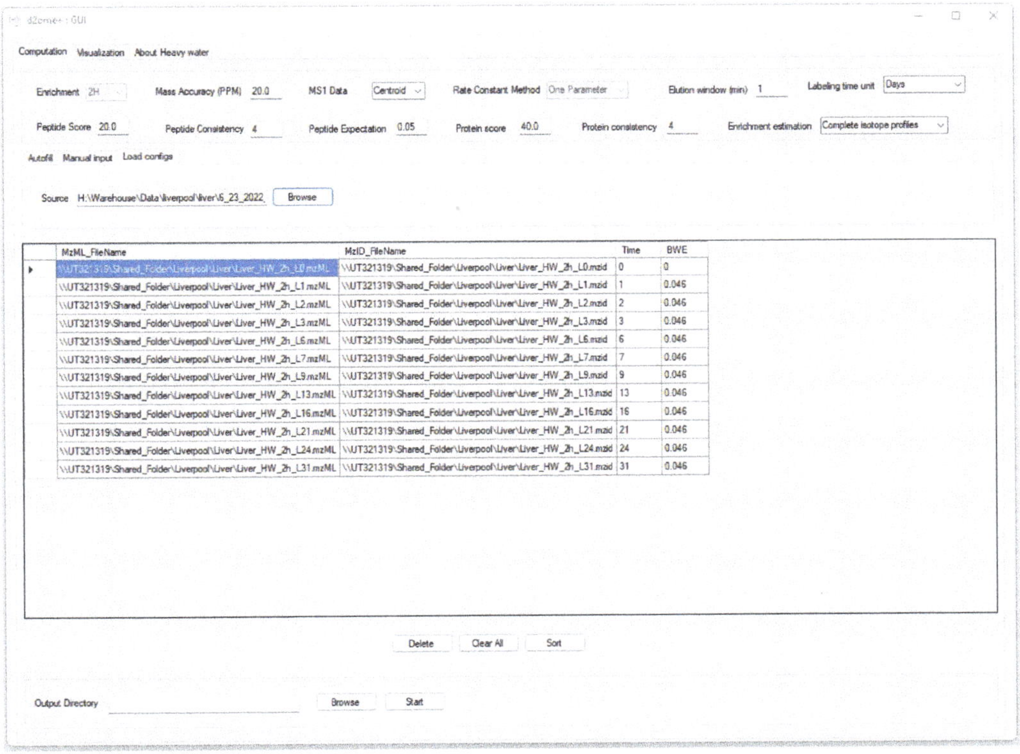

Figure 2. The graphical user interface (GUI) to input data and parameters for the protein turnover rate estimations. Time and BWE are the labeling duration and body water enrichment in deuterium for the corresponding mzML and mzid files (experiments).

Consistent identifications of peptides and proteins from tandem mass spectra are essential in time series experiments. Since DDA is semi-stochastic in the selection of ions to be fragmented, we implemented a match between-runs (MBR) approach to enable the quantification of peptides that are missing in some experiments, but their features (chromatographic elution profile corresponding to m/z and charge state at the allowed chromatographic time elution window) are detectable. The approach implemented the retention time alignment strategy using raw mass spectral profiles [18]. The time window in which the missing peptide features will be searched is adjustable in the GUI. MBR increases proteome coverage across the labeling duration time series.

2.2. The Output of Data Processing

d2ome [10] computes turnover rates for proteins and peptides using the non-linear least squares regression on the monoisotopic RIA. It generates two main outputs: EntryName.RateConst.csv and EntryName.Quant.csv. The first entry in the file names is the Uniprot [19] entry name for a protein. The *.Quant.csv file contains comprehensive information about each peptide of a protein. Each peptide entry is a row of information that contains the amino acid sequence, the charge state of the precursor, theoretical m/z of the peptide sequence, theoretical isotope abundances (natural isotope abundances), precursor m/z, the highest Mascot Ion score, Mascot expectation, mass accuracy (in ppm), scan number, the integrated (from MS1 scans in LC-MS) abundance of the mass isotopomers (six), elution start and end times that were used to calculate the isotopomer abundances, and the monoisotopic peak width in the mass-to-charge domain (used only for data in profile mode).

The rows of *.RateConst.csv file of a protein contain: the peptide sequence, its uniqueness (distinct or shared with other proteins), peptide rate constant and corresponding confidence intervals, the correlation value between theoretical fit and experimental data, RMSE, the absolute deviation between the theoretical and experimental isotope profiles (before the start of labeling), peptide charge, sequence m/z, the number of accessible hydrogens (N_{EH}), the number of data points (NDP), R^2 of the theoretical fit, and the average abundance of the monoisotope.

2.3. Visualization of the Results

The visualization tab of d2ome has two main charts that depict the time series [20] data used for peptides and protein degradation rate computation, Figure 3. It provides easy access to turnover rate estimation results for each protein. For every protein peptide, the monoisotopic RIAs estimated from the isotope profiles in comparison with the theoretical fit can be visualized. This approach visualizes the correspondence between the experimental points and the expected theoretical values, which are computed based on the degradation rate constant.

The estimation of the monoisotopic RIA requires accurate measurements of the abundances of all mass isotopomers of a peptide. Since mammalian samples are complex, peptide species often co-elude and interfere with the mass profile of the target peptide. This GUI enables users to graphically validate the quality of experimental input data (the time series of monoisotopic RIA) in comparison with the theoretical fit. Figure 4 shows the monoisotopic RIAs estimated from the isotope profiles in comparison with the theoretical fit for the peptide sequence, SDEAVKPLGVK^{+2} from FAS_MOUSE protein. For this protein, the experimental isotope distributions of each peptide at every time of labeling are in the FAS_MOUSE.Quant.csv file. The unlabeled and labeled [7] (7 and 31 days) isotope profiles of the SDEAVKPLGVK^{+2} peptide are presented in Figure 5. The monoisotopic RIA was computed as the ratio of monoisotopic abundance to the sum of abundances of all mass isotopomers. The data can be used as additional validation of the label incorporation.

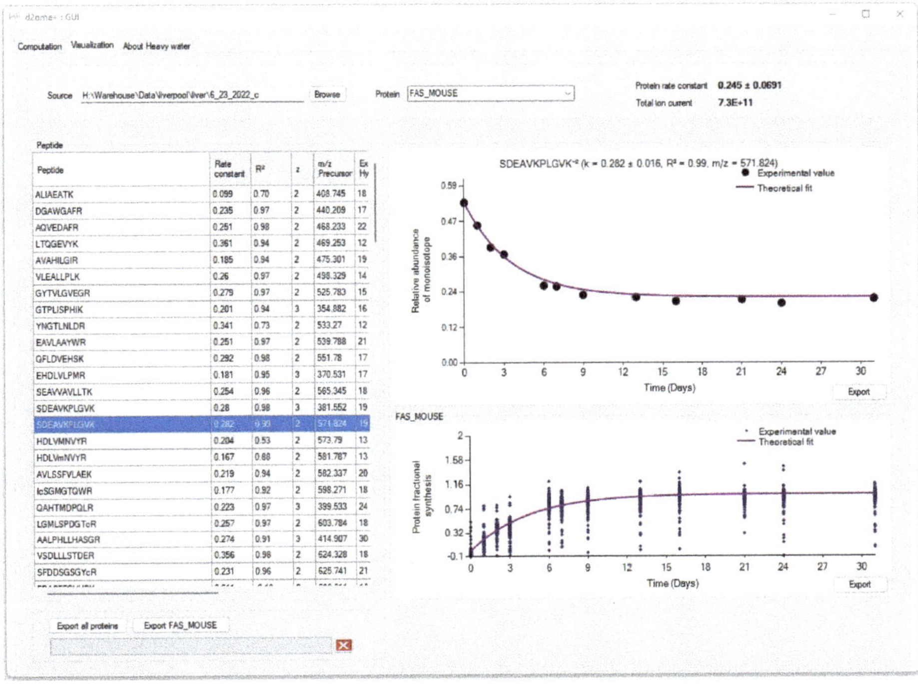

Figure 3. The graphical user interface (GUI) for the results of protein turnover rate estimations. Results for each protein can be accessed by name search. For every peptide of a protein, the experimental time series data and its theoretical fit can be visually examined.

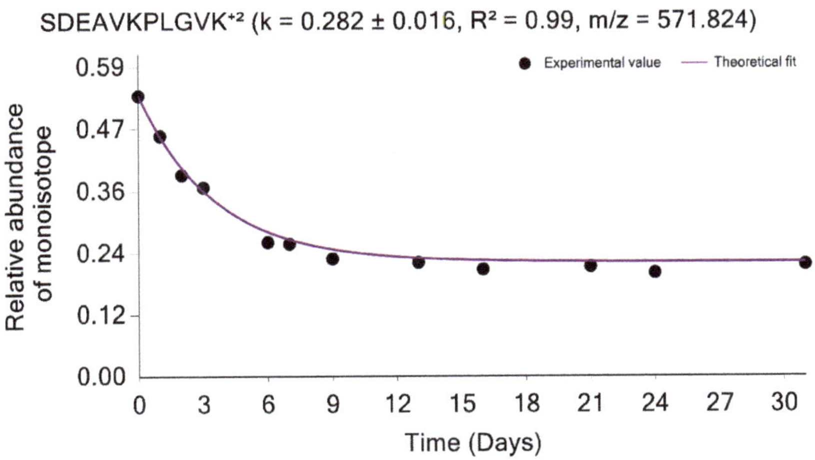

Figure 4. The graphical user interface enables comprehensive visualization of the results of protein turnover studies from metabolic D_2O labeling and LC-MS experiments. Time series of monoisotopic RIAs (y-axis) are shown along the labeling duration (*x*-axis). The solid line shows the fit from the computed degradation constant for SDEAVKPLGVK^{+2} peptide from FAS_MOUSE protein.

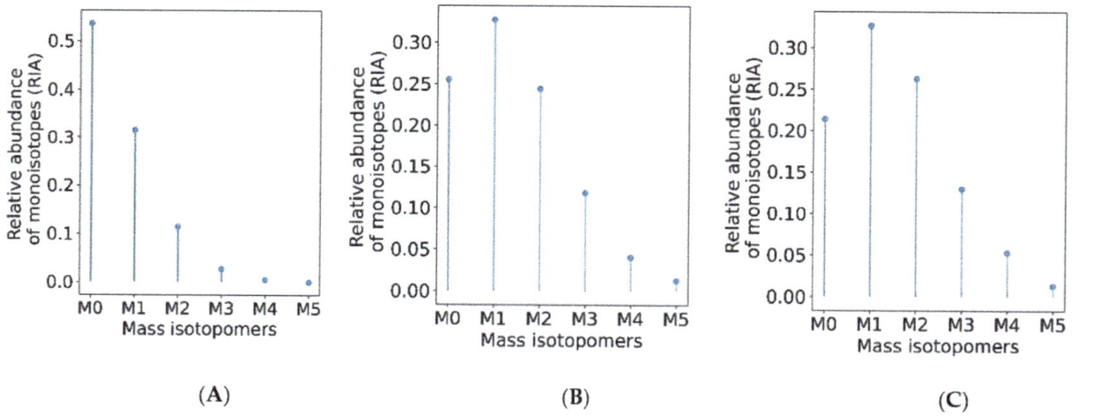

Figure 5. The monoisotopic RIA depletes with the labeling duration. Isotope profiles of SDEAVKPLGVK^{+2} peptide (**A**) from an unlabeled sample (**B**) from a labeled sample (day 7) (**C**) from a labeled sample (day 31).

As mentioned above, the quantifications using MBR transfers are important in stable isotope labeling experiments. The MBR procedure may result in false positive transfers [21]. The GUI provides the opportunity to examine the quality of the label incorporation estimation from the data obtained by using the MBR. Thus, the labeling time points, which were quantified using MBR, can be shown in red; it is demonstrated in Figure 6 for the peptide sequence NLLSVAYK^{+2} from the 1433B_MOUSE protein. Shown in red are the labeling time points (experiments) in which the peptide was not identified from an MS/MS spectrum. Instead, the quantification was performed based on the MBR. The use of MBR increases proteome coverage across the labeling time points. It is helpful to visually examine the MBR quantified time points, and the GUI provides this opportunity.

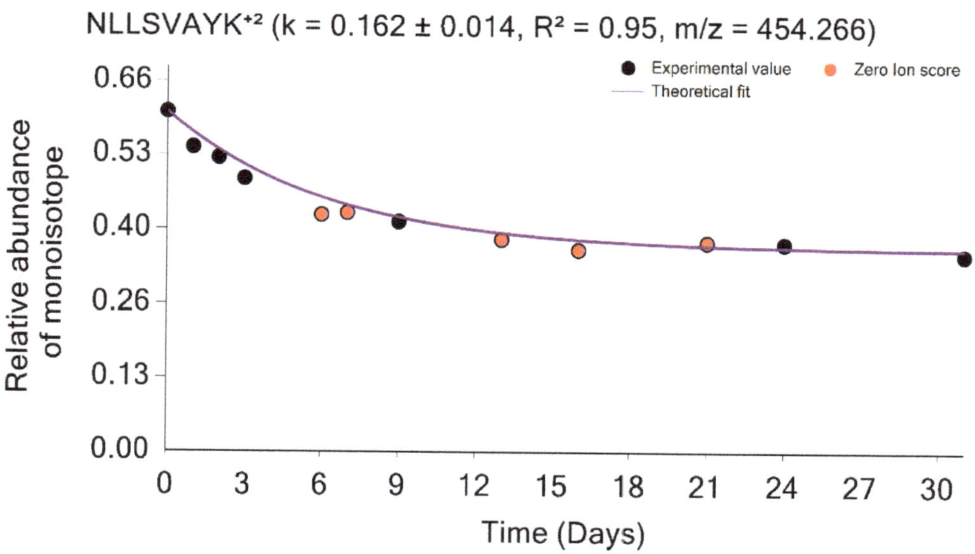

Figure 6. Time course plot of monoisotopic RIAs for NLLSVAYK^{+2} peptide from the 1433B_MOUSE protein. The experimental time points quantified using the match between runs are shown in red.

The GUI also graphically shows the overall label incorporation from all peptides of a protein. It is illustrated in the time series of the FS. Figure 7 shows the protein FS in comparison with the theoretical fit for the FAS_MOUSE protein. The graph presents each peptide's experimental FS as a scatter plot and the theoretical fit based on the protein rate constant as a solid line. Both figures can be exported as high-quality JPEG images. Furthermore, the software enables users to export charts separately or in a batch mode for all identified proteins and peptides. The GUI is a user-friendly application that makes searching and visualizing all proteins and peptides simpler. Users can easily switch between the visualizations of different proteins/peptides.

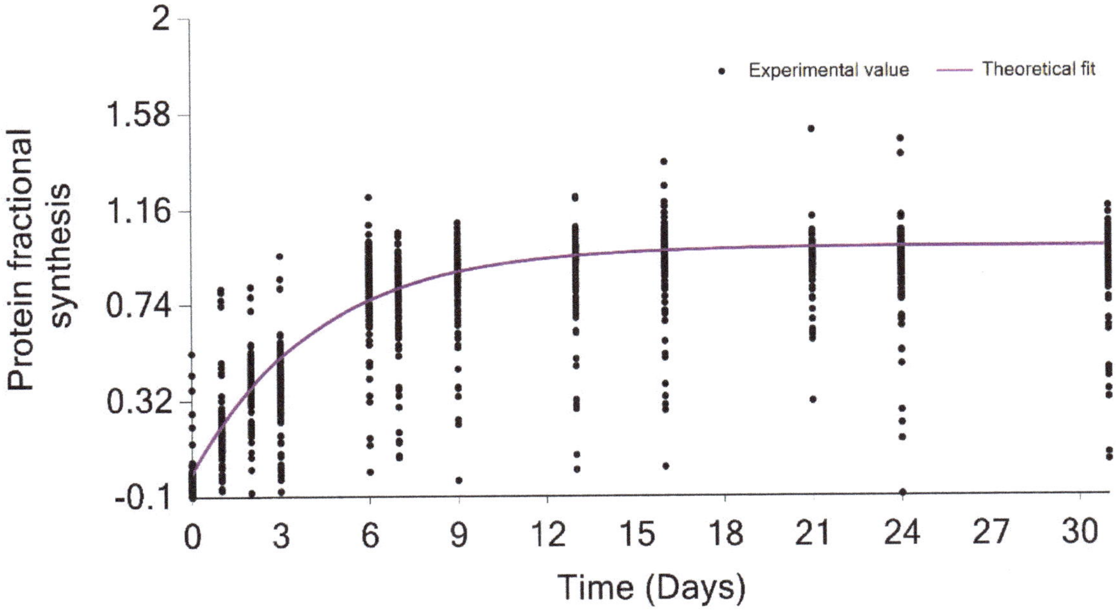

Figure 7. Protein turnover rate is computed from the median of the degradation constant of quantified peptides. The scatter plot indicates the fractional value of peptides in the FAS_MOUSE protein. The solid line shows the fit from the computed turnover rate for the protein.

The visualization window also contains comprehensive information about each peptide of a protein in a tabular format. Each peptide entry is a row of information that contains the amino acid sequence, the charge state of the precursor, the theoretical m/z of the peptide sequence, the correlation between theoretical fit and experimental time series, RMSE, the absolute deviation between the theoretical and experimental isotope profiles, N_{EH}, NDP, R^2 of the theoretical fit, and the monoisotopic average abundance.

The currently available software tools for protein turnover studies from LC-MS-MS/MS data of deuterium-labeled samples (such as DeuterRator [9] and Riana [7]) simplify data analyses also by means of a GUI component. The GUI in DeuteRator [9] simplifies data entry and parameter selection. It plots and saves the FS time series and its theoretical fit for each protein. Output from Riana can be visualized in the R environment using supplied scripts. Our approach to the GUI development was motivated by that of ApplE Turnover [12]. A user can search for each protein, plot the experimental time series and theoretical fit of the monoisotopic RIA for every peptide, display the experimental time series and corresponding theoretical fit for the FS of a protein, export all figures, and view several GOFs of each peptide of a protein. It is possible to review previously processed results. We believe the GUI features address user needs in many cases. Our main goal in developing the GUI was to facilitate the visualization, quality assessment, validation, and dissemination of the turnover rate estimation.

2.4. Future Plans

The GUI developed in this work provides visualization of the theoretical fit to the experimental data points, the GOF measures, and spectral features for a peptide. An additional element for visual validation is the experimental isotope distribution of the peptide at the apex of its elution. This visualization of the distribution would allow us to validate the quality of the monoisotopic RIA estimation. We plan to implement this interactive feature in a future iteration of the GUI. Currently, the GUI interfaces with the database search output from Mascot. We intend to include support for other search engines

3. Methods

In data modeling from metabolic D_2O labeling and LC-MS experiments, the peptide/protein turnover rate is estimated by exponential decay modeling of the time course of the depletion of the monoisotopic RIA, $I_0(t)$, with the labeling duration, t:

$$I_0(t) = I_0^{asymp} + \left(I_0(0) - I_0^{asymp}\right)e^{-kt} \tag{1}$$

where $I_0(0)$ is the monoisotopic RIA of an unlabeled peptide, $I_0{}^{asymp}$ is the monoisotopic RIA at the plateau of labeling, and k is the turnover rate (degradation rate constant) of a peptide. $I_0{}^{asymp}$ is obtained from the body water enrichment in deuterium (p_W) and the number of hydrogens accessible to deuteriums in the water (N_{EH}):

$$I_0^{asymp} = I_0(0)\left(1 - \frac{p_W}{1 - p_H}\right)^{N_{EH}}$$

The turnover rate is obtained from the non-linear regression of the experimental time series data of $I_0(t)$ on the theoretical decay function in Equation (1). The modeling is central to the turnover rate estimation. The GUI depicts experimental time points and the theoretical curve resulting from the regression for every peptide.

Another property used for the analyses of protein turnover is the fractional synthesis. For every peptide, the fractional synthesis (FS) is defined as:

$$FS(t) = \frac{I_0(0) - I_0(t)}{I_0(0) - I_0^{asymp}} = 1 - e^{-kt} \tag{2}$$

In Equation (2), the explicit dependency on the number of exchangeable hydrogens and natural monoisotopic RIA, which are characteristics of each peptide, are removed. The GUI depicts the FSs of all peptides of protein in a single figure.

The GUI in d2ome is a Windows Forms application developed in C# programming language. Windows Forms is a .Net Framework GUI library that provides an interface to develop multipurpose applications. It is composed of controls such as combo boxes, buttons, labels, list boxes, charts, and containers such as panels, group boxes, and others. In the course of the development of the GUI for the d2ome software tool, we had used a tab layout to switch between computation and visualization windows, a data grid view to display detailed peptide information in a tabular format, charts to display peptides time course data, and buttons to execute tasks such as loading data, searching proteins, exporting charts, and others. The GUI can be initiated either from the command line or from the application icon.

The GUI interfaces with d2ome in two stages, Figure 1. In the first stage, the GUI automates data input (matching pairs of mzML [22] and mzid files, body water enrichment, and labeling time course) and the specification of parameters (mass accuracy, the required number of labeling time points, database search scores, etc.). There is no limitation on the number of experiments (input files for processing). The GUI uses the experimental data and parameters to create an input set for d2ome to compute peptide/protein turnover rates. d2ome writes out the results for every protein in .Quant.csv and .RateConts.csv files. The .Quant.csv file of a protein contains information about the amino acid sequence, theoretical isotope distribution of unlabeled peptide, *m*/*z*, charge, scan number of MS/MS identification, mass accuracy, database search score, and mass isotopomer abundances (M_0-M_5) from each experiment for every identified and quantified peptide of the protein. The .Rate.Const.csv file of a protein contains the results of rate constant (turnover rate) calculations for every peptide, GOF to the theoretical model, Equation (1), and statistical properties of the computed rate constant, such as standard deviation (SD), root-mean-squared-error (RMSE), coefficient of determination (R^2), Pearson correlation, averaged (from all experiments) monoisotopic abundance, and protein turnover rate. Normalized (by the median of medians of base peak abundances from each experiment) protein abundance is also reported in the file. These data are depicted by the GUI in the second stage of interfacing with d2ome. All figures (experimental time series and its theoretical fit, FS for a protein) can be exported as high-quality JPEG images. Furthermore, the software enables users to export charts separately or as a batch process for all identified proteins and peptides. The tool is available in the GitHub repository, https://github.com/rgsadygov/d2ome (accessed on 17 November 2022).

Data Used in This Work

The figures and examples shown in the paper were obtained from processing a publicly available data set of mouse liver proteome [7]. Labeling and LC-MS experiments are described in the original publication. In brief, adult male C57BL/6JOlaHsd mice were labeled with deuterium oxide. Murine liver tissues were collected at twelve labeling time points: 0, 1, 2, 3, 6, 7, 9, 13, 16, 21, 24, and 31 days. The body water enrichment in deuterium was determined to be 0.046 in all labeled samples. The mass spectral data were acquired in the data-dependent acquisition mode (DDA) using a Q-Exactive HF quadrupole-Orbitrap mass spectrometer. The raw mass spectral data are available on ProteomeXchange at accession PXD029639.

4. Conclusions

We developed a graphical user interface to facilitate the data analysis of protein turnover studies from time series data of metabolic labeling with D_2O and LC-MS. The turnover rate calculations use a large number of experimental inputs (time series of label enrichment) and parameters (body water enrichment, mass accuracy, peptide/protein identification consistency, etc.). The GUI automates data input and parameter selection.

The validation of the protein turnover results requires information about various spectral features (*m*/*z*, the abundance of the monoisotopic RIA, NDP, etc.) and statistical measures of GOF (R^2, Pearson correlation, SD, RMSE, etc.). The GUI depicts the theoretical fit to the experimental time series data, thus allowing a visual evaluation of the fit. The statistical measures of the model show the quality of the GOF, which also helps to estimate the quality of the theoretical fit. All generated figures for every peptide of a protein can be exported in JPEG format for further dissemination.

Author Contributions: Conceptualization, R.G.S.; methodology, H.M.D.; formal analysis, R.G.S. and H.M.D.; software, H.M.D.; writing-review R.G.S. and H.M.D.; funding acquisition, R.G.S. All authors have read and agreed to the published version of the manuscript.

Funding: The research reported in this publication was supported in part by the NIGMS of the National Institutes of Health (NIH) under Award Number R01GM112044. The content is the authors' responsibility and does not necessarily represent the official views of the NIH.

Institutional Review Board Statement: Not applicable.

Informed Consent Statement: Not applicable.

Data Availability Statement: Tools reported in this paper are available in the GitHub repository: https://github.com/rgsadygov/d2ome (accessed on 17 November 2022).

Acknowledgments: We acknowledge the support by the UTMB.

Conflicts of Interest: The authors declare no conflict of interest.

Abbreviations

DDA—data-dependent acquisition; GOF—goodness of fit; GUI—graphical user interface; FS—fractional synthesis; LC-MS—liquid chromatography coupled with mass spectrometry; MBR—match between runs; NDP—the number of data points; N_{EH}—the number of exchangeable hydrogens; RIA—relative isotope abundance; RMSE—root mean squared error; SD—standard deviation.

References

1. Fornasiero, E.F.; Savas, J.N. Determining and interpreting protein lifetimes in mammalian tissues. *Trends Biochem. Sci.* **2022**. [CrossRef] [PubMed]
2. Sadygov, R.G. Protein turnover models for LC-MS data of heavy water metabolic labeling. *Brief. Bioinform.* **2022**, *23*, bbab598. [CrossRef] [PubMed]
3. Holmes, W.E.; Angel, T.E.; Li, K.W.; Hellerstein, M.K. Dynamic Proteomics: In Vivo Proteome-Wide Measurement of Protein Kinetics Using Metabolic Labeling. *Methods Enzymol.* **2015**, *561*, 219–276. [PubMed]
4. Busch, R.; Kim, Y.K.; Neese, R.A.; Schade-Serin, V.; Collins, M.; Awada, M.; Gardner, J.L.; Beysen, C.; Marino, M.E.; Misell, L.M.; et al. Measurement of protein turnover rates by heavy water labeling of nonessential amino acids. *Biochim. Biophys. Acta* **2006**, *1760*, 730–744. [CrossRef] [PubMed]
5. O'Brien, J.J.; Narayan, V.; Wong, Y.; Seitzer, P.; Sandoval, C.M.; Haste, N.; Smith, M.; Rad, R.; Gaun, A.; Baker, A.; et al. Precise Estimation of In Vivo Protein Turnover Rates. *bioRxiv* **2020**. [CrossRef]
6. Sadygov, R.G. Partial Isotope Profiles Are Sufficient for Protein Turnover Analysis Using Closed-Form Equations of Mass Isotopomer Dynamics. *Anal. Chem.* **2020**, *92*, 14747–14753. [CrossRef]
7. Hammond, D.E.; Simpson, D.M.; Franco, C.; Wright Muelas, M.; Waters, J.; Ludwig, R.W.; Prescott, M.C.; Hurst, J.L.; Beynon, R.J.; Lau, E. Harmonizing Labeling and Analytical Strategies to Obtain Protein Turnover Rates in Intact Adult Animals. *Mol. Cell Proteom.* **2022**, *21*, 100252. [CrossRef]
8. Chepyala, S.R.; Liu, X.; Yang, K.; Wu, Z.; Breuer, A.M.; Cho, J.H.; Li, Y.; Mancieri, A.; Jiao, Y.; Zhang, H.; et al. JUMPt: Comprehensive Protein Turnover Modeling of In Vivo Pulse SILAC Data by Ordinary Differential Equations. *Anal. Chem.* **2021**, *93*, 13495–13504. [CrossRef]
9. Naylor, B.C.; Porter, M.T.; Wilson, E.; Herring, A.; Lofthouse, S.; Hannemann, A.; Piccolo, S.R.; Rockwood, A.L.; Price, J.C. DeuteRater: A tool for quantifying peptide isotope precision and kinetic proteomics. *Bioinformatics* **2017**, *33*, 1514–1520. [CrossRef]
10. Sadygov, R.G.; Avva, J.; Rahman, M.; Lee, K.; Ilchenko, S.; Kasumov, T.; Borzou, A. d2ome, Software for in Vivo Protein Turnover Analysis Using Heavy Water Labeling and LC-MS, Reveals Alterations of Hepatic Proteome Dynamics in a Mouse Model of NAFLD. *J. Proteome Res.* **2018**, *17*, 3740–3748, Correction in *J. Proteome Res.* **2021**, *20*, 4912. [CrossRef]
11. R Core Team. *R: A Language and Environment for Statistical Computing*; R Foundation for Statistical Computing: Vienna, Austria, 2019.
12. Rolfs, Z.; Frey, B.L.; Shi, X.; Kawai, Y.; Smith, L.M.; Welham, N.V. An atlas of protein turnover rates in mouse tissues. *Nat. Commun.* **2021**, *12*, 6778. [CrossRef]
13. Basisty, N.; Shulman, N.; Wehrfritz, C.; Marsh, A.N.; Shah, S.; Rose, J.; Ebert, S.; Miller, M.; Dai, D.F.; Rabinovitch, P.S.; et al. TurnoveR: A Skyline External Tool for Analysis of Protein Turnover in Metabolic Labeling Studies. *J. Proteome Res.* **2022**. [CrossRef]
14. MacLean, B.; Tomazela, D.M.; Shulman, N.; Chambers, M.; Finney, G.L.; Frewen, B.; Kern, R.; Tabb, D.L.; Liebler, D.C.; MacCoss, M.J. Skyline: An open source document editor for creating and analyzing targeted proteomics experiments. *Bioinformatics* **2010**, *26*, 966–968. [CrossRef]
15. Martens, L.; Chambers, M.; Sturm, M.; Kessner, D.; Levander, F.; Shofstahl, J.; Tang, W.H.; Rompp, A.; Neumann, S.; Pizarro, A.D.; et al. mzML—A community standard for mass spectrometry data. *Mol. Cell Proteomics* **2011**, *10*, R110000133. [CrossRef]
16. Jones, A.R.; Eisenacher, M.; Mayer, G.; Kohlbacher, O.; Siepen, J.; Hubbard, S.J.; Selley, J.N.; Searle, B.C.; Shofstahl, J.; Seymour, S.L.; et al. The mzIdentML data standard for mass spectrometry-based proteomics results. *Mol. Cell Proteom.* **2012**, *11*, M111-014381. [CrossRef]
17. Perkins, D.N.; Pappin, D.J.; Creasy, D.M.; Cottrell, J.S. Probability-based protein identification by searching sequence databases using mass spectrometry data. *Electrophoresis* **1999**, *20*, 3551–3567. [CrossRef]
18. Sadygov, R.G.; Maroto, F.M.; Huhmer, A.F. ChromAlign: A two-step algorithmic procedure for time alignment of three-dimensional LC-MS chromatographic surfaces. *Anal. Chem.* **2006**, *78*, 8207–8217. [CrossRef]
19. UniProt, C. UniProt: The universal protein knowledgebase in 2021. *Nucleic Acids Res.* **2021**, *49*, D480–D489.

20. Sadygov, V.R.; Zhang, W.; Sadygov, R.G. Timepoint Selection Strategy for In Vivo Proteome Dynamics from Heavy Water Metabolic Labeling and LC-MS. *J. Proteome Res.* **2020**, *19*, 2105–2112. [CrossRef]
21. Lim, M.Y.; Paulo, J.A.; Gygi, S.P. Evaluating False Transfer Rates from the Match-between-Runs Algorithm with a Two-Proteome Model. *J. Proteome Res.* **2019**, *18*, 4020–4026. [CrossRef]
22. Gilski, M.J.; Sadygov, R.G. Comparison of Programmatic Approaches for Efficient Accessing to mzML Files. *J. Data Min. Genom. Proteom.* **2011**, *2*, 103–109. [CrossRef] [PubMed]

International Journal of
Molecular Sciences

MDPI

Article

Proteomics Unveils Post-Mortem Changes in Beef Muscle Proteins and Provides Insight into Variations in Meat Quality Traits of Crossbred Young Steers and Heifers Raised in Feedlot

Mariane Severino [1], Mohammed Gagaoua [2,3,*], Welder Baldassini [1,4], Richard Ribeiro [4], Juliana Torrecilhas [4], Guilherme Pereira [1,4], Rogério Curi [1,3], Luis Artur Chardulo [1,4], Pedro Padilha [5] and Otávio Machado Neto [1,4,*]

1 College of Agriculture and Veterinary Science (FCAV), São Paulo State University (UNESP), Jaboticabal, Sao Paulo 14884-900, Brazil
2 Food Quality and Sensory Science Department, Teagasc Food Research Centre, Ashtown, Dublin 15, D15 DY05 Dublin, Ireland
3 Physiologie, Environnement et Génétique Pour l'Animal et les Systèmes d'Élevage (PEGASE), INRAE, Institut Agro, 35590 Saint-Gilles, France
4 College of Veterinary Medicine and Animal Science (FMVZ), São Paulo State University (UNESP), Botucatu, Sao Paulo 18618-681, Brazil
5 Institute of Bioscience (IB), São Paulo State University (UNESP), Botucatu, Sao Paulo 18618-681, Brazil
* Correspondence: mohammed.gagaoua@inrae.fr or gmber2001@yahoo.fr (M.G.); otavio.machado@unesp.br (O.M.N.)

Abstract: Proteomics has been widely used to study muscle biology and meat quality traits from different species including beef. Beef proteomics studies allow a better understanding of the biological processes related to meat quality trait determination. This study aimed to decipher by means of two-dimensional electrophoresis (2D-PAGE), mass spectrometry and bioinformatics the changes in post-mortem muscle with a focus on proteins differentially expressed in the *Longissimus thoracis* (LT) muscle of immunocastrated young heifers and steers. Carcass traits, chemical composition, pH, instrumental color (L*, a*, b*), cooking loss and Warner-Bratzler shear force (WBSF) of meat from F1 Montana-Nellore cattle were also evaluated. Backfat thickness (BFT) and intramuscular fat content (IMF) were 46.8% and 63.6% higher in heifers ($p < 0.05$), respectively, while evaporation losses (EL) were 10.22% lower compared to steers. No differences ($p > 0.05$) were observed for tenderness evaluated by WBSF (3, 10, and 17 days post-mortem), pH, and color traits (L*, a* and b*) between the experimental groups. The study revealed several proteins to be differentially expressed proteins in heifers compared steers ($p < 0.05$). In heifers, proteins involved in nutrient transport (TF, ALB, and MB), energy metabolism (ALDOA, GAPDH, and PKM), and oxidative stress and response to stress (HSPA8 and CA3) were associated with a greater BFT and IMF deposition. The higher expression of these proteins indicated greater oxidative capacity and lower glycolytic activity in the LT muscle of heifers. In steers, there was greater abundance of protein expression related to muscle contraction and proteins of structure (ACTA1, TPM2 and TNNT3), energy metabolism (ENO1, ENO3, PYGM, PGM1 and TPI1) and ATP metabolism (ATP5F1B, PEBP1 and AK1), indicating greater glycogenolysis in LT muscle, suggesting a shift in the glycolytic/oxidative fibers of steers.

Keywords: beef quality; proteome; carcass properties; gender; mass spectrometry; 2D-PAGE

Citation: Severino, M.; Gagaoua, M.; Baldassini, W.; Ribeiro, R.; Torrecilhas, J.; Pereira, G.; Curi, R.; Chardulo, L.A.; Padilha, P.; Neto, O.M. Proteomics Unveils Post-Mortem Changes in Beef Muscle Proteins and Provides Insight into Variations in Meat Quality Traits of Crossbred Young Steers and Heifers Raised in Feedlot. *Int. J. Mol. Sci.* **2022**, *23*, 12259. https://doi.org/10.3390/ijms232012259

Academic Editor: Paolo Iadarola

Received: 26 September 2022
Accepted: 11 October 2022
Published: 14 October 2022

1. Introduction

Gender or the sexual condition of cattle is known to be an important factor affecting animal performances and growth, carcass properties and meat quality [1]. Such differences are related to tissue growth as well as their distribution in the carcass. Likewise, castration can improve body fat deposition in beef cattle, and limited sexual behavior leads to easier rearing practices and less carcass damage, which improves carcass fatness and, consequently, meat quality traits [2]. Alternatively, immunological castration (immunocastration) is a relatively new approach, and some studies have described the effects on meat quality of beef cattle [3,4]. However, there is scarcity in the studies comparing young heifers and steers produced according to the "super early-maturing" system that allocate animals to feedlot feeding at 8 months of age (slaughtered up to 15 months of age, [5]). Although immunocastration of female cattle is not a common practice in tropical beef production systems like Brazil, it has the potential to be an easily applied tool to improve the welfare of heifers and cows used for beef production, as reviewed [6]. Moreover, no studies have reported so far reported the relationships between meat quality traits and muscle tissue proteome and which pathways are impacted in crossbred young heifers and steers immunocastrated and feedlot finished.

Proteomics is a powerful tool in meat science due to its technical performance in understanding the effect of the factors influencing meat quality as well as in deciphering the biochemical and metabolic mechanisms occurring to muscle-to-meat conversion in the post-mortem period [7–10]. It further allows for the study each quality trait, such as pH decline, muscle proteins degradation during ageing, oxidation and post-translational modification of proteins in an in-depth manner [11–13]. Thus, omics-related analytical technologies and bioinformatics tools have been significantly applied in the last two decades in the field of meat research to identify proteins related to several meat quality traits, since they are the main constituents of muscle tissue and are responsible for the regulation of main metabolic pathways [11]. In this context, meat quality research in beef cattle and other livestock species used the proteomic approach to evaluate tenderness [7,14,15], marbling [10], color [16,17], water-holding capacity [18,19], and dark-cutting beef [20] among other meat quality traits [11].

The hypothesis of the present study was that differences might exist in the carcass properties and meat quality traits of heifers and steers, particularly the deposition of subcutaneous and intramuscular fat, which can be a consequence of protein changes and their expression. In this context, the aim of this study using a proteomics approach was to identify proteins differentially expressed in the *Longissimus thoracis* (LT) muscle, and how they are related with the physicochemical differences of meat produced by immunocastrated F1 Montana-Nellore young heifers and steers feedlot finished. We further aimed to reveal the molecular pathways and mechanisms behind such mechanisms using advanced bioinformatics analyses.

2. Results

2.1. Carcass Traits, Chemical Composition and Meat Quality

Backfat thickness (BFT) of heifers' carcasses was 46.8% higher ($p < 0.05$) compared to steers' (Table 1). The variables FBW and CY were not altered between genders, while HCW and REA tended to be greater in steers ($p < 0.10$). Intramuscular fat (IMF) content of heifers was 63.6% higher and moisture content was 2.22% lower than in steers (Table S1). Evaporation loss (EL) was 10.22% higher in steers. On the other hand, DL was 22.3% higher in heifers, while CL, WBSF, pH and lightness (L*) were influenced by the ageing period (Table 2).

Table 1. Live body weight and carcass traits of immunocastrated F1 Montana-Nellore young heifers and steers feedlot finished.

Variables [1]	Heifers	Steers	SEM	p-Value
IBW (kg)	289.40	284.00	9.57	0.78
FBW (kg)	373.75	391.13	10.41	0.57
HCW (kg)	202.18	219.25	5.97	0.09
CY (%)	54.81	55.82	0.00	0.53
BFT (mm)	14.90	10.15	1.10	0.04

Table 1. *Cont.*

Variables [1]	Heifers	Steers	SEM	*p*-Value
REA (cm^2)	75.99	90.48	1.66	0.06

[1] IBW and FBW: initial and final body weight, respectively; HCW: hot carcass weight; CY: Carcass yield; BFT: Backfat thickness; REA: Ribeye area.

Table 2. Meat quality traits of immunocastrated F1 Montana-Nellore young steers and heifers feedlot finished.

Variables [1]	Heifers Ageing (Days)			Steers Ageing (Days)			SEM	*p*-Value Gender	*p*-Value Ageing Time
	3	10	17	3	10	17			
L*	33.76	34.78	36.54	32.94	34.50	34.84	0.49	0.24	0.06
a*	16.26	16.55	16.73	15.94	16.73	15.99	0.14	0.45	0.53
b*	6.08	6.18	6.53	5.93	6.21	6.05	0.08	0.41	0.63
pH	5.60	5.71	5.73	5.60	5.68	5.75	0.02	0.65	<0.01
WBSF (N)	55.70	40.40	31.28	54.42	40.79	37.26	3.62	0.42	<0.01
EL (%)	24.27	20.43	17.10	21.46	17.47	17.14	0.01	0.01	<0.01
DL (%)	6.69	4.49	4.28	7.13	6.54	5.22	0.00	0.02	0.003
CL (%)	30.96	24.92	21.38	28.59	24.01	22.36	0.01	0.25	<0.01

[1] L*: lightness, a*: redness, b*: yellowness, pH = meat pH, WBSF: Warner-Bratzler shear force, EL: evaporation loss, DP: drip loss, CL: cooking loss.

2.2. Muscle Tissue Proteome

The comparison of the 2DE gels allowed us to see that the mean number of protein spots per treatment was 119 ± 25 for steers and 115 ± 32 for heifers. The number of spots of the reference gels were 150 for steers and 160 for heifers. Of this total, after the imaging analysis investigation, 50 spots were selected as differentially expressed ($p < 0.05$) in heifers (Figure 1) and steers (Figure 2).

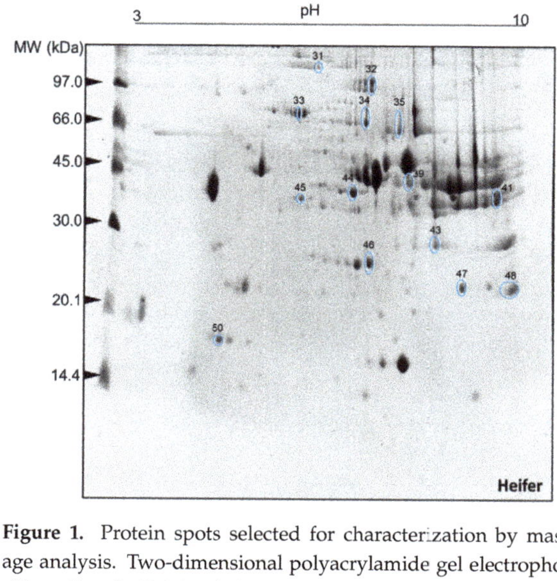

Figure 1. Protein spots selected for characterization by mass spectrometry (ESI-MS) after image analysis. Two-dimensional polyacrylamide gel electrophoresis (2D-PAGE): 12.5% (*w*/*v*) and pH gradient 3–10. Muscle tissue samples (*Longissimus thoracis*) from immunocastrated F1 Montana-Nellore young heifers feedlot finished. The information on the number IDs of differentially abundant spot proteins are given in Table 3.

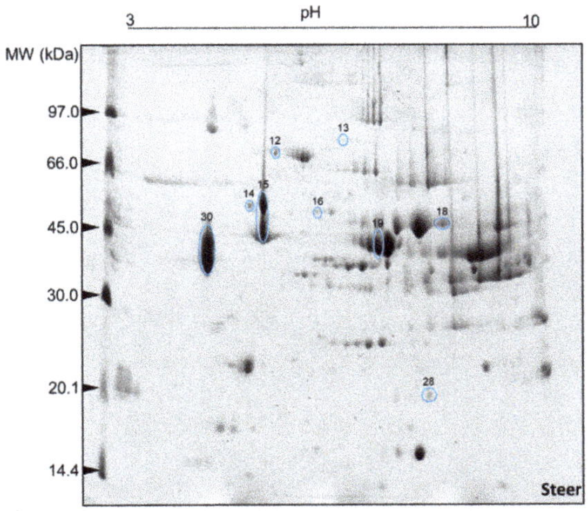

Figure 2. Protein spots selected for characterization by mass spectrometry (ESI-MS) after image analysis. Two-dimensional polyacrylamide gel electrophoresis (2D-PAGE): 12.5% (*w/v*) and pH gradient 3–10. Muscle tissue samples (*Longissimus thoracis*) from immunocastrated F1 Montana-Nellore young steers feedlot finished. The information on the number IDs of differentially abundant spot proteins are given in Table 3.

Several proteins were identified that were abundantly expressed in the muscle of steers and heifers (Table 3). The proteins identified perform functions of muscle contraction/regulation, carbohydrate metabolism, ATP activity, cytoprotection, cellular defense, energy metabolism, and binding. Three main groups (clusters) were distinguished in heifers: oxidative stress and cell defense proteins (HSPA8 and CA3), proteins related to muscle contraction (MYBPC1, TNNT1 and TNNI2), and energy metabolism proteins (CKM, PKM, ALDOA and GAPDH). Regarding the proteins identified in steers, there were groups (clusters) of proteins related to energy metabolism (ATPIF1, ENO1, ENO3, PEBP1, PYGM, PGM1 and TPI1) and muscle contraction (TPM2, TNNT3 and ACTA1).

Table 3. Proteins separated by two-dimensional electrophoresis (2D-PAGE) and identified by mass spectrometry (ESI-MS/MS) in samples of *Longissimus thoracis* muscle from immunocastrated F1 Nellore-Montana young heifers and steers feedlot finished.

Spot ID	Gene Symbol	Full Protein Names	Uniprot ID	Mascot Score	Coverage (%)	pI/MW Experimental	pI/MW Theoretical	Expression [1]
				Energy metabolism				
16	ENO1	Phosphopyruvate hydratase	A0A3Q1MXQ0	5261.196	47.00	6.17/49,703	6.48/54,808	1.95 (UP in steers)
18	ENO3	Beta-enolase	Q3ZC09	10,199.91	57.60	8.19/44,261	7.55/47,438	5.51 (UP in steers)
32	PYGM	Alpha 1–4 glucan phosphorylase	F1MJ28	7318.068	63.42	7.09/97,750	7.05/97,735	1.13 (UP in steers)
34	PGM1	Phosphoglucomutase-1	Q08DP0	9944.941	78.11	6.99/65,655	7.21/61,874	1.89 (UP in steers)
46	TPI1	Triosephosphate isomerase	Q5E956	21,146.33	88.35	7.29/25,236	7.27/26,917	1.28 (UP in steers)
47	AK1	Adenylate kinase isoenzyme 1	P00570	12,755.28	58.76	8.71/22,375	8.35/21,778	1.73 (UP in steers)
35	PKM	Pyruvate kinase	A5D984	12,176.24	63.09	7.7/60,091	7.1/58,519	1.37 (UP in heifers)
39	ALDOA	Fructose-bisphosphate aldolase	A6QLL8	7590.983	70.33	7.73/42,863	7.94/39,949	2.20 (UP in heifers)
41	GAPDH	Glyceraldehyde-3-phosphate dehydrogenase	P10096	30,163.61	47.75	9.19/34,988	8.97/36,096	4.01 (UP in heifers)
19	CKM	Creatine kinase M-type	Q9XSC6	16,598.07	70.60	7.02/43,080	6.8/43,217	1.89 (UP in heifers)
14	ATP5F1B	ATP synthase subunit beta, mitochondrial	P00829	12,307.51	56.44	5.05/51,780	5.04/56,283	2.07 (UP in steers)
				Muscle structure				
31	MYBPC1	Myosin binding protein C1	A6QP89	1502.969	33.97	6.23/113,152	6.5/135,007	4.07 (UP in heifers)
30	TPM2	Tropomyosin beta chain	Q5KR48	5317.557	44.01	4.33/40,213	4.55/32,950	1.44 (UP in steers)
45	TNNT1	Troponin T, slow skeletal muscle	Q8MKH6	3035.667	19.39	5.88/37,343	5.67/31,284	2.49 (UP in heifers)
48	TNNI2	Troponin I2, fast skeletal muscle	F6QIC1	3650.82	54.49	9.44/21,323	9.78/21,141	2.66 (UP in heifers)
44	TNNT3	Troponin T, fast skeletal muscle	Q8MKI4	1410.723	36.9	6.84/39,524	6.85/32,124	1.37 (UP in steers)
15	ACTA1	Actin, alpha skeletal muscle	P68138	29,409.83	81.96	5.28/42,998	5.23/42,393	2.50 (UP in steers)
				Binding proteins				
13	TF	Serotransferrin	Q29443	270.5198	17.19	6.58/82,905	6.6/79,920	1.74 (UP in heifers)
33	ALB	Albumin	P02769	13261.69	64.91	5.97/69,477	5.87/71289	1.19 (UP in heifers)
50	MB	Myoglobin	A0A1K0FUF3	11,507.29	56.49	4.49/17,121	7.80/17077	3.65 (UP in heifers)
28	PEBP1	Phosphatidylethanolamine binding protein 1	P13696	6676.657	64.17	7.94/21,444	7.82/21099	2.08 (UP in steers)

Table 3. *Cont.*

Spot ID	Gene Symbol	Full Protein Names	Uniprot ID	Mascot Score	Coverage (%)	pI/MW Experimental	pI/MW Theoretical	Expression [1]
				Energy metabolism				
43	CA3	Carbonic Anhydrase 3	Q3SZX4	10,936.17	60.00	8.37/27,806	8.1/29655	5.24 (UP in heifers)
				Heat shock proteins				
12	HSPA8	Heat shock cognate 71 kDa protein	P19120	19,088.86	44.92	5.46/73,533	5.25/71468	2.59 (UP in heifers)

[1] Protein spots expression in experimental groups (steers versus heifers). For 2D gel image comparisons, a reference gel per treatment was listed, which contained the highest number and best definition of spots, and the reference gel of a treatment was contrasted with each gel of another treatment, totaling 15 comparisons.The number of BP, MF and CC of the identified proteins differed between heifers (Figure 3) and steers (Figure 4). Considering the distribution of the top 20 levels (Gene Ontology), BPs were identified in greater quantities in steers, among which cellular metabolism (19), nitrogen compounds (16), small molecules and biosynthesis (16) stand out. Divergent processes were observed in smaller amounts in heifers, the main ones being related to establishment (8), vesicle-mediated transport (6) and response to oxidative stress (5). Regarding MF, a predominance of membrane transport (14) and ionic binding (8) proteins was observed in steers. Ionic binding sequences (5) were also verified in heifers, as well as with hydrolysis activity (3). The CC identified in both treatments was similar between the groups, with emphasis on several proteins associated with the organization of cell structure, cytoplasmic and those present in organelles.

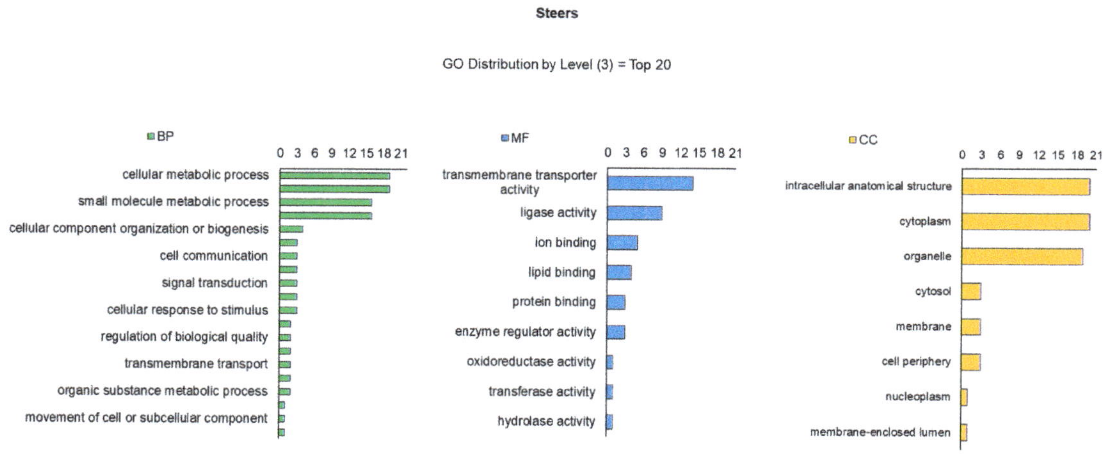

Figure 3. Classification of proteins identified in tissue samples (*Longissimus thoracis*) from immunocastrated F1 Montana-Nellore young heifers feedlot finished. The OMICSBOX software was used to classify the proteins according to biological process (BP), molecular function (MF), and cellular component (CC).

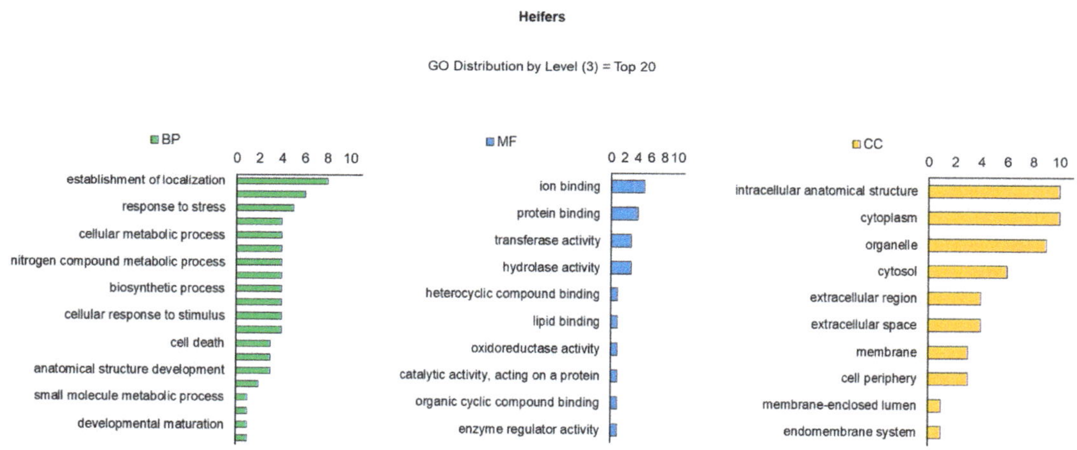

Figure 4. Classification of proteins identified in tissue samples (*Longissimus thoracis*) from immunocastrated F1 Montana-Nellore young steers feedlot finished. The OMICSBOX software was used to classify the proteins according to biological process (BP), molecular function (MF), and cellular component (CC).

The main enriched terms and pathways identified in this study using the differentially expressed proteins for heifers and steers are summarized in Figure 5. Based on gene ontology terms, the ADP metabolic process through "striated muscle thin filament" was highly and significantly up-regulated in steers compared to heifers. Cluster pathways related with cellular and developmental processes were more enriched in heifers. Such pathways of generation of metabolites and energy in heifers help to explain the higher IMF found in the meat of these animals.

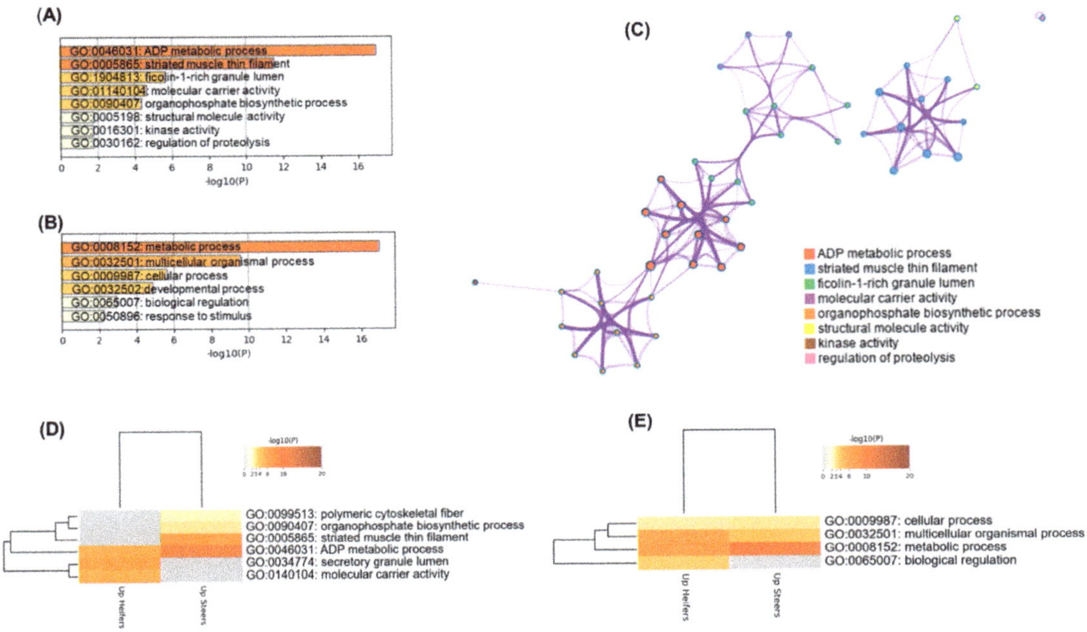

Figure 5. Bioinformatics analyses based on the differentially expressed proteins identified in this experiment. (**A**) Enriched ontology clusters based on the significantly enriched gene ontology (GO) terms obtained using the protein lists of steers (n = 11) and heifers (n = 12) identified in *Longissimus thoracis* muscle tissue of feedlot-finished immunocastrated F1 Montana-Nellore. (**B**) Enriched biological process. The graphs highlight all the enriched terms across the protein lists with the importance of energy metabolism (metabolic and ADP processes), metabolic, cellular and developmental processes according to −Log p-values. (**C**) Enriched network related to the previous terms of (**A**) highlighting the degree of interconnectedness. (**D**,**E**) Hierarchical heatmap clustering comparing the enriched GO terms within steers and heifers as well the main biological process in each condition. The heatmaps colored by the p-values are indicated by color, where grey cells indicate the lack of significant enrichment, palest brown indicates a low p-value and darkest brown indicates a high p-value.

The current GO analysis (Figure 5D) suggests that "secretory granule lumen and "molecular carrier activity" are associated specific to heifers and can be associated with greater fatness (BFT in carcass and IMF in meat). Additionally, "biological regulation" was specifically associated with greater IMF found in heifers. Other GO terms were common to both protein lists, some that were more significant for steers such as metabolic process (Figure 5E). The protein-protein interactions were analyzed (Figure 6). There were two main groups (clusters): proteins related to energy metabolism (PKM, PYGM, GAPDH, CKM, TPI1, ENO1, ENO3, PGM1, AK1, ALDOA, and ATPSB) and proteins related to muscle contraction (TNNT1, TNNT3, TNNI2, TPM2, ACTA1 and MYBPC1). Furthermore, a small interaction network involving binding proteins (ALB, TF, MB, PEBP1 and CA3) and heat shock protein (HSPA8) was identified.

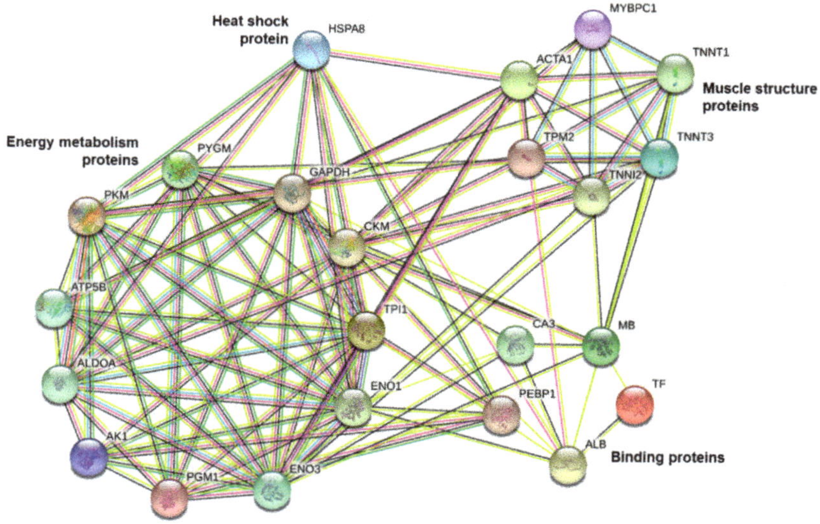

Figure 6. Analysis of protein-protein interactions using the differentially expressed proteins in muscle tissue (*Longissimus thoracis*) of feedlot-finished immunocastrated heifers and steers F1 Montana-Nellore.

Several Quantitative trait loci (QTL) were found to be related with carcass and meat quality traits (n = 12). Overall, 10 chromosomes grouped the 21 proteins (gene names) and, among the major QTLs, most of the proteins were related to the energy metabolism pathway, followed by signaling and transport. Several of the proteins found to change in this study were biomarkers of marbling degree, IMF, beef tenderness and QTLs at the same time (Table 4).

Table 4. List of the Quantitative trait loci (QTL) of carcass and meat quality traits and their chromosomes (Chr.) obtained using the list of the proteins from immunocastrated F1 Nellore-Montana young heifers and steers feedlot finished.

QTL Linked to QTLdb [1]	Gene Symboles	UniProtID (Bovine)	Chr.
Marbling score (n = 4)	HSPA8; MYBPC1; PGM1; PYGM	A6QP89; P19120; P79334; Q08DP0	Chr.15; Chr.5; Chr.3; Chr.29
Fat thickness at the 12th rib (n = 4)	ATP5F1B; MB; MYBPC1; PGM1	P00829; P02192; A6QP89; Q08DP0	Chr.5; Chr.5; Chr.5; Chr.3
Intramuscular fat (n = 1)	ATP5F1B	P00829	Chr.5;
Juiciness (n = 2)	ENO1; PYGM	P79334; Q9XSJ4	Chr.16; Chr.29
Shear force (n = 5)	ALB; ALDOA; HSPA8; PEBP1; PYGM	P02769; A6QLL8; P19120; P13696; P79334	Chr.6; Chr.25; Chr.15; Chr.17; Chr.29
Tenderness score (n = 2)	ALDOA; PYGM	A6QLL8; P79334	Chr.25; Chr.29
Adhesion (n = 1)	TPM2	Q5KR48	Chr.8
Muscle pH (n = 1)	PKM	A5D984	Chr.10

[1] ProteQTL tool included in ProteINSIDE (http://www.proteinside.org/, accessed on 20 August 2022) interrogates a public library of published QTL in the Animal QTL Database (https://www.animalgenome.org/QTLdb/, accessed on 20 August 2022) that contains cattle QTL and association data curated from published scientific articles.

3. Discussion

In tropical countries, such as Brazil, more than 80% of the beef come from pasture systems distributed over 170 million hectares of land. Feedlot systems are used as an alternative to ensure beef supply mainly during the dry season, with better quality to regional and international markets. According to a recent survey, animals on these feedlots included mostly Nellore (75%), a *Bos indicus* cattle, but also some European × Nellore crossbreeds, and other Zebu breeds [21]. Although Nellore cattle is by far the most prevalent breed in tropical regions of Brazil, breeds like Aberdeen Angus and Bonsmara and, also, composite programs (e.g., Montana Tropical) are increasingly growing in the last decades [22].

This is the first study to report that differences in the meat quality of F1 Nellore-Montana young steers and heifers feedlot finished, are related with the expression of several proteins from different pathways such as ALB, MB, CA3, ALDOA, GAPDH, PKM, PYGM, PGM1, HSPA8, and CKM in the post-mortem LT muscle. Some of these proteins were reported as biomarkers of IMF deposition in other studies [11,23–25].

3.1. Proteins with Possible Roles in Intramuscular Fat Content and Meat Quality

In the present study, the greater IMF of heifers can be linked with proteins involved in nutrient transport (ALB and MB), energy metabolism (ALDOA, GAPDH, PKM), and cell protection and response to stress (HSPA8). In fact, these proteins are known to be associated with greater IMF deposition capacity, which consequently can be the reason for the generation of lower EL. Proteins differentially expressed in the LT muscle of heifers also revealed the enrichment of the oxidative activity pathways (ALB, MB and CA3 proteins). Moreover, glycolytic/gluconeogenesis enzymes (ALDOA, GAPDH and PKM) may characterize the use of non-carbohydrate substrates in the production of energy by the tissue [26], thus helping to explain the higher proportion of IMF in the meat of immunocastrated heifers when compared to immunocastrated steers. Further studies using shotgun proteomics approaches on a higher number of animals are needed to validate these proteins as potential biomarkers.

Proteins such as ALB and MB indicated oxidative activity in the muscle of heifers. Albumin regulates the colloidal osmotic pressure of the blood, and binds and transports fatty acids, cholesterol and some ions (copper, zinc and calcium) via the bloodstream [27]. A previous study by Baldassini et al. [28], identified the higher expression of ALB in Nellore bulls with higher IMF and lower WBSF (more tender meat). This protein was also described in another study using a similar gel-based proteomic approach [29], whereby ALB was associated with IMF content. The uptake of fatty acids by ALB indicates the use of lipids as muscle oxidative substrate, used in slow-twitch fibers (type I) which can support the oxidative activity in the muscle of heifers. In addition, myoglobin carries the oxygen necessary for oxidative metabolism, a characteristic of slow-twitch (type I) fibers [30,31] in the muscle of heifers.

The proteins ALDOA, GAPDH and PKM are enzymes of the glycolysis and gluconeogenesis pathways. ALDOA reversibly converts fructose-1,6-bisphosphate to glyceraldehyde-3-phosphate and dihydroxyacetone phosphate. Picard et al. [32] reported similar results to the present study for European cattle, in which ALDOA was more expressed in cows compared to castrated males. The enzyme GAPDH also reversibly converts glyceraldehyde 3-phosphate to 1,3-bisphosphoglycerate. In this sense, an earlier study found higher expression of the enzymes GAPDH and ALDOA in the proteome of steers, with higher IMF content, when compared to Nellore bulls [24]. In the present study, these enzymes were also more abundant in heifers due to the higher IMF content in the meat.

The allosteric enzyme PKM catalyzes the irreversible transfer of a phosphate group from phosphoenolpyruvate to an ADP molecule with final production, at the substrate level, of pyruvate and ATP [33]. In the present study, the proteome of heifers indicates greater glucose degradation by the oxidative pathway, suggesting greater mitochondrial activity with greater synthesis of fatty acids for IMF deposition.

Carbonic anhydrase (CA3) has the function of reversible hydration of carbon dioxide (CO_2) and is closely related to increased oxidative metabolism. In proteomic studies conducted using steers of Asian origin, Qinchuan [34] and Hanwoo [35], CA3 was less expressed in animals with high marbling, leading to divergent results to the present study. The protein CA3 is present in higher concentrations in the cytoplasm of skeletal muscle cells with a predominance of type I fibers, stimulating ATP synthesis by the rapid conversion of glycolytic intermediates into oxaloacetate and citrate [36]. Again, the greater expression of CA3, together with PKM, suggest greater oxidative metabolic activity in the muscle tissue of heifers compared to steers. Although characterization of muscle fibre type was not performed in the current study, further biochemical approaches are needed to validate this hypothesis.

In addition to that, creatine kinase Type M (CKM), converting ADP to ATP from phosphocreatine, provides energy for muscle and reported to affect meat quality and its variability [7,37,38]. The greater expression of CKM in the muscle of heifers may be indicative of a greater energy demand of ATP to maintain muscle functions immediately after slaughter.

The proteins PYGM and PGM1 are key enzymes in the use of glycogen as a substrate for the synthesis of glucose-6-phosphate and glycogenolysis [39]. Researchers reported that PYGM was overabundant in muscle from low growth rate than high growth rate crossbred steers at the time of harvest [29]. After slaughter, this metabolic pathway is extremely important in the anaerobic synthesis of ATP with the final accumulation of intracellular lactate. The greater expression of these proteins in the muscle of steers indicates a greater gluconeogenesis, probably due to the greater accumulation of glycogen ramifications, typical of muscles with a greater number of glycolytic/oxidative fibers (Type IIA) [31,40], fibers with greater hypertrophic capacity, and water accumulation.

The chemical composition of the meat from steers showed higher moisture compared to heifers, which showed a higher amount of IMF. Each gram of glycogen retains four grams of water that is released in the cooking process [41], a fact that could explain the lower water-holding capacity (or higher EL) observed in beef from steers. Moreover, researchers [42] analyzed proteins expression related to both IMF and visceral fat in cows and bulls, and reported that TPM2 expression in adipose tissues were lower in bulls compared to cows, suggesting that TPM2 is positively associated with marbling score and quality grade. Our data also have demonstrated that TPM2 were differentially expressed depending on sex, which indicates that sex hormones are key factors affecting the TPM2 expression and, consequently, lipid accumulation in meat. Therefore, when there is more IMF there is less water content in the heifers' meat.

3.2. *Key Roles of Oxidative Stress and Cell Defense*

The cognate heat shock protein (HSPA8) has a key role in cellular cell death such as apoptosis and autophagy, conferring greater selectivity to degrading proteins in the lysosome [43]. They are also involved in the cytosolic export of nuclear proteins [43]. This is a well-known biomarker of beef quality (tenderness) from the heat shock proteins pathways revealed by Gagaoua and co-workers in their integromics meta-analysis [7]. The HSP70 was also shown to play a role in osteogenesis by upregulating the expression of osteogenic genes [44]. Therefore, HSP70 could be associated with carcass traits through an involvement in muscle and skeletal development as previously evidenced by Gagaoua et al. [45]. Working with pigs, Di Luca et al. [19] reported the increased abundance of HSPs in samples with low post-mortem muscle exudate. The greater expression of HSPA8 in the muscle of heifers may be indicative of a more effective establishment of *rigor mortis*, suggesting a greater proteolytic efficiency under stress conditions (hypoxia).

3.3. *Muscle Structure, Contractile and Associated Proteins*

In steers, the greater abundance of proteins related to muscle contraction and structure (ACTA1 and TNNT3) allow for the explanation of the results of HCW and REA, which trend to be greater in these animals compared to heifers. Some of these proteins identified have been related to muscle growth in other previous studies [46,47], which agree with the results observed in the current study. Moreover, an earlier proteomic study on feedlot finished lambs [48] reported greater expression of proteins TNNT3 and MYL1 in LT muscle. Both were related to the regulation of myosins and, consequently, muscle growth. Thus, the main factors involved in muscle contraction, as observed in the present study, may be affected by gender status.

Similarly, steers and heifers slaughtered at the same final body weight (FBW) differ in carcass traits and meat quality, as reported in previous studies [1]. Such differences can be related to a greater abundance of protein expression related to muscle contraction and proteins of structure (ACTA1 and TNNT3), tissue growth, as well as carcass fatness. Moreover, as reported in the current study, whereby contractile and associated proteins were highly and significantly up-regulated in steers, researchers [29] found that ACTA1 and TNNT3, involved in biological pathways such as glycolysis/gluconeogenesis and muscle contraction, were upregulated in crossbreed feedlot finished steers, regardless of meat aging period. These authors also reported that the growth rate (feedlot versus pasture finishing systems) affected proteins expression in LT muscle and led to an overabundance of ACTA1 and TNNT3, which are good biomarkers of beef tenderness, as reported in other studies [7,17]. However, in the current study, despite the effects on the LT proteome, tenderness evaluated by WBSF was not affected by gender.

3.4. Limitations

The inclusion of one additional biological type in the experimental design of the study, for example, a third group ("control" bulls), would have been an advantage for this study for robust comparisons of differences in skeletal muscle proteome and meat quality traits. Such an experimental group may help to better describe the protein expression in LT muscle to further dissect the contribution of the individual changing proteins belonging to energy metabolism, nutrient transport and signaling pathways, as well as proteins of muscle contraction in response to sexual condition or gender, as reported in the literature [1,2,24].

Although several molecular mechanisms are affected after slaughter according to the literature [49,50], such differences sometimes do not reflect changes in final beef quality, as observed in the current study for tenderness (WBSF), and the meat color of crossbred young heifers and steers raised on feedlot. The WBSF and color variables did not reach statistical significance, which may be due to the small number of animals used for the meat quality. Taken together, these results suggest that a greater number of animals should be evaluated in the future.

4. Materials and Methods

4.1. Animals, Carcass Traits and Muscle/Meat Sampling

Sixteen-immunocastrated F1 Montana-Nellore animals (eight heifers and eight steers), half siblings, were fattened in an experimental feedlot at of the São Paulo State University "Júlio de Mesquita Filho"–UNESP (Botucatu, São Paulo, Brazil) from December 2018 to April 2019. The animals were housed in collective pens and separated by gender. All animals received the same diet and three doses of the immunocastration vaccine (Bopriva®) throughout the finishing period. The first, second and third doses of vaccine were applied at 30, 60 and 90 days after weaning (8 months of age), respectively. The immunocastration of the females was carried out aiming to submit these animals to the same rearing conditions as the males. Additionally, this condition causes temporary immune suppression of ovarian function, reducing estrogen and progesterone levels, thus decreasing ovarian and uterine weights. In addition, this practice was also applied to prevent mounting behaviors and their associated injuries, which help to improve heifer welfare [6].

The diet (% dry matter, DM) was composed of 15% forage (sugarcane bagasse) and 85% concentrate (64% corn, 17% soybean meal, and 4% mineral mixture). Males and females started the feedlot with an initial body weight (BW) of 284.00 ± 45.26 kg and 289.40 ± 17.40 kg, respectively. The animals were weighed at the beginning and at the end of the experimental period, which lasted 110 days. The experimental groups were slaughtered at 15 months of age in a commercial slaughterhouse in the city of Boituva, São Paulo, located 120 km from the place where the experiment was conducted. This followed the state inspection procedures and was preceded by 16-h water and feed fasting. After slaughter, the carcasses were individually weighed to record the hot carcass weight (HCW) and carcass yield (CY), which was calculated using final BW and HCW (CY = HCW/BW × 100).

Individual samples (approximately 2 g) of LT muscle were collected from the right half carcass between the 12th and 13th thoracic vertebrae in the hot carcass (pre rigor mortis) and frozen in liquid nitrogen.

Subsequently, the LT samples were stored in a freezer (−80 °C) until they were used in proteomic analyses. The carcasses were cooled for approximately 48 h at 1 °C. During deboning (48 h of cooling), the backfat thickness (BFT) was then measured with a caliper between the 12th and 13th thoracic vertebra in the LT muscle. The rib eye area (REA) at the 12th/13th rib interface was also measured. Subsequently, beef samples were collected and later sectioned into 2.54 cm steaks for physicochemical analyses. The first steak was taken between the 11th and 13th ribs and the others were in the cranio-caudal direction.

4.2. Chemical Composition of Meat

To evaluate the proximate composition, the samples were thawed in a refrigerator at 4 °C for 24 h and the subcutaneous fat was removed with the aid of a scalpel. The steak was then ground in a multiprocessor for five minutes, using approximately 180 g of sample [51]. The analyses were carried out by infrared spectroscopy in a FoodScan™ equipment (FOSS, Hillerød, Denmark), in which the average levels of moisture, protein, fat and total collagen were determined. The averages of moisture, protein, fat and ash were obtained through three readings per sample, and at each reading the sample was removed from the plate, homogenized again and returned to the plate for the next reading.

4.3. pH and Meat Color

The pH, instrumental color, cooking loss and shear force were measured according to the ageing periods (3, 10, and 17 days post-mortem). Ageing was carried out in a refrigerated BOD incubator (TE-371, TECNAL, Piracicaba, Brazil) at a temperature of 0 to 2 °C, in polyethylene packaging bags (20 × 30 cm; Bemis Company, São Paulo, Brazil) for high vacuum and low oxygen permeability. The pH was determined with a Hanna digital pH meter (Model HI 99163, Hanna Instruments, Woonsocket, RI, USA) equipped with a penetration electrode. Standard buffers (pH 4.0 and 7.0) were used in calibration procedures.

Meat color (L* = lightness; a* = redness; b* = yellowness) was obtained from the average value of three readings for each variable (L*, a* and b*), after 30 min of oxygenation. The CIELab system of the CR-400 colorimeter (light source A, absorbance angle 10°, display Y: 0.01% to 160% reflectance, Konica Minolta Sensing, Inc., Tokyo, Japan) was used. The colorimeter calibration was performed with a standard black and white plate.

4.4. Cooking Loss and Shear Force

To assess cooking loss (CL) and Warner-Bratzler shear force (WBSF), the procedure proposed by Wheeler et al. [52] was adopted and the recommendations of the American Meat Science Association were followed [53]. The samples were placed on a grid coupled to a glass refractory. A thermocouple connected to a digital thermometer (DT-612, ATP Instrumentation, Ashby-de-la-Zouch, England) was used, which was inserted in the center of each sample to monitor the internal end-point temperature.

The samples were cooked in an industrial electric oven (Feri90 Venâncio, Venâncio Aires, Rio Grande do Sul, Brazil) preheated to 170 °C and equipped with a thermostat to minimize temperature variations. Once the internal temperature of the steaks reached 40 °C, they were turned over and remained in the oven until the final temperature reached 71 °C. The samples were then kept at room temperature for 15 min, weighed, and refrigerated at 4 °C for 24 h. The CL was divided into evaporation loss (EL) and drip loss (DL), determined as percentage. The DL was obtained by weighing only the refractory before and after cooking the sample. The EL was obtained by weighing the sample before and after cooking.

For the determination of WBSF, eight cylinders with a diameter of 1.27 cm were sectioned in a Brookfield CT-3 Texture Analyzer (AMETEK Brookfield, Middleborough, MA, USA), equipped with a stainless steel 3.07-mm-thick Warner–Bratzler blade with a vee-shaped (60° angle) cutting edge. Results were reported as the average of eight values per sample, in Newton (N).

4.5. Proteomics

The extraction, precipitation, separation, imaging and protein identification procedures were carried out according to the literature [7,28], with minimal adaptations.

4.5.1. Extraction and Precipitation of Proteins

Individual samples of LT muscle collected during the pre-rigor mortis period were individually processed to obtain two-dimensional electrophoresis gels (2D-PAGE). For each treatment (sex class), eight individual samples (biological replicates) were used (eight steer gels and eight heifer gels) with three technical replicates (three gels for each sample), totaling 48 gels. For each sample, approximately 0.2 g of the muscle was ground in 1.0 mL lysis buffer using an Ultra-Turrax high shear mixer (Marconi–MA102/E, Piracicaba, São Paulo, Brazil) at 20,000 rpm twice for 30 s. The protein extracts were separated from the solid part by 15 min of centrifugation at 10,000× *g* rpm at 4 °C. The protein content of these extracts was placed in 80% (*v*/*v*) acetone solution and kept at 5 °C for 2–3 h to ensure that the procedure occurred for a sufficient time. The centrifugation process was then repeated for 25 min at 10.000× *g* rpm to obtain protein pellets for quantification and 2D-PAGE. The obtained pellets were washed beforehand to quantify proteins and were used in electrophoretic runs. One portion of the protein pellets was re-solubilized in 0.5 M NaOH to quantify total protein. The total protein concentration of the bovine muscle tissue samples was quantified by the Biuret method [54].

4.5.2. Protein Separation by Two-Dimensional Electrophoresis (2D-PAGE)

An additional protein fraction was re-solubilized in a specific buffer containing 0.07 M urea, 0.02 M thiourea, 2% 3-[(3-cholaminopropyl)-dimethylammonium]-1-propanesulfonate–CHAPS (*m/v*), 10% ampholyte (pH range 3–10), and 0.002% bromophenol blue. Moreover, 2.8 mg of 1,4-dithiothreitol was added and this solution was used in electrophoretic separations.

Briefly, approximately 375 µg of protein extracts (1.5 µg/µL) were loaded into first dimension strips (13 cm) and hydrated for 12 h. Subsequently, the protein extracts were separated on pH 3–10 with isoelectric focus in an Ettan IPGphor 3 device (GE Healthcare, United States), in which the proteins were fractionated by the isoelectric point, pH value when the net charge total protein is zero.

The strips were placed in equilibrium solutions for reduction, alkylation and weresubjected to the second dimension (2D) of electrophoresis in a 12.5% (*m/v*) polyacrylamide gel. At the end of the 2D run, approximately 500 mL of colloidal Coomassie stain was used to mark the protein spots of the gels for 72 h. These gels were subsequently destained with ultrapure water.

The gels were scanned, and the images were imported into ImageMaster Platinum software, version 7.0, for comparisons (contrasts) of images between treatments and obtaining information such as number of spots per gel, percentage of matching (correspondence between the spots proteins in the gels), isoelectric point (p*I*), molecular weight (MW), and volume of the spots. The correspondence (matching) of the gels within each sample (three technical repetitions) was greater than 95%, demonstrating that 95% of the spots were present in the technical replicates, indicating good reproducibility. For image comparisons, a reference gel per treatment was listed [55], which contained the highest number and best definition of spots, and the reference gel of a treatment was contrasted with each gel of another treatment, totaling 15 comparisons.

4.5.3. Tryptic Digestion of Protein Spots and Identification of Proteins by ESI-MS/MS

Protein spots from experimental groups (heifers versus steers) were selected based on the molecular weight (MW) and isoelectric point (p*I*) obtained by image analysis and then cut out (fragments of approximately 1 mm^3), and prepared according to the method of [56]. The sediments were transferred to microtubes and submitted to the following four steps: The first step involved the removal of the dye with 25 mM ammonium bicarbonate (Ambic)/acetonitrile (50:50, v/v) Step two was the reduction and alkylation in which the gel fragments were rehydrated in a reducing solution and incubated for 40 min at 56 °C.

After removal of the reducing solution, an alkylating solution was added and the fragments were incubated in the dark for 30 min at room temperature. Step three involved the digestion consisting of overnight incubation at 37 °C with 10 ng·μL^{-1} trypsin in 25 mM Ambic for 15 min (Trypsin Gold Mass Spectrometry, Promega, Madison, WI, United States). The final step saw the elution of peptides extracted from the gel using three steps: (A) 50% ACN with 1% formic acid incubated for 15 min at 40 °C under sonication, and the supernatant was collected and transferred to a new tube; (B) 60% methanol with 1% formic acid incubated for 15 min at 40 °C under sonication, and the supernatant was collected and transferred to a new tube; and (C) 100% ACN; the extracts were dried in a vacuum centrifuge and peptides were dissolved in 10 µL of 3% ACN with 0.1% formic acid.

The mass spectra of the peptides were obtained by analyzing aliquots of the solutions in a nanoACQUITY UPLC-Xevo TQ-MS System (Waters, Manchester, UK). The proteins were identified in the UniProt database (UniProtKB/Swiss-Prot database) for the *Bos taurus* genome.

4.5.4. Bioinformatics

Bioinformatics analyses were conducted for the classification of differentially expressed proteins in muscle tissues from animals (heifers versus steers) in terms of biological processes (BP), molecular function (MF), and cellular components (CC). For this purpose, the accession number of the proteins identified by ESI/MS/MS were entered into the UniProt database (www.uniprot.org, accessed on 5 August 2022) and their FASTA sequences were extracted. After this step, the proteins were analyzed using the OMICSBOX v.2.0 (https://www.biobam.com/omicsbox/, accessed on 5 August 2022) and Blast2GO tools [57].

Additionally, the interactions between the proteins identified in the treatments were analyzed using the open source STRING 11.0 platform (https://string-db.org/, accessed on 20 August 2022). The same list of proteins whose expression differed between the experimental groups were used in these analyses [58].

Subsequently, further bioinformatics analyses were performed following the procedures described by Gagaoua et al. [7] using Metascape® platform. Briefly, gene identifiers were converted using Uniprot Retrieve/ID mapping. Thus, key information regarding the proteins (gene names [GN]) and their relationships with carcass and meat quality traits described in the current study were annotated for each gender. These procedures aiming to compare the two protein lists to better understand the common and divergent molecular signatures. Hierarchical heatmap clustering was also created using enriched GO terms analyzed by Metascape® (https://metascape.org/, accessed on 20 August 2022).

Additionally, the ProteQTL tool included in ProteINSIDE (http://www.proteinside.org/, accessed on 20 August 2022) was used for rapid searching of carcass and meat quality quantitive trait loci (QTL) among the list of putative biomarkers following the details described by Gagaoua et al. [59]. ProteQTL interrogates a public library of published QTL in the Animal QTL Database (https://www.animalgenome.org/QTLdb, accessed on 20 August 2022) that contains cattle QTL and association data curated from published scientific articles [7].

4.5.5. Statistical Analysis

Carcass and meat quality variables were analyzed regarding the homogeneity and normality of the residues, which expressed by means and their respective errors. Data were submitted to analysis of variance (ANOVA) using the F Test and using the SAS GLM procedure (version 9.1, Cary, NC, USA). Comparisons between means were made using a

Tukey's Test and the p value < 0.05 was adopted as the critical level of probability [60–62]. The design was completely randomized according to the following model:

$$Yij = \mu + ti + \epsilon ij,$$

where, Yij is the observed value for the experimental unit referring to treatment i in repetition j; μ is the general effect of the mean; t is the treatment effect (gender) and ϵ is the experimental error.

The protein spot volumes data were imported into ImageMaster Platinum (v. 7.0) software and mean and standard deviation were calculated for selected spots. The images were compared between treatments by means of the matching of the spots regarding their distribution, volume, relative intensity, p*I*s and MW. The program generated an ANOVA within the comparisons between treatments, and the divergent protein spots ($p < 0.05$) were selected. Additionally, the Mann-Whitney test (Wilcoxon rank-sum test) was used when the normality criteria was violated in any of the treatments. For both tests (Student or Mann-Whitney), significance was detected at the 0.05 level. For all data, trends were considered at $0.05 < p \leq 0.10$.

5. Conclusions

The results obtained showed higher IMF content in meat from heifers compared to steers, with a better response to stress in the muscle of heifers. Even with immunocastration, males did not increase lipid synthesis in muscle tissue in the same proportion as females. The expression pattern of transport proteins, energy metabolism, cellular defense, and glycogenolysis differ among heifers and steers, suggesting greater oxidative capacity and lower glycolytic activity in the muscle of heifers. Such proteome changes in LT muscle help to explain the differences found in meat quality traits, particularly marbling.

Supplementary Materials: The following supporting information can be downloaded at: https://www.mdpi.com/article/10.3390/ijms232012259/s1.

Author Contributions: Conceptualization, W.B. and O.M.N.; methodology, M.S., M.G., R.R. and J.T.; software, M.G., W.B. and G.P.; validation, W.B., M.G., P.P. and L.A.C.; formal analysis, W.B., M.G., M.S., R.R. and J.T.; investigation, M.S., W.B., M.G. and R.R.; resources, L.A.C., P.P. and O.M.N.; data curation, W.B. and M.G.; writing—original draft preparation, M.S., R.C., M.G. and W.B.; writing—review and editing, M.G., R.C., G.P. and J.T.; visualization, W.B. and M.G.; supervision, O.M.N., L.A.C. and M.G.; project administration, O.M.N.; funding acquisition, O.M.N. All authors have read and agreed to the published version of the manuscript.

Funding: This research was funded by São Paulo Research Foundation (FAPESP), grant numbers 2019/11028-0; 2018/00981-5 and 2016/04478-0. This study was also financed in part by CAPES-Finance Code 001.

Institutional Review Board Statement: The animal study protocol was approved by Ethics Committee (CEUA) of the São Paulo State University "Júlio de Mesquita Filho"–UNESP, Campus of Botucatu (protocol number 0107/2018).

Informed Consent Statement: Not applicable.

Data Availability Statement: All relevant data are within the paper.

Acknowledgments: Mohammed Gagaoua acknowledges the support of the Marie Sklodowska-Curie grant agreement No. 713654 under the project number MF20180029.

Conflicts of Interest: The authors declare that they have no conflict of interest. The funders had no role in the design of the study; in the collection, analyses, or interpretation of data; in the writing of the manuscript; or in the decision to publish the results.

References

1. Blanco, M.; Ripoll, G.; Delavaud, C.; Casasús, I. Performance, carcass and meat quality of young bulls, steers and heifers slaughtered at a common body weight. *Livest. Sci.* **2020**, *240*, 104156. [CrossRef]

2. Mueller, L.F.; Balieiro, J.C.D.C.; Ferrinho, A.; Martins, T.; Corte, R.R.P.D.S.; De Amorim, T.R.; Furlan, J.D.J.M.; Baldi, F.; Pereira, A.S.C. Gender status effect on carcass and meat quality traits of feedlot Angus × Nellore cattle. *Anim. Sci. J.* **2019**, *90*, 1078–1089 [CrossRef] [PubMed]
3. Andreo, N.; Bridi, A.M.; Soares, A.L.; Prohmann, P.E.F.; Peres, L.M.; Tarsitano, M.A.; Giangareli, B.D.L.; Takabayashi, A.A. Fatty acid profile of beef from immunocastrated (BOPRIVA ®) Nellore bulls. *Meat Sci.* **2016**, *117*, 12–17. [CrossRef]
4. Miguel, G.Z.; Faria, M.H.; Roca, R.; Santos, C.T.; Suman, S.; Faitarone, A.B.; Delbem, N.L.; Girao, L.; Homem, J.M.; Barbosa, E.K.; et al. Immunocastration improves carcass traits and beef color attributes in Nellore and Nellore×Aberdeen Angus crossbred animals finished in feedlot. *Meat Sci.* **2014**, *96*, 884–891. [CrossRef] [PubMed]
5. Silveira, A.C. Novilho superprecoce: Técnicas de nutrição e manejo. Simpósio Goiano sobre Manejo e Nutrição de bovinos de corte e leite. *Goiânia Colégio Bras. De Nutr. Anim.* **2003**, *5*, 153–166.
6. Needham, T.; Lambrechts, H.; Hoffman, L. Castration of male livestock and the potential of immunocastration to improve animal we. *S. Afr. J. Anim. Sci.* **2017**, *47*, 731. [CrossRef]
7. Gagaoua, M.; Terlouw, E.C.; Mullen, A.M.; Franco, D.; Warner, R.D.; Lorenzo, J.M.; Purslow, P.P.; Gerrard, D.; Hopkins, D.L.; Troy, D.; et al. Molecular signatures of beef tenderness: Underlying mechanisms based on integromics of protein biomarkers from multi-platform proteomics studies. *Meat Sci.* **2021**, *172*, 108311. [CrossRef] [PubMed]
8. Gagaoua, M.; Zhu, Y. Proteomics Advances in Beef Production. In *Food Proteomics*; Academic Press: Cambridge, MA, USA, 2022; pp. 151–182. [CrossRef]
9. Gagaoua, M.; Terlouw, C.; Richardson, I.; Hocquette, J.-F.; Picard, B. The associations between proteomic biomarkers and beef tenderness depend on the end-point cooking temperature, the country origin of the panelists and breed. *Meat Sci.* **2019**, *157*, e107871. [CrossRef]
10. Picard, B.; Gagaoua, M.; Micol, D.; Cassar-Malek, I.; Hocquette, J.F.; Terlouw, C.E. Inverse relationships between biomarkers and beef tenderness according to contractile and metabolic properties of the muscle. *J. Agric. Food Chem.* **2014**, *62*, 9808–9818. [CrossRef] [PubMed]
11. Gagaoua, M.; Picard, B. *Proteomics to Explain and Predict Meat Quality*; Woodhead Publishing: UK, 2022; pp. 393–431. [CrossRef]
12. Munekata, P.E.; Pateiro, M.; López-Pedrouso, M.; Gagaoua, M.; Lorenzo, J.M. Foodomics in meat quality. *Curr. Opin. Food Sci.* **2021**, *38*, 79–85. [CrossRef]
13. Purslow, P.P.; Gagaoua, M.; Warner, R.D. Insights on meat quality from combining traditional studies and proteomics. *Meat Sci.* **2020**, *174*, 108423. [CrossRef] [PubMed]
14. Picard, B.; Gagaoua, M. Meta-proteomics for the discovery of protein biomarkers of beef tenderness: An overview of integrated studies. *Food Res. Int.* **2020**, *127*, 108739. [CrossRef] [PubMed]
15. Picard, B.; Gagaoua, M. Proteomic Investigations of Beef Tenderness. In *Proteomics in Food Science: From Farm to Fork*; Elsevier Inc.: Amsterdam, The Netherlands, 2017. [CrossRef]
16. Joseph, P.; Nair, M.N.; Suman, S.P. Application of proteomics to characterize and improve color and oxidative stability of muscle foods. *Food Res. Int.* **2015**, *76*, 938–945. [CrossRef]
17. Gagaoua, M.; Hughes, J.; Terlouw, E.C.; Warner, R.D.; Purslow, P.P.; Lorenzo, J.M.; Picard, B. Proteomic biomarkers of beef colour. *Trends Food Sci. Technol.* **2020**, *101*, 234–252. [CrossRef]
18. Yang, B.; Liu, X. Application of proteomics to understand the molecular mechanisms determining meat quality of beef muscles during postmortem aging. *PLoS ONE* **2021**, *16*, e0246955. [CrossRef]
19. Di Luca, A.; Mullen, A.M.; Elia, G.; Davey, G.; Hamill, R. Centrifugal drip is an accessible source for protein indicators of pork ageing and water-holding capacity. *Meat Sci.* **2011**, *88*, 261–270. [CrossRef]
20. Gagaoua, M.; Warner, R.D.; Purslow, P.; Ramanathan, R.; Mullen, A.M.; López-Pedrouso, M.; Franco, D.; Lorenzo, J.M.; Tomasevic, I.; Picard, B.; et al. Dark-cutting beef: A brief review and an integromics meta-analysis at the proteome level to decipher the underlying pathways. *Meat Sci.* **2021**, *181*, 108611. [CrossRef]
21. De Andrade, T.S.; Albertini, T.Z.; Barioni, L.G.; De Medeiros, S.R.; Millen, D.D.; Dos Santos, A.C.R.; Goulart, R.S.; Lanna, D.P.D. Perception of consultants, feedlot owners, and packers regarding management and marketing decisions on feedlots: A national survey in Brazil (Part II). *Can. J. Anim. Sci.* **2020**, *100*, 759–770. [CrossRef]
22. Ferraz, J.B.S.; de Felício, P.E. Production systems—An example from Brazil. *Meat Sci.* **2010**, *84*, 238–243. [CrossRef]
23. Bazile, J.; Picard, B.; Chambon, C.; Valais, A.; Bonnet, M. Pathways and biomarkers of marbling and carcass fat deposition in bovine revealed by a combination of gel-based and gel-free proteomic analyses. *Meat Sci.* **2019**, *156*, 146–155. [CrossRef]
24. Silva, L.; Rodrigues, R.; Assis, D.E.; Benedeti, P.D.; Duarte, M.S.; Chizzotti, M. Explaining meat quality of bulls and steers by differential proteome and phosphoproteome analysis of skeletal muscle. *J. Proteom.* **2019**, *199*, 51–66. [CrossRef] [PubMed]
25. Poleti, M.D.; Regitano, L.C.A.; Souza, G.H.M.F.; Cesar, A.S.M.; Simas, R.C.; Silva-Vignato, B.; Oliveira, G.B.; Andrade, S.C.S.; Cameron, L.C.; Coutinho, L.L.; et al. Longissimus dorsi muscle label-free quantitative proteomic reveals biological mechanisms associated with intramuscular fat deposition. *J. Proteom.* **2018**, *179*, 30–41. [CrossRef] [PubMed]
26. Dashty, M. A quick look at biochemistry: Carbohydrate metabolism. *Clin. Biochem.* **2013**, *46*, 1339–1352. [CrossRef] [PubMed]
27. Evans, T.W. Review article: Albumin as a drug—Biological effects of albumin unrelated to oncotic pressure. *Aliment. Pharmacol. Ther.* **2002**, *16*, 6–11. [CrossRef]

28. Baldassini, W.A.; Braga, C.P.; Chardulo, L.A.L.; Silva, J.A.I.V.; Malheiros, J.M.; de Albuquerque, L.G.; Fernandes, T.T.; Padilha, P.D.M. Bioanalytical methods for the metalloproteomics study of bovine *longissimus thoracis* muscle tissue with different grades of meat tenderness in the Nellore breed (*Bos indicus*). *Food Chem.* **2015**, *57*, 363–370. [CrossRef]
29. Antonelo, D.S.; Gómez, J.F.; Silva, S.L.; Beline, M.; Zhang, X.; Wang, Y.; Pavan, B.; Koulicoff, L.A.; Rosa, A.F.; Goulart, R.S.; et al. Proteome basis for the biological variations in color and tenderness of longissimus thoracis muscle from beef cattle differing in growth rate and feeding regime. *Food Res. Int.* **2022**, *153*, 110947. [CrossRef]
30. Faustman, C.; Sun, Q.; Mancini, R.; Suman, S.P. Myoglobin and lipid oxidation interactions: Mechanistic bases and control. *Meat Sci.* **2010**, *86*, 86–94. [CrossRef]
31. Picard, B.; Gagaoua, M. Muscle Fiber Properties in Cattle and Their Relationships with Meat Qualities: An Overview. *J. Agric. Food Chem.* **2020**, *68*, 6021–6039. [CrossRef]
32. Picard, B.; Gagaoua, M.; Al-Jammas, M.; De Koning, L.; Valais, A.; Bonnet, M. Beef tenderness and intramuscular fat proteomic biomarkers: Muscle type effect. *PeerJ* **2018**, *6*, e4891. [CrossRef]
33. Wu, W.; Gao, X.-G.; Dai, Y.; Fu, Y.; Li, X.-M.; Dai, R.-T. Post-mortem changes in sarcoplasmic proteome and its relationship to meat color traits in M. semitendinosus of Chinese Luxi yellow cattle. *Food Res. Int.* **2015**, *72*, 98–105. [CrossRef]
34. Zhang, Y.Y.; Zan LSen Wang, H.B.; Xin, Y.P.; Adoligbe, C.M.; Ujan, J.A. Effect of sex on meat quality characteristics of Qinchuan cattle. *Afr. J. Biotechnol.* **2010**, *9*, 28. [CrossRef]
35. Shen, Y.N.; Kim, S.H.; Yoon, D.H.; Lee, H.G.; Kang, H.S.; Seo, K.S. Proteome Analysis of Bovine Longissimus dorsi Muscle Associated with the Marbling Score. *Asian-Australasian J. Anim. Sci.* **2012**, *25*, 1083–1088. [CrossRef] [PubMed]
36. Gagaoua, M.; Couvreur, S.; Le Bec, G.; Aminot, G.; Picard, B. Associations among Protein Biomarkers and pH and Color Traits in Longissimus thoracis and Rectus abdominis Muscles in Protected Designation of Origin Maine-Anjou Cull Cows. *J Agric Food Chem* **2017**, *65*, 3569–3580. [CrossRef] [PubMed]
37. Zou, B.; Zhao, D.; He, G.; Nian, Y.; Da, D.; Yan, J.; Li, C. Acetylation and Phosphorylation of Proteins Affect Energy Metabolism and Pork Quality. *J. Agric. Food Chem.* **2020**, *68*, 7259–7268. [CrossRef] [PubMed]
38. della Malva, A.; Gagaoua, M.; Santillo, A.; De Palo, P.; Sevi, A.; Albenzio, M. First insights about the underlying mechanisms of Martina Franca donkey meat tenderization during aging: A proteomic approach. *Meat Sci.* **2022**, *193*, 108925. [CrossRef]
39. Nelson, D.; Cox, M. *Lehninger—Principios de Bioquimica*, 5th ed.; Freeman, W.H., Ed.; MACMILLAN: New York Basingstoke, NY, USA, 2009.
40. Ryu, Y.; Kim, B. The relationship between muscle fiber characteristics, postmortem metabolic rate, and meat quality of pig longissimus dorsi muscle. *Meat Sci.* **2005**, *71*, 351–357. [CrossRef]
41. Traore, S.; Aubry, L.; Gatellier, P.; Przybylski, W.; Jaworska, D.; Kajak-Siemaszko, K.; Santé-Lhoutellier, V. Higher drip loss is associated with protein oxidation. *Meat Sci.* **2012**, *90*, 917–924. [CrossRef]
42. Cho, J.H.; Jeong, J.Y.; Lee, R.H.; Na Park, M.; Kim, S.-H.; Park, S.-M.; Shin, J.-C.; Jeon, Y.-J.; Shim, J.-H.; Choi, N.-J.; et al. Regional Differences of Proteins Expressing in Adipose Depots Isolated from Cows, Steers and Bulls as Identified by a Proteomic Approach. *Asian-Australasian J. Anim. Sci.* **2016**, *29*, 1197–1206. [CrossRef]
43. Stricher, F.; Macri, C.; Ruff, M.; Muller, S. HSPA8/HSC70 chaperone protein. *Autophagy* **2013**, *9*, 1937–1954. [CrossRef]
44. Chen, E.; Xue, D.; Zhang, W.; Lin, F.; Pan, Z. Extracellular heat shock protein 70 promotes osteogenesis of human mesenchymal stem cells through activation of the ERK signaling pathway. *FEBS Lett.* **2015**, *589*, 4088–4096. [CrossRef]
45. Gagaoua, M.; Monteils, V.; Couvreur, S.; Picard, B. Identification of Biomarkers Associated with the Rearing Practices, Carcass Characteristics, and Beef Quality: An Integrative Approach. *J. Agric. Food Chem.* **2017**, *65*, 8264–8278. [CrossRef] [PubMed]
46. D'Alessandro, A.; Zolla, L. Meat science: From proteomics to integrated omics towards system biology. *J. Proteom.* **2013**, *78*, 558–577. [CrossRef] [PubMed]
47. Lefaucheur, L.; Lebret, B.; Ecolan, P.; Louveau, I.; Damon, M.; Prunier, A.; Billon, Y.; Sellier, P.; Gilbert, H. Muscle characteristics and meat quality traits are affected by divergent selection on residual feed intake in pigs1. *J. Anim. Sci.* **2011**, *89*, 996–1010. [CrossRef]
48. Paim, T.D.P.; Viana, P.; Van Tilburg, M.F.; Moura, A.D.A.; De Souza, J.R.; McManus, C.; Abdalla, A.L.; Louvandini, H. Feeding effects of cottonseed and its co-products on the meat proteome from ram lambs. *Sci. Agricola* **2019**, *76*, 463–472. [CrossRef]
49. Lana, A.; Zolla, L. Proteolysis in meat tenderization from the point of view of each single protein: A proteomic perspective. *J. Proteom.* **2016**, *147*, 85–97. [CrossRef]
50. Gagaoua, M.; Terlouw, E.M.C.; Boudjellal, A.; Picard, B. Coherent correlation networks among protein biomarkers of beef tenderness: What they reveal. *J. Proteom.* **2015**, *128*, 365–374. [CrossRef] [PubMed]
51. Anderson, S.; Aldana, S.; Beggs, M.; Birkey, J.; Conquest, A.; Conway, R.; Hemminger, T.; Herrick, J.; Hurley, C.; Ionita, C.; et al. Determination of Fat, Moisture, and Protein in Meat and Meat Products by Using the FOSS FoodScan Near-Infrared Spectrophotometer with FOSS Artificial Neural Network Calibration Model and Associated Database: Collaborative Study. *J. AOAC Int.* **2007**, *90*, 1073–1083. [CrossRef] [PubMed]
52. Wheeler, T.L.; Koohmaraie, M.; Cundiff, L.V.; Dikeman, M.E. Effects of cooking and shearing methodology on variation in Warner-Bratzler shear force values in beef2. *J. Anim. Sci.* **1994**, *72*, 2325–2330. [CrossRef]
53. AMSA. *Research Guidelines for Cookery, Sensory Evaluation, and Instrumental Tenderness Measurements of Meat*; American Meat Science Association Educational Foundation: Savoy, IL, USA, 2015.

54. Gornall, A.G.; Bardawill, C.J.; David, M.M. Determination of serum proteins by means of the biuret reaction. *J. Biol. Chem.* **1949**, *177*, 751–766. [CrossRef]
55. Paredi, G.; Mori, F.; de Marino, M.G.; Raboni, S.; Marchi, L.; Galati, S.; Buschini, A.; Fiego, D.P.L.; Mozzarelli, A. Is the protein profile of pig Longissimus dorsi affected by gender and diet? *J. Proteom.* **2019**, *206*, 103437. [CrossRef]
56. Shevchenko, A.; Tomas, H.; Havlis, J.; Olsen, J.V.; Mann, M. In-gel digestion for mass spectrometric characterization of proteins and proteomes. *Nat. Protoc.* **2006**, *1*, 2856–2860. [CrossRef] [PubMed]
57. Götz, S.; Garcia-Gomez, J.M.; Terol, J.; Williams, T.D.; Nagaraj, S.H.; Nueda, M.J.; Robles, M.; Talón, M.; Dopazo, J.; Conesa, A. High-throughput functional annotation and data mining with the Blast2GO suite. *Nucleic Acids Res.* **2008**, *36*, 3420–3435. [CrossRef] [PubMed]
58. Franceschini, A.; Szklarczyk, D.; Frankild, S.; Kuhn, M.; Simonovic, M.; Roth, A.; Lin, J.; Minguez, P.; Bork, P.; von Mering, C.; et al. STRING v9.1: Protein-protein interaction networks, with increased coverage and integration. *Nucleic Acids Res.* **2012**, *41*, D808–D815. [CrossRef] [PubMed]
59. Gagaoua, M.; Troy, D.; Mullen, A.M. The Extent and Rate of the Appearance of the Major 110 and 30 kDa Proteolytic Fragments during Post-Mortem Aging of Beef Depend on the Glycolysing Rate of the Muscle and Aging Time: An LC–MS/MS Approach to Decipher Their Proteome and Associated Pathways. *J. Agric. Food Chem.* **2021**, *69*, 602–614. [CrossRef]
60. Picard, B.; Gagaoua, M.; Al Jammas, M.; Bonnet, M. Beef tenderness and intramuscular fat proteomic biomarkers: Effect of gender and rearing practices. *J. Proteom.* **2019**, *200*, 1–10. [CrossRef]
61. Gagaoua, M.; Bonnet, M.; Ellies-Oury, M.-P.; De Koning, L.; Picard, B. Reverse phase protein arrays for the identification/validation of biomarkers of beef texture and their use for early classification of carcasses. *Food Chem.* **2018**, *250*, 245–252. [CrossRef]
62. Gagaoua, M.; Bonnet, M.; De Koning, L.; Picard, B. Reverse Phase Protein array for the quantification and validation of protein biomarkers of beef qualities: The case of meat color from Charolais breed. *Meat Sci.* **2018**, *145*, 308–319. [CrossRef]

International Journal of
Molecular Sciences

MDPI

Article

Revealing the Hidden Diagnostic Clues of Male Infertility from Human Seminal Plasma by Dispersive Solid Phase Extraction and MALDI-TOF MS

Serena Correnti [1,†], Mariaimmacolata Preianò [1,†], Pierpaolo Murfone [2], Annalisa Fregola [3], Massimo Bitonti [3], Rocco Savino [4] and Rosa Terracciano [2,*]

1 Department of Health Sciences, Magna Græcia University, 88100 Catanzaro, Italy
2 Department of Experimental and Clinical Medicine, Magna Græcia University, 88100 Catanzaro, Italy
3 Biopepticom S.R.L., 88100 Catanzaro, Italy
4 Department of Medical and Surgical Sciences, Magna Græcia University, 88100 Catanzaro, Italy
* Correspondence: terracciano@unicz.it
† These authors contributed equally to this work.

Abstract: Seminal plasma (SP) mirrors the local pathophysiology of the male reproductive system and represents a non-invasive fluid for the study of infertility. Matrix-Assisted Laser Desorption/Ionization-Time-of-Flight Mass Spectrometry (MALDI-TOF-MS) provides a high-throughput platform to rapidly extrapolate the diagnostic profiles of information-rich patterns. In this study, dispersive solid phase extraction (*d*-SPE) combined with MALDI-TOF-MS was applied for the first time to the human SP, with the aim of revealing a diagnostic signature for male infertility. Commercially available octadecyl (C_{18})-, octyl (C_8)-bonded silica sorbents and hexagonal mesoporous silica (HMS) were tested and the robustness of MALDI-TOF peptide profiling was evaluated. Best performances were obtained for C_{18}-bonded silica with the highest detection of peaks and the lowest variation of spectral features. To assess the diagnostic potential of the method, C_{18}-bonded silica *d*-SPE and MALDI-TOF-MS were used to generate enriched endogenous peptide profiles of SP from 15 fertile and 15 non-fertile donors. Principal component analysis (PCA) successfully separated fertile from non-fertile men into two different clusters. An array of seven semenogelin-derived peptides was found to distinguish the two groups, with high statistical significance. These findings, while providing a rapid and convenient route to selectively enrich native components of SP peptidome, strongly reinforce the prominent role of semenogelins in male infertility.

Keywords: MALDI-TOF/TOF; mass spectrometry; proteome; seminal plasma; male infertility; seminal fluid; biomarker; semenogelins; dispersive solid phase extraction

Citation: Correnti, S.; Preianò, M.; Murfone, P.; Fregola, A.; Bitonti, M.; Savino, R.; Terracciano, R. Revealing the Hidden Diagnostic Clues of Male Infertility from Human Seminal Plasma by Dispersive Solid Phase Extraction and MALDI-TOF MS. *Int. J. Mol. Sci.* **2022**, *23*, 10786. https://doi.org/10.3390/ijms231810786

Academic Editor: Paolo Iadarola

Received: 27 July 2022
Accepted: 12 September 2022
Published: 15 September 2022

1. Introduction

Human seminal plasma (SP) is an acellular fluid, easily obtained after semen centrifugation, which contains specific proteins and endogenous peptides originating from the testis, epididymis, and male accessory glands [1,2]. Due to its intrinsic nature, being much closer to the male reproductive tract, it is significantly more enriched than serum and urine in secreted and/or shed proteins relevant for the study of infertility, male disorders, and other related pathologies [3]. SP has therefore captured a growing interest as a clinical sample for noninvasive diagnostics, with special attention being paid to the highly complex mixtures of small peptides present therein, which may offer a great potential for information rich patterns for clinical diagnosis. Indeed, several investigations have identified SP peptides involved in sperm motility and fertility [4,5]. Other studies have recognized the importance of SP peptides not only as putative markers for primary prostate cancer diagnosis [6] but also for their potential role in bactericidal activity [7,8]. Finally, recent studies have highlighted the pivotal role of specific peptides released by physiological cleavage of semen coagulum proteins (Semenogelins and Prostatic acid phosphatase) in forming amyloidogenic fragments, which are abundant in SP and which boost semen-mediated enhancement of HIV infection [9,10].

The delineation of new omics platforms, which accurately and rapidly provide a wide array of molecular entities from clinical specimens, including tissues and bodily fluids, may be of interest in the field of diagnostic and personalized medicine. With the rapid development of mass spectrometry (MS) technology, Matrix-Assisted Laser Desorption/Ionization-Time-of-Flight Mass Spectrometry (MALDI-TOF MS) has gradually increased its applications in biomedicine and clinical diagnostics, owing to its low acquisition times, ease of use, ruggedness, very high automation and throughput, and good sensitivity [11,12]. In particular, in this direction, MALDI-TOF MS provides an accessible and high-throughput platform to rapidly extrapolate by comparative analysis information-rich patterns arising from altered expression of specific peptides between normal and pathological conditions [13–16].

However, until now, in the case of human SP only a few investigations used MALDI-TOF MS either to assess fingerprints of SP lipids [17] or to assess the N-glycome profile for human SP [18]. Liu et al., by isobaric tags for relative and absolute quantitation (iTRAQ) labeling and liquid chromatography (LC)-MALDI strategy discovered infertility-related SP proteins in men with normal semen parameters and clinical pregnancy by R-ICSI after IVF failure [19]. However, MALDI-TOF peptidic profiles of non-digested human SP is reported only in one study. Fung and colleagues described direct analysis of unfractionated human SP showing peptide peaks features in a m/z range from 500 to 10,000 [20]. Additionally, only one investigation based on Surface-enhanced laser desorption/ionization (SELDI)-TOF MS reported differential protein expression in SP from fertile and infertile males [21]. In view of the above reported considerations, acquisition of rich and informative MALDI-TOF spectra preserving valuable information about endogenous and naturally occurring peptides of SP might be an attractive strategy to deliver fast and sensitive clinical diagnostic assays for pathologies and disorders of the male reproductive tract. In order to reach this goal, a practical and efficient SP peptide enrichment strategy is required before MALDI-TOF MS analysis. Currently, solid-phase extraction (SPE) prior to MALDI-TOF MS analysis is a rapid and convenient tool for profiling of clinical specimens [22]. We previously developed a procedure based on dispersive-SPE (*d*-SPE) in which the sorbent phase is suspended in a specific bio-fluid, providing more effective interaction between the sorbent and the peptides [23,24]. In the present study, for the first time, we applied a *d*-SPE coupled to MALDI-TOF MS to the enrichment and detection of human SP peptide patterns. Commercially available octadecyl (C_{18})- and octyl (C_8)-bonded silica sorbents and hexagonal mesoporous silica (HMS) were used to enrich and selectively harvest peptidic components of SP. We systematically optimized a convenient procedure based on different sorbents performances, assessing the ability of the method to generate highly stable and reproducible spectra. Finally, as proof-of-concept a key peptide-pattern within spectra was extrapolated by differential display statistical analysis that was able to distinguish between fertile and unfertile groups.

2. Results and Discussion

2.1. Pre-Analytical Assessment of Residual Proteolytic Activity in SP

It is well accepted that the presence in SP of proteins as well as tissue-specific mediators should provide new insights in the origin of male infertility [25]. In particular, the expression in SP of naturally occurring peptides, mainly arising from proteolysis of abundant and larger proteins, might be correlated to specific function of male reproductive organs [26–28]. As a result, the complexity and large variety of the size and the charge of peptides make peptidomic analysis of SP quite challenging. Therefore, in this study, we firstly assessed pre-analytical and analytical variables, aiming to establish an optimized protocol based on *d*-SPE coupled to MALDI-TOF/TOF MS for peptide pattern analysis of human SP in the low molecular weight mass spectral range. It is well known, that sample collections and processing procedures may significantly affect mass spectra profile. In particular, since semen liquefaction depends on a complex proteolytic cascade [26–28], a major challenge with SP low molecular weight profiling study is to check and to monitor if residual proteolytic activity is still present in SP and if it might influence the spectral readouts. The addition of a protease inhibitor cocktail (PIC) containing 4-(2-aminoethyl) benzenesulfonyl fluoride hydrochloride (AEBSF), a serine protease inhibitor that blocks PSA activity, was already described by Robert et al. [28]. However, in order to set up a valid pre-analytical processing protocol for our study, the stability of SP peptide profile obtained from a normozoospermic healthy volunteer was firstly investigated. PIC containing AEBSF was immediately added after the liquefaction of coagulum as described in the Material and Methods section. Specifically, the sample was prepared in three independent experiments and the variation of the total peak numbers between samples with and without PIC was assessed at several time points, more precisely at 0 and after 60, 90, 120, and 150 min at room temperature and after 1 and 120 days of storage at −80 °C. The total number of m/z peaks with a signal-to-noise (S/N) $\geq$ 10 detected in the MALDI-TOF mass spectra is shown in Supplementary Figure S1 as function of the time. No statistically significant variation in peak number was observed neither up to 2.5 h at room temperature nor up to 120 days of storage at −80 °C. Moreover, no statistically significant difference in peak number was observed at any time point between samples with and without the use of PIC. Definitively, these data demonstrate that SP is stable for at least 2.5 h at room temperature and for at least 4 months when stored at −80 °C. Additionally, any potential variation that might be observed between different cohorts of subjects would hardly be an artifact due to storage conditions, provided that the analysis is performed at any time within 4 months from the collection if the sample is stored at −80 °C within 2.5 h. Therefore, it might be tempting to speculate that no significant, residual proteolytic activity is still present in SP after the liquefaction of coagulum because no statistically significant difference in peak number was observed at any time point between samples with and without the use of PIC.

2.2. *d*-SPE Enrichment with Hydrophobic C_{18} and C_8 silica Sorbents and with HMS

Selective enrichment strategies before MALDI-TOF MS analysis may represent an important tool to finely modulate the low molecular weight profiling of clinical samples [22,29]. In fact, depending on the efficiency of the extraction method used, poor, medium, or rich peptide fingerprints can be obtained. A molecular profile rich in peaks is highly desirable, as more informative pattern diagnostic features could be achieved for discriminant peaks analysis [15,30]. In this investigation, our efforts were mainly devoted to identifying an optimal sample preparation method suitable for desalting and efficiently enriching peptides from human SP specimens in order to generate rich informative and robust MALDI-TOF screens.

Ideally, MALDI-TOF analysis of a clinical specimen should be accomplished by as few as possible processing steps in order to reduce sample loss. Additionally, it would be necessary to remove interferences and contaminants, which might generate high background and ambiguous peaks in the spectra. Another important requirement is to enrich the analytes of interest, in our case SP peptides, for increased sensitivity. SPE is a well known separation procedure based on the property of a chromatographic medium (the solid phase) to sequester one analyte or a subset of them from a specific solution of interest and is normally used not only for food matrices but also for clinical specimens. In *d*-SPE, the higher contact surface between the solid phase and the analytes allows reducing time needed to reach the extraction equilibrium in comparison to SPE. Consequently, *d*-SPE requires a shorter extraction time, providing a more effective capture of analytes. Owing to its outstanding rapidity, to low solvent consumption compared to standard SPE, to high efficiency, and to its wide applicability, this procedure is attracting growing interest [24,31,32].

Based on this rationale, in this study we explored for the first time commercially available C_8- and C_{18}-bonded silica beads and HMS as harvesting sorbents for human SP peptide enrichment by *d*-SPE. Optimal conditions were assessed for adsorption, washing, and elution steps, in order to ensure the best MALDI-TOF mass spectra in terms of number of detected peaks, repeatability on peak heights, area, and signal-to-noise ratio (see Supplementary Materials and Supplementary Tables S1–S4, Supplementary Figures S2–S5). Other critical parameters, such as the number of absorbents, the pH, the type of washing solvent, and the elution time, were established on the basis of our previous results obtained with clinical biological matrices for the enrichment/extraction of peptides [24].

Silica bonded C_{18} and C_8 sorbents were tested owing to their well-known interactions for hydrophobic peptides. In general, slight differences in selectivity are observed between C_{18} and C_8 sorbents [33]. Instead, in the case of HMS a larger spectrum of peptides might be extracted because the textural properties of these silica sorbents coupled to the characteristic wormhole mesostructured ensure very satisfactory results for integrating size selectivity with effective adsorptive mechanism [24]. Representative examples of the SP peptidic profiles obtained from the same sample before and after *d*-SPE processing with the different sorbents used in our study are shown in Figures 1 and 2.

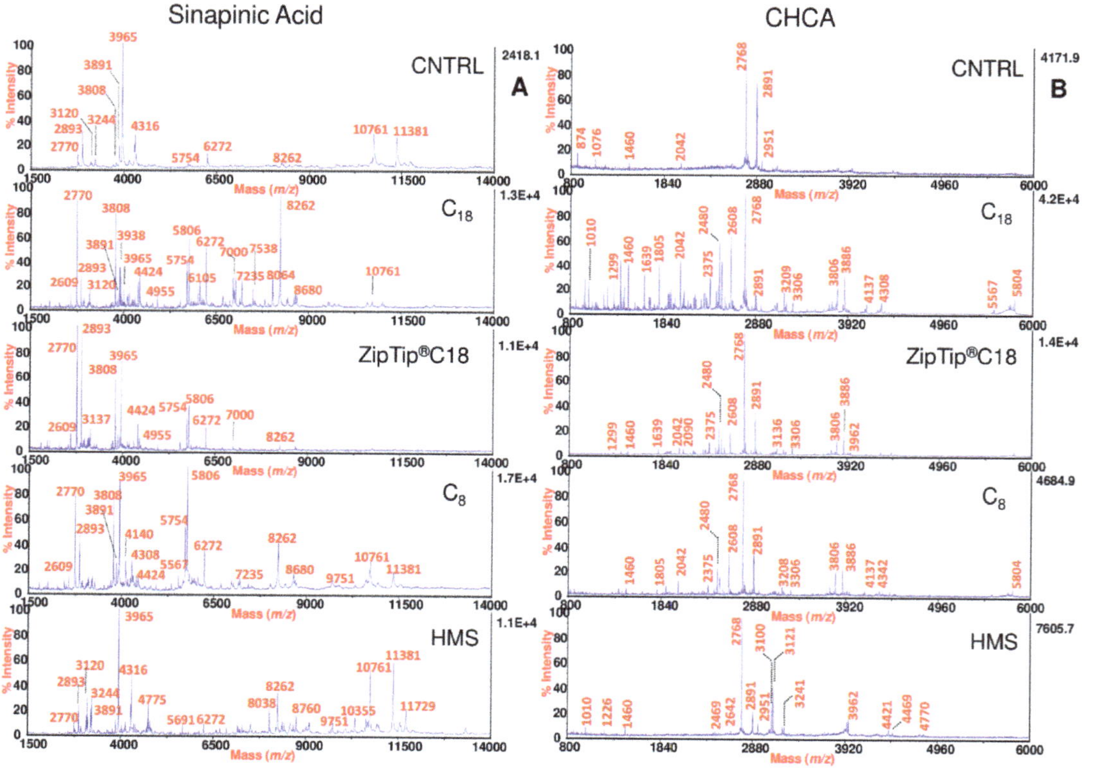

Figure 1. MALDI-TOF MS full view of SP using SA and CHCA matrices. MALDI-TOF mass spectra of SP obtained from the same fertile normozoospermic man using SA (**A**) and CHCA (**B**) as matrix, before and after processing by C_{18}, ZipTip®C18, C_8, and HMS. For SA, the spectra are shown in the *m*/*z* range from 1500 to 14,000. For CHCA, the spectra are shown in the *m*/*z* range from 800 to 6000, with labeled monoisotopic peaks.

In particular, the spectra were acquired both in sinapinic acid (SA) and α-cyano-4-hydroxycynnamic acid (CHCA), which are the most common MALDI matrices used for peptide profiling. Different *m*/*z* ranges are shown, in order to better highlight both in "full" (Figure 1) and "zoom" (Figure 2) views, the high peaks enrichment provided by the use of the *d*-SPE processing.

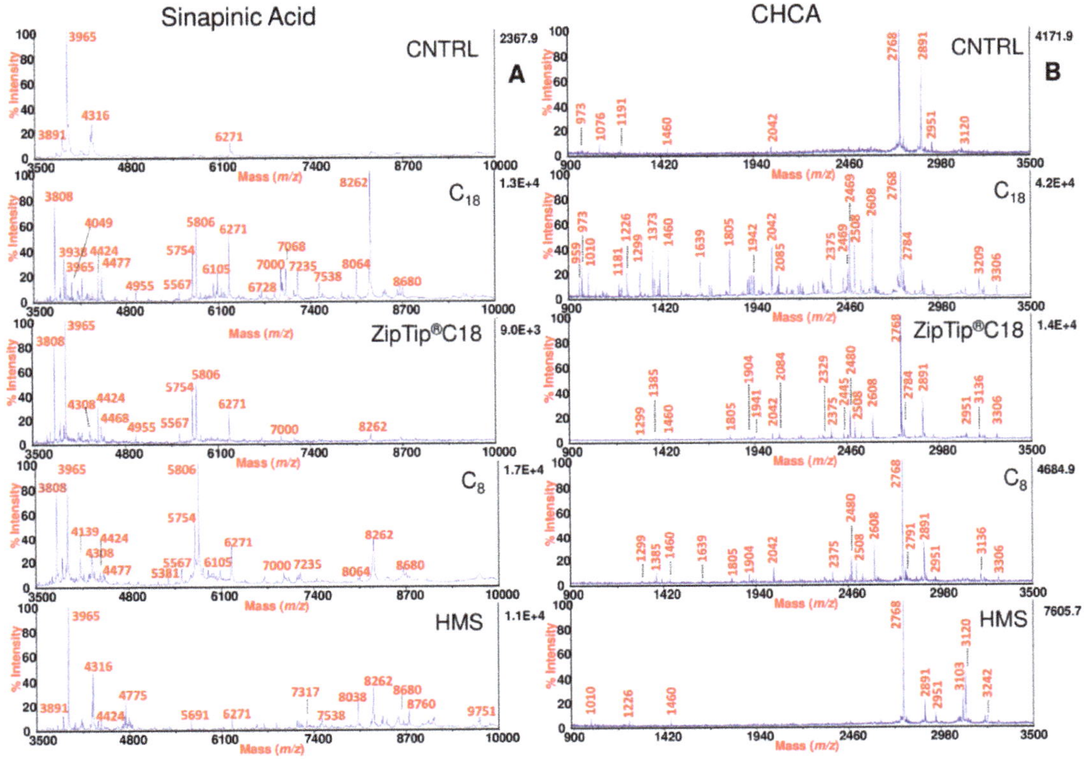

Figure 2. MALDI-TOF MS zoom view of SP using SA and CHCA matrices. MALDI-TOF mass spectra of SP obtained from one fertile normozoospermic subject using SA (**A**) and CHCA (**B**), before and after processing by C_{18}, ZipTip®C18, C_8, and HMS. For SA, the spectra are shown in the m/z range from 3500 to 10,000. For CHCA, the spectra are shown in the m/z range from 900 to 3500, with labeled monoisotopic peaks.

Commercial ZipTip Pipette Tips equipped with C_{18} chromatographic media (conventionally used SPE approach) were also included in this study for comparative purposes. As expected, the use of these pre-treatment methods, removing salts, interferences and other contaminants, which generate signal suppression in the MALDI-TOF analysis, significantly amplified the peptidic repertoire of SP compared to the same untreated sample (Figures 1 and 2). Comparing each of the spectral view of *d*-SPE processed samples (Figures 1 and 2) to the same untreated sample, the fine tuning in the peptide profiling through the different sorbents allowed the detection of specific subset of SP peptides, which were missing without pre-treatment. Moreover, in the described dispersive procedure (see Section 3), since the contact area between the sorbents and the analytes is amplified in comparison to the classical SPE, a more effective interaction reduces the times of the protocol and the volumes of the solvent required for the elution of the peptides retained on the sorbents. In fact, only 10/15 min are required for the adsorption step and few microliters (15/25) for the elution of peptides adsorbed on the sorbents (Supplementary Table S1–S3). Dispersive methods appear to generate MALDI-TOF spectral portraits with an increased number of peaks and with higher intensity, area, and S/N after the enrichment (see Figures 1,2 and 3A,B). Generally, for all sorbent tested, in the case of SA, the best detection of the SP peptide peaks was observed in a m/z range from 1500 to 14,000 (Figure 1A). For CHCA, the m/z range showing the best detection was lower than that observed for SA, displaying best peaks fingerprints in the range between 800 and 6000 (Figure 1B).

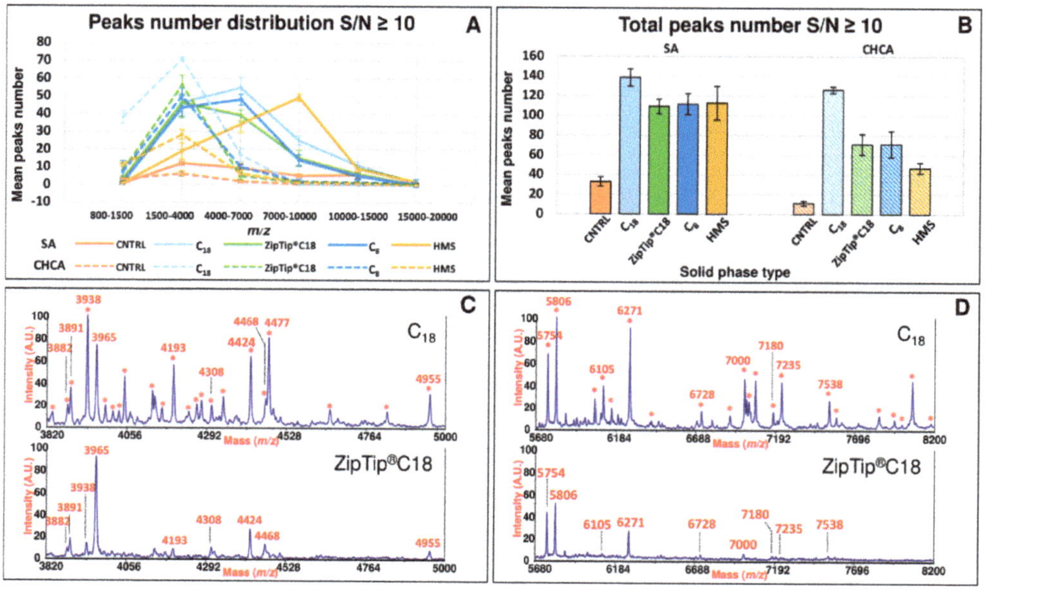

Figure 3. MALDI-TOF peaks number distribution using different *d*-SPE sorbents and peptides enrichment comparison between C_{18} *d*-SPE and ZipTip®C18. Panel (**A**) shows the mean peaks number distribution for SP samples over different *m*/*z* ranges using SA and CHCA before and after processing by C_{18}, ZipTip®C18, C_8 and HMS. Panel (**B**) displays the total peaks number detected for SP samples using SA and CHCA before and after processing by C_{18}, ZipTip®C18, C_8, and HMS. Panels (**C**,**D**) display the comparison of spectral portraits in SA of SP between C_{18} and ZipTip®C18 in the *m*/*z* range from 3820 to 5000 and from 5680 to 8200, respectively. Asterisks indicate peaks selectively extracted by C_{18} sorbent, which are absent or detected with a lower signal intensity in ZipTip®C18 treated sample.

A more detailed study of *d*-SPE performances was obtained through the distribution of peaks eluted by the diverse sorbents in different *m*/*z* regions of the spectrum (Figure 3A). An increase of the peak number was observed from 800 to 4000 in CHCA for all kinds of sorbents. Among all stationary phases, pre-processing using C_{18} silica sorbent showed the best performance (in term of extracted peaks) in the low mass range (1500–4000) for spectra acquired in CHCA and in the medium mass range (4000–7000) for spectra acquired in SA (Figure 3A). C_8 silica sorbent (and ZipTip) showed a peak distributions similar to C_{18} (Figure 3A). Specifically, C_8 sorbent paralleled the C_{18} peak distribution behavior although running in a slightly lower region of the y-axis. The similar trend in peak distribution is explained by the quite similar adsorption behavior based on hydrophobic interactions. A different peaks distribution of the eluted peaks was observed when HMS particles were used, in particular for the spectra acquired in SA. In fact, as shown in Figure 3A, the maximum of the distribution for HMS was shifted to 7000–10,000 *m*/*z* region of mass spectra acquired in SA. While, for the spectra acquired in CHCA, the maximum was observed in the same range (1500–4000) of the other sorbents. The HMS shows hydrophilic silica features and mesoporous wormhole structure that warrants not only an exceptional dispersibility but also an optimal enrichment ability for peptides based on a cut-off mechanism (depending on the mesopore diameter) and a variety of interactions [24]. Therefore, it is reasonable to observe a different shape in the distribution of peaks in the different regions of the mass spectra. The distribution of peptides for HMS mirrors mainly the size exclusion effect and, as a consequence, the majority of peaks were detected in the m/z range between 7000 and 10,000 (Figure 3A), while a low efficiency was observed for higher molecular weight species.

In the case of the CHCA, the shorter m/z ranges analyzed do not allow highlighting differences in the maximum, although a different slope is observed from 1500 to 4000. Considering the total number of peaks detected with an $S/N \geq 10$, best results were obtained for C_{18} sorbent, with ~140 peaks detected over the entire m/z range from 800 to 20,000 when the spectra were acquired in SA, and ~120 peaks when the spectra were acquired in CHCA (Figure 3B). Although the total peak number was slightly lower, also C_8 and HMS showed an optimal efficiency in SP peptide capture as illustrated in Figure 3B with comparable total peaks detected (approximately 110) in SA. In the C_{18} sorbent, the presence of a great coverage of octadecyl chain bound to the silica surface, associated to the great methylene selectivity, while reducing silanols interactions, allows for very hydrophobic interactions between the SP peptides and the stationary phase. For the C_8, the shorter octyl chain provides less hydrophobic interactions, and maybe this provides a possible explanation for the slightly lower number of peaks extracted for this sorbent in comparison to C_{18} (Figure 3B). A lower number of peaks was detected for HMS in CHCA in comparison to the other sorbents (Figure 3B). As already stated above, the adsorptive mechanism for this mesoporous sorbent is quite different from the C_8 and C_{18}; moreover, we do not exclude that during our *d*-SPE procedure with HMS some low mass proteins might have been removed during the washing step. Furthermore, it might be also possible that a part of the low molecular weight SP peptides are bound to larger SP proteins as it happens in the case of plasma or serum peptides for the presence of carrier proteins, such as human serum albumin [34]. In this case, due to the cutoff mechanism of HMS, the large carrier proteins with bound LMWP cannot be adsorbed into the nanometric porous network wormhole channels of the HMS. This phenomenon might explain the lower number of peptides extracted by HMS in the low molecular m/z range in comparison to the other sorbents (Figure 3A,B).

Figure 3C,D show the comparison of spectral portraits in absolute intensity units of SP between C_{18} and Ziptip C_{18} in the m/z range from 3820 to 5000 and 5680 to 8200, respectively. In particular, several peaks (indicated by asterisks) extracted by C_{18} sorbent are not observed or show a lower intensity in the case of ZipTip C18. These data highlight the higher performance of *d*-SPE in comparison to SPE.

Although inter-individual variations of protein composition in SP might influence differential proteomic/peptidomics analysis [35,36], our effort was to standardize and optimize the experimental procedure with the aim to provide a robust tool for biomarker discovery. In order to warrant a robust analytical tool, repeatability and reproducibility of the method were assessed for spot-to-spot and within-spot reproducibility as reported in the Materials and Method section and in the Supplementary Materials. The results are summarized in Supplementary Figures S2–S5. Mean CVs% obtained for peak heights, area and S/N ranged between 10% and 15% both for inter and intra spot for all spectra obtained after pre-treatment, while in the case of the untreated sample, the CV% was comprised between 20% and 25% (Supplementary Figure S5). Among all *d*-SPE sorbents and ZipTip, the lowest variation in signal intensity, area, and S/N was observed for C_{18} *d*-SPE (mean CVs less than 10%-see also Supplementary Figure S5). Overall, the data showed a well adequate analytical robustness, suggesting that these SP peptidic fingerprints might be part of a diagnostic profile for properly weighing substantial alterations between fertile and infertile groups, likely reflecting the physiological or pathological state of the male reproductive tract.

2.3. Differential Comparative Analysis between Fertile and Infertile Groups

To assess the diagnostic potential of this platform, 15 SP specimens from normozoospermic fertile subjects and 15 from non-fertile subjects were analyzed and peptide enriched fingerprints by C_{18} *d*-SPE were compared by statistical data analysis and statistical assessment model (principal component analysis-PCA). Table 1 reports the clinical characteristics of the cohort subjects enrolled in this study including all semen parameters for semen subtype classification according to the World Health Organization (WHO) 2010 guidelines [37]. Further details of the study groups are listed in Supplementary Table S5.

Table 1. Clinical characteristics of the subjects enrolled in the study.

Patient Features (Mean ± SD)	NZ [1] n = 15	AT [2] n = 2	OAT [3] n = 5	TZ [4] n = 6	OZ [5] n = 2
Age	28.7 ± 7.3	43 ± 2.8	31.2 ± 5.6	30.8 ± 7.9	33 ± 14.1
Ejaculated volume (mL)	3.3 ± 1.6	3.4 ± 1.3	2.3 ± 0.8	4.3 ± 2.1	3.15 ± 0.2
pH	7.7 ± 0.3	7.5 ± 0	7.7 ± 0.4	7.5 ± 0.4	8.25 ± 0.2
Sperm count (million)	209.2 ± 125.7	155.9 ± 62.3	3.4 ± 2.4	269.1 ± 72.9	20 ± 5.3
Progressive motility (%)	49.3 ± 7.3	23 ± 7.1	10.6 ± 12.3	41.2 ± 6.5	40.5 ± 7.8
Total motility (%)	61.6 ± 7	45 ± 4.2	16.8 ± 10.1	57.2 ± 8.5	53.5 ± 2.1
Normal morphology (%)	7.4 ± 3.4	2.5 ± 0.7	0.7 ± 0.5	2.4 ± 0.9	4.25 ± 0.3

1. Normozoospermic subject, fertile man. In this category all semen parameters are normal within the acceptable reference values provided by the World Health Organization (WHO) 2010 guidelines: total number of spermatozoa, and percentages of progressively motile and morphologically normal spermatozoa, equal to or above the lower reference limits (total sperm number $\geq 39 \times 10^6$ spermatozoa per ejaculate, sperm progressive motility $\geq 32\%$ and normal sperm morphology $\geq 4\%$). 2. Asthenoteratozoospermic patient, an infertile man with percentages of both progressively motile and morphologically normal spermatozoa below the lower reference limits. 3. Oligoasthenoteratozoospermic patient, an infertile man with total number of spermatozoa and percentages of both progressively motile and morphologically normal spermatozoa below the lower reference limits. 4. Teratozoospermic patient, an infertile man with percentage of morphologically normal spermatozoa below the lower reference limit. 5. Oligozoospermic patient, an infertile man with total number of spermatozoa below the lower reference limit.

As an explorative effort, we assessed by PCA if SP C_{18} sorbent enriched peptide MALDI signatures from the 30 donors (15 fertile vs. 15 infertile) allow any possible clustering in an unsupervised modality. PCA is a processing tool that analyzes the variance of a dataset in an unsupervised manner, considering the expected and unexpected variance, with high dimensionality data sets [38]. In particular, PCA, performed with the use of MarkerView™ software, reduces the complexity of a mass spectral dataset. Based on an unbiased assessment of inherent spectral differences, this tool shows any intrinsic clustering of examined samples. The results are shown in Figure 4: two separate clusters were visualized indicating that unsupervised PCA satisfactorily segregates infertile patients from fertile normozoospermic controls. Infertile individuals (red dots) appear quite tightly clustered and quite cleanly separated in the plot from the fertile individuals (green dots). Samples clustering and differentiation confirm data homogeneity (Figure 4) and strongly suggest that the platform described might provide a useful baseline resource for future fertility biomarker studies.

Table 2 lists the peaks differentially expressed between the fertile and infertile groups in a statistically significant manner.

Table 2. Discriminant peaks significantly different between fertile (n. 15) and infertile (n. 15) men.

m/z [1]	Uniprot ID (Accession Number)	Protein Identity	Peptide Sequence	Study Groups and Peptide Expression Level	*p* Value [2]
2331	P04279	SEM I; Fragm:330–349	ITIPSQEQEHSQKANKISYQ	↑Fertile/↓Infertile	0.007
2362	Q02383	SEM II; Fragm:248–267	HGPKDIFTTQDELLVYNKNQ	↑Fertile/↓Infertile	0.001
2482	P04279	SEM I; Fragm:195–215	VLQTEELVANKQQRETKNSHQ	↑Fertile/↓Infertile	0.00003
2893	P04279	SEM I; Fragm:428–453	HGSHGGLDIVIIEQEDDSDRHLAQHL	↑Fertile/↓Infertile	0.0000006
3059	P04279	SEM I; Fragm:248–273	HGSKDIFSTQDELLVYNKNQHQTKNL	↑Fertile/↓Infertile	0.001
3083	Q02383	SEM II; Fragm:248–273	HGPKDIFTTQDELLVYNKNQHQTKNL	↑Fertile/↓Infertile	0.010
3938	Q02383	SEM II; Fragm:549–582	ESSESHNIVITEHEVAQDDHLTQQYNEDRNPIST	↑Fertile/↓Infertile	0.001

1. The m/z values of precursor ions refer to average masses MH^+. 2. The p values were calculated with unpaired t-test on peak intensity; significance was set at a p value less than 0.05.

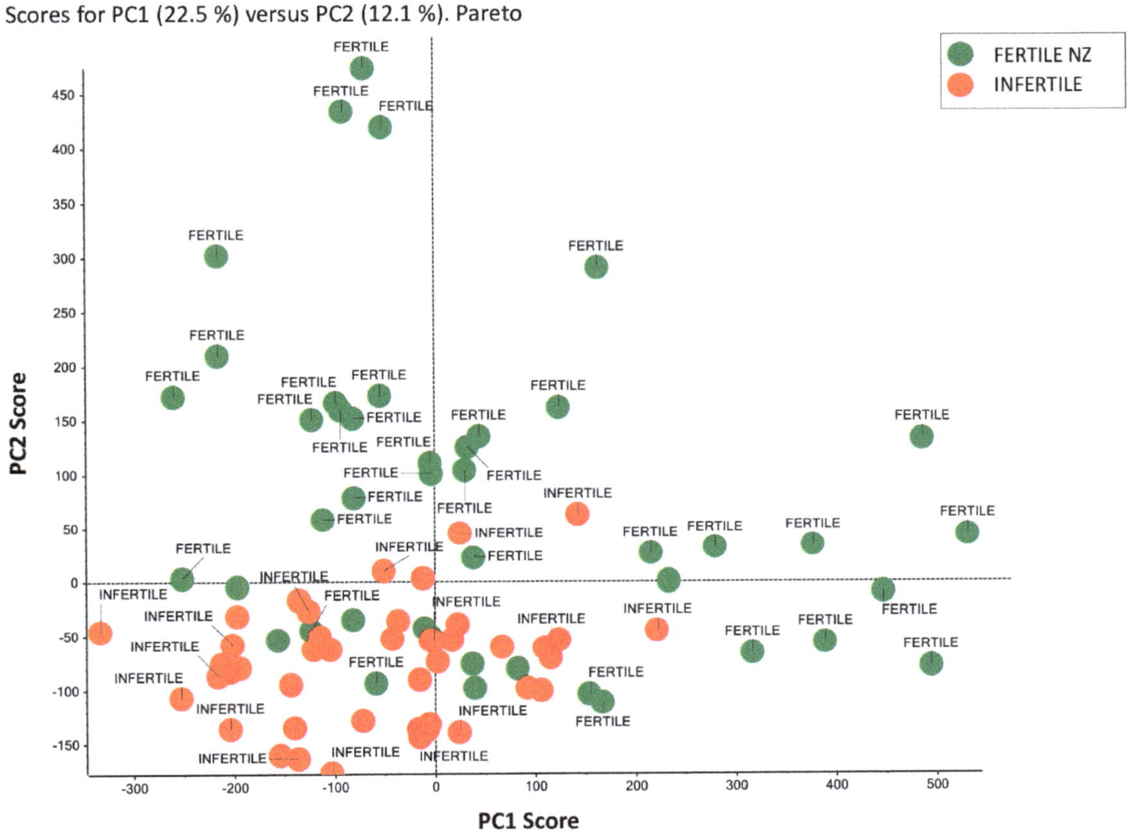

Figure 4. Unsupervised PCA of SP from fertile and infertile men enrolled in this study. Unsupervised PCA scores plot acquired by MarkerView™ software represents SP clustering of fertile (green dots) vs. infertile men (red dots). Pareto scaling was applied on the MALDI-MS data set.

The graphical comparison of the normalized peak heights between the two groups, illustrated in the box plots reported in Figure 5, shows that each main discriminant m/z peak is downregulated in infertile men with very significant p-value for the peaks 2893 (p = 0.0000006) and 2482 (p = 0.00003). Highly significant p-values were also found for the peaks 2362, 3059, and 3938 with p-values < 0.005. The other peaks 2331 and 3083 show a p-value $\leq$ 0.01 (Figure 5).

In order to better mark differential peak expressions, the overlaid mass traces from the 15 fertile (green profiles) and 15 infertile men (red profiles) are shown in the m/z regions from 2300 to 2400 (Figure 6A) for the peaks 2331 and 2362. The other m/z peaks are shown in the spectral ranges from 2440 to 2530 (Figure 6B), from 2860 to 2930 (Figure 6C), from 3050 to 3090 (Figure 6D), and in the m/z range from 3930 to 3950 (Figure 6E).

The seven differentially expressed peptides were then identified by direct sequencing by MALDI-TOF/TOF mass analysis, as shown in Table 2 and Supplementary Figure S6 in which the primary sequences, the identities, and the MS/MS mass spectra are shown. These peaks resulted as array of seven Semenogelins-derived peptides. In particular, four peaks (m/z = 2331, 2482, 2893, 3059) originated from Semenogelin-1 (SEM I) and three peaks (m/z = 2362, 3083, 3938) were fragments semenogelin-2 (SEM II). Interestingly, the SEM I fragments with m/z 2362 and 3059 (see Table 2) shared a common aminoacidic sequence 248–267. More intriguingly, the SEM II fragment with m/z = 3083 covers the sequence 248–273, which shows very high sequence similarity with SEM I fragment 248–273.

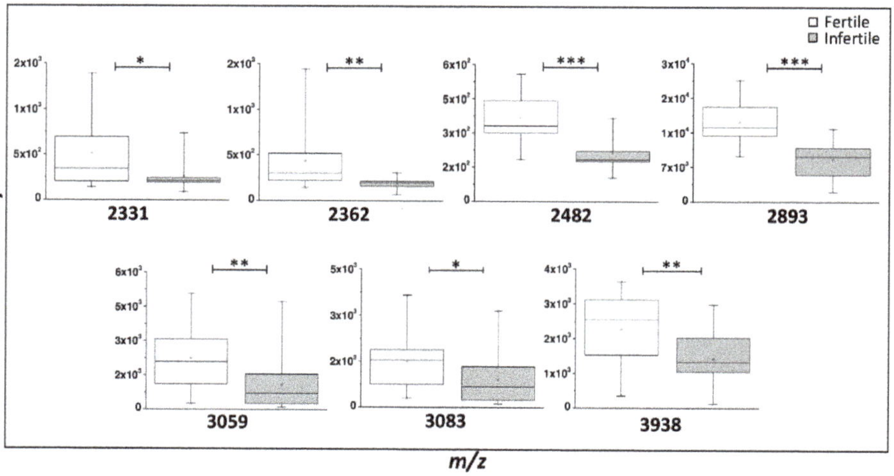

Figure 5. Main discriminant peaks box plot analysis. Box plot of the peak intensities for statistically discriminant m/z signals between 15 fertile and 15 infertile men performed by OriginLab® software. The p values were calculated with unpaired t-test on normalized peak intensity and the asterisks show the level of significance between the two groups; * p values < 0.01, ** p values < 0.005, *** p values < 0.0001.

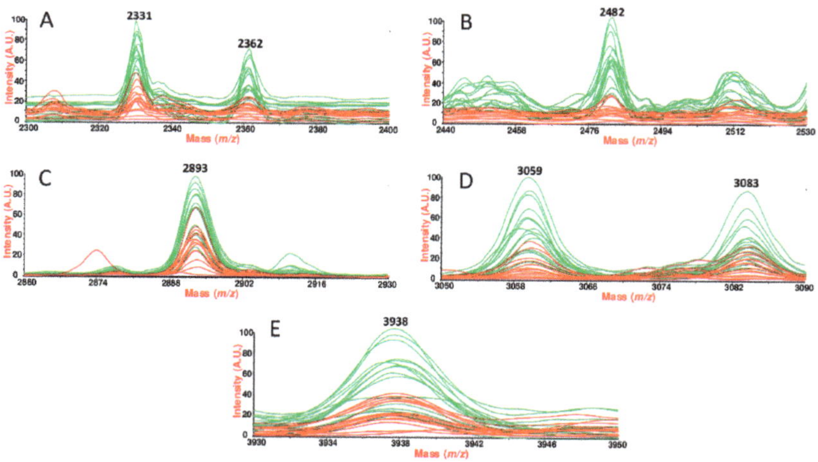

Figure 6. Traces overlay for discriminant peaks between fertile and infertile men. MALDI-TOF fingerprints comparison of SP samples between fertile (green) subjects and infertile (red) patients, obtained using Data Explorer software. Spectra overlay in absolute units highlights the peak intensity variation resulted statistically significant for the m/z peaks 2331 and 2362 (**A**), 2482 (**B**), 2893 (**C**), 3059 and 3083 (**D**), 3938 (**E**).

Semenogelins are abundant proteins in SP; as a consequence, the amount of these fragments may reflect the balance between the activity of specific enzymes (including proteases) and their inhibitors [6,39]. In fact, immediately after ejaculation, the spermatozoa are immobilized in the semen coagulum matrix. The semenogelins, which are the main proteins forming the semen coagulum, are proteolytically cleaved by prostate-specific antigen (PSA also known as KLK3). Owing to its proteolytic activity, PSA breaks up the coagulum bound to the spermatozoa surface, which increases their motility [40]. The resulting liquefaction of the clot after semenogelins fragmentation, warrants the physiological motility required by spermatozoa to reach the female reproductive tract [40,41]. Interestingly, PSA expression is decreased in infertile subjects with reduced sperm motility [42,43]. Moreover, in a recent study, comparative and quantitative proteome analysis between two groups of men with oligoasthenozoospermia (total number of spermatozoa and percentages of progressively motile spermatozoa below the lower reference limits) and normozoospermia (total number of spermatozoa and percentages of progressively motile and morphologically normal spermatozoa are equal to or above the lower reference limits), PSA was found decreased in SP of men with oligoasthenozoospermia [44]. Therefore, it is tempting to speculate that the decreased level of expression of semenogelins-derived peptides in the SP of infertile men may be associated to a decreased expression of PSA in the unfertile group.

SEM I was found increased in oligoteratozoospermia (total number of spermatozoa and percentages of morphologically normal spermatozoa are below the lower reference limits) [45], while in the recent proteomics study by Martins et al. [46] both SEM I and SEM II resulted under-expressed in primary and secondary infertile individuals in comparison to the control group. However, western blot validation experiments confirmed only SEM II decreased in primary infertility, while any change in the expression of SEM I and SEM II by western blot analysis was observed in secondary infertility group [46].

These contrasting findings on semenogelins change of expression may arise from the intrinsic limitations of bottom up approaches and the presence in SP of both intact semenogelins and peptide-derived semenogelins. In fact, in a bottom up approach all the proteins are digested before LC-MS/MS analysis, therefore it is not possible to discriminate if the peptides that contribute to the identification of SEM I and II originate from intact precursors or from a fragment derived peptide. It is important to underline that, as top-down strategy, our method does not require the use of trypsin or more in general proteolytic digestion in comparison to bottom-up strategy. Therefore, the spectral readouts acquired by this platform reveal SP peptides in their native and biologically active forms.

Currently, mechanism underlying the physiological roles of the semenogelins, including their proteolytically-derived peptides, are still unclear; however, their implication and clinical relevance in male infertility is well described in literature [47–49]. Considering the important role of semenogelins in coagulation and liquefaction process and their great impact on fertility, our platform provides a new rapid and robust tool to reveal new potential diagnostic semenogelin-derived signature of male infertility hidden in SP. The data obtained in this pilot study will be further extended to a larger cohort in order to validate these preliminary findings. Additionally, the use of other sorbents, such as C4 and mesoporous silica, are currently under investigation in our group with the aim to test their potential to amplify the peptidomic repertoire of SP.

3. Materials and Methods

3.1. dSPE Sorbents

Silica gel C_8-Reversed phase (60759) and Discovery® DSC-18 SPE Bulk Packing (52600-U) were purchased from Supelco (Merck, Darmstadt, Germany).

HMS (code 541036 wormhole silica mesostructured) was purchased from Sigma-Aldrich (St. Louis, MO, USA). HMS (code 541036 wormhole silica mesostructured). ZipTip®C18 was purchased from Merk Millipore Corporation (Darmstadt, Germany).

3.2. Reagents

All the chemicals and reagents were analytical grade. Water was purchased from Sigma Aldrich. Acetonitrile (ACN) (HPLC grade) and trifluoroacetic acid (TFA) (ACS grade) were obtained from Merck (Darmstadt, Germany). The MALDI matrices Sinapinic Acid (SA) and alpha-cyano-4-hydroxycinnamic acid (CHCA) were purchased from Fluka (St. Louis, MO, USA). Peptide mass standards kit for calibration of AB SCIEX MALDI-TOFTM instruments (AB Sciex, Framingham, MA, USA). Protease inhibitor cocktail (PIC; P8340, Sigma, St. Louis, MO, USA) and Pierce™ BCA Protein Assay Kit (23225, Thermo

Scientific™, Rockford, IL, USA) were used, according to the manufacturer's instructions. Ammonium bicarbonate (09830) was purchased from Sigma Aldrich (St. Louis, MO, USA).

3.3. Patient Recruitment

The study on the volunteer subjects was conducted in accordance with the Declaration of Helsinki and was approved by the Ethics Committee of MAGNA GRAECIA UNIVERSITY and MATER DOMINI HOSPITAL (protocol code 2014.39, date of approval 16 April 2014). All subjects were enrolled after the informed written consent was obtained from each participant. Only the first sperm evaluation was used in this analysis. Samples from patients who had vasectomy or history of orchitis, testicular trauma, sexually transmitted disease, varicocele, inguinal hernia operation, and cryptorchism were excluded. A questionnaire was distributed to obtain information on age, smoking habits, alcohol use (regular, irregular, or total abstinence), and use or abuse of other substances and drugs (yes or no). Semen from fertile men and infertile patients was obtained by masturbation into sterile containers after (3–5) days of sexual abstinence. All specimens were processed and analyzed anonymously.

The detailed clinical characteristics of the subjects enrolled in this study are summarized in Table 1.

3.4. Preparation of Seminal Plasma

Each collected ejaculated sample was allowed to liquefy for 15–30 min at 37 °C. Semen parameters were assessed according to the World Health Organization guidelines 2010 [37]. After the complete liquefaction of coagulum, liquefied sample from each donor, was divided into two aliquots. In one aliquot a protease inhibitor cocktail (PIC) was immediately added in a 1:100 v/v ratio. Then, both aliquots were processed to obtain SP. In particular, each clinical sample was centrifuged at 15,000× g for 15 min at 4 °C. The supernatant (the SP) resulted as a clear and fluid phase that was separated from pellets (debris) and cellular components, aliquoted, and stored at −80° C until use. Protein concentration of SP was determined by the bicinchoninic acid (BCA) assay according to the manufacturer's instructions. Supplementary Table S5 summarizes for each SP sample the determined total protein content.

3.5. Assessment of Protease Activity in SP

To assess stability of SP peptidic profile the freshly obtained semen from a normozoospermic fertile donor was used. After the complete liquefaction of coagulum, liquefied sample was split in two and PIC was added to one of the samples. Then, both samples were processed to obtain SP. In particular, each clinical sample was centrifuged at 15,000× g for 15 min at 4 °C. The supernatant (the SP) resulted as a clear and fluid phase that was separated from pellets (debris) and cellular components. Aliquots were analyzed immediately and after 60, 90, 120, and 150 min at room temperature; then, aliquots were frozen at −80°C and analyzed after 1 and 120 days of storage. The number of peaks were comparatively analyzed (Supplementary Figure S1).

3.6. Seminal Plasma Samples Normalization

The total protein content of SP specimens obtained from each study participant was determined by BCA assay (Supplementary Table S5). The concentrations of SP specimens collected from each donor patient, ranged from 30.7 to 82.0 mg/mL. Therefore, an adequate volume of each clinical sample was either concentrated (by vacuum centrifugation) or diluted (by the addition of deionized water) to 50 µL, in order to obtain a final concentration of 50 mg/mL and a total protein content of 2.5 mg

3.7. C_{18} and C_8 d-SPE Optimized Procedure

A total of 10 mg of C_{18} or C_8 silica sorbents were mixed with 100 µL of SP sample (50 µL of normalized SP sample in 50 µL of deionized water). The slurry was gently vortexed at room temperature for 10 min, then it was centrifuged at 2000× g for 2 min. The sorbent was then separated from the supernatant and was washed twice with 20 μL of 0.1% TFA. After the last wash, peptides bound to solid phase were eluted with 15 μL of a 1:1 (v/v) solution of ACN/0.1% TFA. Eluates were in part aliquoted and stored at −80 °C and, in part, immediately used for MALDI-TOF MS analysis.

Detailed experimental procedures for optimization of final protocols are described in the Supplementary Materials.

3.8. d-SPE HMS Optimized Procedure

A total of 10 mg of HMS were mixed with 100 μL of SP sample (50 μL of normalized SP sample in 50 μL of deionized water). The slurry was shaken at room temperature for 15 min. The suspension was centrifuged at 2000 g for 2 min, then HMS particles were separated from the supernatant and washed twice with 20 μL 0.1% TFA. After the last wash, species adsorbed on HMS were extracted with 25 μL of a 1:1 (v/v) solution of ACN/0.1% TFA. Eluates were in part aliquoted and stored at −80 °C and in part immediately used for MALDI-TOF MS analysis.

Detailed experimental procedures for optimization of final protocols are described in the Supplementary Materials.

3.9. ZipTip C18

After protein quantification, each SP sample was diluted (by the addition of deionized water) to a final concentration of 0.4 mg/mL, as suggested by the manufacturer's instructions of ZipTip®C18 pipette tips.

3.10. MALDI-TOF MS Analysis

For MALDI-TOF MS sample preparation, a saturated matrix solution of SA in 35% ACN/0.1% TFA was prepared. The CHCA matrix was prepared by dissolving 4 mg in 1 mL of a solution prepared with 50% of ACN in 0.1% TFA. The solutions were then sonicated for 1 min. SP samples were prepared by a dry-droplet method. A total of 1 μL of non-treated SP or previously treated with HMS, C_{18}, C_8 sorbent, or ZipTip®C18 was mixed with 4 μL of SA or CHCA solution prepared as described above and 1 μL of the resulting mixture was spotted on the MALDI target plate (Opti-TOF 384-Well Insert, ABSciex, Framingham, MA, USA).

MALDI-TOF mass spectra were acquired on AB SCIEX MALDI-TOF/TOF 5800 mass spectrometer (ABSciex, Framingham, MA, USA), equipped with a diode-pumped, Nd:YLF laser with λ = 345 nm wavelength. Each sample was run in triplicate. For MALDI MS measurements in SA the following settings were applied: bin size was set at 4 ns, final detector voltage was 2.070 kV with multiplier value at 0.75; 3000 laser shots were accumulated for each spectrum. MS data were calibrated via external calibration using the 5800 Mass Standards kit (AB SCIEX, Framingham, MA, USA) containing insulin bovine (MH^+ 5734.59), thioredoxin (MH^+ 11,674.48), and horse apomyoglobin (MH^+ 16,952.56). For MALDI MS measurements in CHCA the following settings were applied: bin size was set at 1 ns, final detector voltage was 1.980 kV with multiplier value at 0.66; 3000 laser shots were accumulated for each spectrum. MS data were calibrated via external calibration using the 5800 Mass Standards kit (AB SCIEX, Framingham, MA, USA) containing des-Arg^1-Bradykinin (MH^+ 904.4681), Angiotensin I (MH^+ 1296.6853), Glu-Fibrinopeptide B (MH^+ 1570.6774), ACTH (clip 1–17) (MH^+ 2093.0867), ACTH (clip 18–39) (MH^+ 2465.1989), ACTH (clip 7–38) (MH^+ 3657.9294).

Data Explorer version 4.11 software (AB SCIEX, Framingham, MA, USA) was used for data acquisition and data processing.

3.11. Repeatability Assessment

The SP sample from one normozoospermic donor was used to assess the repeatability of the MALDI analysis by three independent experiments performed on non-treated (control) and treated (C_{18}, C_8, HMS,) samples. For each experiment, 6 spectra were acquired with a total of 18 replicates.

3.12. Spot-to-Spot and within Spot Reproducibility

A total of 10 samples were prepared for MALDI-TOF MS analysis. Two samples included control SP samples in CHCA and SA collected from one normozoospermic donor. The other eight samples were obtained from four different SP enrichment preparations, three using *d*-SPE with three different sorbents (C_8, C_{18} and HMS), one using SPE by ZipTip collected from the same normozoospermic donor using control SP sample.

For each of the ten MALDI samples three spots were loaded on the MALDI target plate. Six MALDI-TOF mass spectra were acquired from each spot. Eighteen spectra for each sample were acquired for intra and inter spot MALDI-TOF reproducibility assessment. A total of 180 mass spectra were acquired for experiments in SA. At the same manner, a total of 180 mass spectra were acquired for experiments in CHCA. S/N, peak intensity and peak area from 30 selected peaks both in CHCA and in SA acquired spectra were used for calculating spot-to-spot and within spot reproducibility (Supplementary Figure S5).

3.13. Differential Statistical Peptide Pattern Analysis

MALDI-TOF mass spectra were first acquired and processed using Data Explorer version 4.11 software (AB SCIEX, Framingham, MA, USA) and subsequently analyzed for differential peptide patterns. Mass t2d data files were uploaded into MarkerView™ software 1.2.1.1 (AB Sciex, Foster City, CA, USA) and differential peptide profiling was assessed by unpaired, two-tailed Student's t-test with aligned MALDI-TOF mass spectra and the normalized peak height for each m/z value. The list of differentially expressed peaks was then filtered by manual inspection and each peak was verified as described in a previous report [50]. Box plot analysis between fertile and infertile men for statistically discriminant m/z signals was performed by OriginLab® software (version 7.0, OriginLab Corporation, Northampton, MA, USA).

3.14. MALDI-TOF/TOF Sequencing Experiments

The differentially expressed peptides were directly subjected to the MALDI-TOF/TOF analysis for acquiring sequence information (Supplementary Figure S6). For the MS/MS measurements, 1 μL of SP sample was mixed with 4 μL of matrix solution (4 mg/mL of CHCA in 50% ACN and 0.1% TFA), and 1 μL of the obtained solution was spotted on the MALDI target plate. The voltage settings were 8.0 kV and 15.0 kV for the ion source 1 and source 2, respectively. Air was used as the collision gas and MS/MS spectra were acquired by accumulating twenty spectra (1000 shots each) at 1000 Hz pulse rate and laser energy setting of 5000–6000. The experimental collision-induced dissociation (CID)-MS/MS ion spectra were compared to theoretical MS/MS spectra generated from Protein Prospector (http://prospector.ucsf.edu/, accessed on 18 July 2022). In Supplementary Figure S6, the MALDI-TOF/TOF mass spectra with b and y ion series and the assigned peptide sequences of identified species are reported.

3.15. PCA

MS data were exported from the 5800 MALDI ABScieX as t2d files and were then processed with MarkerView™ software 1.2.1.1 (AB Sciex, Foster City, CA, USA), with well-defined mass tolerance limits imposed for PCA. Specifically, unsupervised PCA was performed in order to visualize samples clustering. PCA results are commonly plotted in two- or three-dimensional plots that reflect the behavior of the samples (scores plot) or variables (loadings plot). Pareto scaling and no weighting was applied on the MALDI-MS data set comprising the normalized m/z peak intensities from SP peptides enriched with C_{18} sorbent using three replicate spectra for each subject. In brief, Pareto data set processing performs mean centering and scaling using square root of standard deviation of peak intensities.

Supplementary Materials: The following supporting information can be downloaded at: https://www.mdpi.com/article/10.3390/ijms231810786/s1.

Author Contributions: Conceived and designed the experiments, R.T. and R.S.; performed the experiments, S.C., M.P. and P.M.; analyzed the data, R.T., M.P. and R.S.; software, M.P.; contributed reagents/materials/analysis tools: S.C., M.P., A.F., P.M. and R.T.; collected samples, A.F., M.B. and S.C.; wrote the paper: R.T.; supervision, R.T. and R.S. All authors have read and agreed to the published version of the manuscript.

Funding: This research received no external funding.

Institutional Review Board Statement: The study on the volunteer subjects was conducted in accordance with the Declaration of Helsinki and was approved by the Ethics Committee of MAGNA GRAECIA UNIVERSITY and MATER DOMINI HOSPITAL (protocol code 2014.39, date of approval 16 April 2014).

Informed Consent Statement: Informed consent was obtained from all subjects involved in the study. Written informed consent has been obtained from the patients to publish this paper.

Acknowledgments: S.C. is supported by a fellowship from the Ph.D. Program in Life Sciences (XXXVI cycle), University "Magna Græcia" of Catanzaro, Italy.

Conflicts of Interest: The authors declare no conflict of interest.

References

1. Pilch, B.; Mann, M. Large-scale and high-confidence proteomic analysis of human seminal plasma. *Genome Biol.* **2006**, *7*, R40. [CrossRef]
2. Gilany, K.; Minai-Tehrani, A.; Savadi-Shiraz, E.; Rezadoost, H.; Lakpour, N. Exploring the human seminal plasma proteome: An unexplored gold mine of biomarker for male infertility and male reproduction disorder. *J. Reprod. Infertil.* **2015**, *16*, 61–71.
3. Drabovich, A.P.; Saraon, P.; Jarvi, K.; Diamandis, E.P. Seminal plasma as a diagnostic fluid for male reproductive system disorders. *Nat. Rev. Urol.* **2014**, *11*, 278–288. [CrossRef]
4. Bosler, J.S.; Davies, K.P.; Neal-Perry, G.S. Peptides in seminal fluid and their role in infertility: A potential role for opiorphin inhibition of neutral endopeptidase activity as a clinically relevant modulator of sperm motility: A review. *Reprod. Sci.* **2014**, *21*, 1334–1340. [CrossRef]
5. Fritz, R.; Mukherjee, A.; Zaghi, S.; Agalliu, I.; Jindal, S.; Tashima, A.K.; Lloyd, D.; Davies, F.; Davies, K.P. Identification and characterization of RSIY-11, a novel seminal peptide derived from semenogelin-1, which acts as a neutral endopeptidase inhibitor modulating sperm motility. *J. Assist. Reprod. Genet.* **2019**, *36*, 1891–1900. [CrossRef]
6. Neuhaus, J.; Schiffer, E.; von Wilcke, P.; Bauer, H.W.; Leung, H.; Siwy, J.; Ulrici, W.; Paasch, U.; Horn, L.; Stolzenburg, J. Seminal plasma as a source of prostate cancer peptide biomarker candidates for detection of indolent and advanced disease. *PLoS ONE* **2013**, *8*, e67514. [CrossRef]
7. Edström, A.M.; Malm, J.; Frohm, B.; Martellini, J.A.; Giwercman, A.; Mörgelin, M.; Cole, A.M.; Sørensen, O.E. The major bactericidal activity of human seminal plasma is zinc-dependent and derived from fragmentation of the semenogelins. *J. Immunol.* **2008**, *18*, 3413–3421. [CrossRef]
8. Malm, J.; Sørensen, O.; Persson, T.; Frohm-Nilsson, M.; Johansson, B.; Bjartell, A.; Lilja, H.; Ståhle-Bäckdahl, M.; Borregaard, N.; Egesten, A. The human cationic antimicrobial protein (hCAP-18) is expressed in the epithelium of human epididymis, is present in seminal plasma at high concentrations, and is attached to spermatozoa. *Infect. Immun.* **2000**, *68*, 4297–4302. [CrossRef]
9. Münch, J.; Rücker, E.; Ständker, L.; Adermann, K.; Goffinet, C.; Schindler, M.; Wildum, S.; Chinnadurai, R.; Rajan, D.; Specht, A.; et al. Semen-derived amyloid fibrils drastically enhance HIV infection. *Cell* **2007**, *131*, 1059–1071. [CrossRef]
10. Roan, N.R.; Müller, J.A.; Liu, H.; Chu, S.; Arnold, F.; Stürzel, C.M.; Walther, P.; Dong, M.; Witkowska, H.E.; Kirchhoff, F.; et al. Peptides released by physiological cleavage of semen coagulum proteins form amyloids that enhance HIV infection. *Cell Host Microbe* **2011**, *10*, 541–550. [CrossRef]
11. Palmblad, M.; Tiss, A.; Cramer, R. Mass spectrometry in clinical proteomics—From the present to the future. *Proteom. Clin. Appl.* **2009**, *3*, 6–17. [CrossRef]
12. Duncan, M.W.; Nedelkov, D.; Walsh, R.; Hattan, S.J. Applications of MALDI Mass Spectrometry in Clinical Chemistry. *Clin. Chem.* **2016**, *62*, 134–143. [CrossRef]
13. Klupczynska, A.; Swiatly, A.; Hajduk, J.; Matysiak, J.; Dyszkiewicz, W.; Pawlak, K.; Kokot, Z.J. Identification of Serum Peptidome Signatures of Non-Small Cell Lung Cancer. *Int. J. Mol. Sci.* **2016**, *17*, 410. [CrossRef]
14. Prodan, A.; Brand, H.; Imangaliyev, S.; Tsivtsivadze, E.; van der Weijden, F.; de Jong, A.; Paauw, A.; Crieelard, W.; Keijser, B.; Veerman, E. A Study of the Variation in the Salivary Peptide Profiles of Young Healthy Adults Acquired Using MALDI-TOF MS. *PLoS ONE* **2016**, *11*, e0156707. [CrossRef]

15. Chinello, C.; Cazzaniga, M.; De Sio, G.; Smith, A.J.; Gianazza, E.; Grasso, A.; Rocco, F.; Signorini, S.; Grasso, M.; Bosari, S.; et al. Urinary signatures of Renal Cell Carcinoma investigated by peptidomic approaches. *PLoS ONE* **2014**, *9*, e106684. [CrossRef]
16. Preianò, M.; Savino, R.; Villella, C.; Pelaia, C.; Terracciano, R. Gingival Crevicular Fluid Peptidome Profiling in Healthy and in Periodontal Diseases. *Int. J. Mol. Sci.* **2020**, *21*, 5270. [CrossRef]
17. Camargo, M.; Intasqui, P.; de Lima, C.B.; Montani, D.A.; Nichi, M.; Pilau, E.J.; Gozzo, F.C.; Lo Turco, E.G.; Bertolla, R.P. Maldi-tof fingerprinting of seminal plasma lipids in the study of human male infertility. *Lipids* **2014**, *49*, 943–956. [CrossRef]
18. Kałuża, A.; Jarząb, A.; Gamian, A.; Kratz, E.M.; Zimmer, M.; Ferens-Sieczkowska, M. Preliminary MALDI-TOF-MS analysis of seminal plasma N-glycome of infertile men. *Carbohydr. Res.* **2016**, *435*, 19–25. [CrossRef]
19. Liu, X.; Liu, G.; Zhu, P.; Wang, Y.; Wang, J.; Zhang, W.; Wang, W.; Li, N.; Wang, X.; Zhang, C.; et al. Characterization of seminal plasma proteomic alterations associated with the IVF and rescue-ICSI pregnancy in assisted reproduction. *Andrology* **2020**, *8*, 407–420. [CrossRef]
20. Fung, K.Y.; Glode, L.M.; Green, S.; Duncan, M.W. A comprehensive characterization of the peptide and protein constituents of human seminal fluid. *Prostate* **2004**, *61*, 171–181. [CrossRef]
21. Cadavid, J.A.P.; Alvarez, A.; Markert, U.R.; Cardona Maya, W. Differential protein expression in seminal plasma from fertile and infertile males. *J. Hum. Reprod. Sci.* **2014**, *7*, 206–211.
22. Greco, V.; Piras, C.; Pieroni, L.; Ronci, M.; Putignani, L.; Roncada, P.; Urbani, A. Applications of MALDI-TOF mass spectrometry in clinical proteomics. *Expert Rev. Proteom.* **2018**, *15*, 683–696. [CrossRef]
23. Terracciano, R.; Casadonte, F.; Pasqua, L.; Candeloro, P.; Di Fabrizio, E.; Urbani, A.; Savino, R. Enhancing plasma peptide MALDI-TOF-MS profiling by mesoporous silica assisted crystallization. *Talanta* **2010**, *80*, 1532–1538. [CrossRef]
24. Terracciano, R.; Preianò, M.; Maggisano, G.; Pelaia, C.; Savino, R. Hexagonal Mesoporous Silica as a Rapid, Efficient and Versatile Tool for MALDI-TOF MS Sample Preparation in Clinical Peptidomics Analysis: A Pilot Study. *Molecules* **2019**, *24*, 2311. [CrossRef]
25. Candenas, L.; Chianese, R. Exosome Composition and Seminal Plasma Proteome: A Promising Source of Biomarkers of Male Infertility. *Int. J. Mol. Sci.* **2020**, *21*, 7022. [CrossRef]
26. Neuhaus, J.; Schiffer, E.; Mannello, F.; Horn, L.C.; Ganzer, R.; Stolzenburg, J.U. Protease Expression Levels in Prostate Cancer Tissue Can Explain Prostate Cancer-Associated Seminal Biomarkers-An Explorative Concept Study. *Int. J. Mol. Sci.* **2017**, *18*, 976. [CrossRef]
27. Robert, M.; Gagnon, C. Purification and characterization of the active precursor of a human sperm motility inhibitor secreted by the seminal vesicles: Identity with semenogelin. *Biol. Reprod.* **1996**, *55*, 813–821. [CrossRef]
28. Robert, M.; Gibbs, B.F.; Jacobson, E.; Gagnon, C. Characterization of prostate-specific antigen proteolytic activity on its major physiological substrate, the sperm motility inhibitor precursor/semenogelin I. *Biochemistry* **1997**, *36*, 3811–3819. [CrossRef]
29. Preianò, M.; Maggisano, G.; Murfuni, M.S.; Villella, C.; Colica, C.; Fregola, A.; Pelaia, C.; Lombardo, N.; Pelaia, G.; Savino, R.; et al. Rapid Detection and Identification of Antimicrobial Peptide Fingerprints of Nasal Fluid by Mesoporous Silica Particles and MALDI-TOF/TOF Mass Spectrometry: From the Analytical Approach to the Diagnostic Applicability in Precision Medicine. *Int. J. Mol. Sci.* **2018**, *19*, 4005. [CrossRef]
30. Corigliano, A.; Preianò, M.; Terracciano, R.; Savino, R.; De Gori, M.; Galasso, O.; Gasparini, G. C3f is a potential tool for the staging of osteoarthritis. *J. Biol. Regul. Homeost. Agents* **2017**, *31*, 29–35.
31. Ścigalski, P.; Kosobucki, P. Recent Materials Developed for Dispersive Solid Phase Extraction. *Molecules* **2020**, *25*, 4869. [CrossRef]
32. Casado, N.; Morante-Zarcero, S.; Pérez-Quintanilla, D.; Câmara, J.S.; Sierra, I. Dispersive Solid-Phase Extraction of Polyphenols from Juice and Smoothie Samples Using Hybrid Mesostructured Silica Followed by Ultra-high-Performance Liquid Chromatography-Ion-Trap Tandem Mass Spectrometry. *J. Agric. Food Chem.* **2019**, *67*, 955–967. [CrossRef]
33. Herraiz, T.; Casal, V. Evaluation of solid-phase extraction procedures in peptide analysis. *J. Chromatogr. A* **1995**, *708*, 209–221. [CrossRef]
34. Tirumalai, R.S.; Chan, K.C.; Prieto, D.A.; Issaq, H.J.; Conrads, T.P.; Veenstra, T.D. Characterization of the low molecular weight human serum proteome. *Mol. Cell. Proteom.* **2003**, *2*, 1096–1103. [CrossRef] [PubMed]
35. Giacomini, E.; Ura, B.; Giolo, E.; Luppi, S.; Martinelli, M.; Garcia, R.C.; Ricci, G. Comparative analysis of the seminal plasma proteomes of oligoasthenozoospermic and normozoospermic men. *Reprod. BioMed. Online* **2015**, *30*, 522–531. [CrossRef] [PubMed]
36. Kovac, J.R.; Pastuszak, A.W.; Lamb, D.J. The use of genomics, proteomics, and metabolomics in identifying biomarkers of male infertility. *Fertil. Steril.* **2013**, *99*, 998–1007. [CrossRef]
37. Cooper, T.G.; Noonan, E.; von Eckardstein, S.; Auger, J.; Baker, H.W.; Behre, H.M.; Haugen, T.B.; Kruger, T.; Wang, C.; Mbizvo, M.T.; et al. World Health Organization reference values for human semen characteristics. *Hum. Reprod. Update* **2010**, *16*, 231–245. [CrossRef]
38. Ivosev, G.; Burton, L.; Bonner, R. Dimensionality reduction and visualization in principal component analysis. *Anal. Chem.* **2008**, *80*, 4933–4944. [CrossRef]
39. Emami, N.; Deperthes, D.; Malm, J.; Diamandis, E.P. Major role of human KLK14 in seminal clot liquefaction. *J. Biol. Chem.* **2008**, *283*, 19561–19569. [CrossRef] [PubMed]
40. Veveris-Lowe, T.L.; Kruger, S.J.; Walsh, T.; Gardiner, R.A.; Clements, J.A. Seminal fluid characterization for male fertility and prostate cancer: Kallikrein-related serine proteases and whole proteome approaches. *Semin. Thromb. Hemost.* **2007**, *33*, 87–99. [CrossRef] [PubMed]

41. de Lamirande, E. Semenogelin, the main protein of the human semen coagulum, regulates sperm function. *Semin. Thromb. Hemost.* **2007**, *33*, 60–68. [CrossRef] [PubMed]
42. Elzanaty, S.; Richthoff, J.; Malm, J.; Giwercman, A. The impact of epididymal and accessory sex gland function on sperm motility. *Hum. Reprod.* **2002**, *17*, 2904–2911. [CrossRef] [PubMed]
43. Ahlgren, G.; Rannevik, G.; Lilja, H. Impaired secretory function of the prostate in men with oligo-asthenozoospermia. *J. Androl.* **1995**, *16*, 491–498. [PubMed]
44. Liu, X.; Wang, W.; Zhu, P.; Wang, J.; Wang, Y.; Wang, X.; Liu, J.; Li, N.; Wang, X.; Lin, C.; et al. In-depth quantitative proteome analysis of seminal plasma from men with oligoasthenozoospermia and normozoospermia. *Reprod. BioMed. Online* **2018**, *37*, 467–479. [CrossRef]
45. Sharma, R.; Agarwal, A.; Mohanty, G.; Jesudasan, R.; Gopalan, B.; Willard, B.; Yadav, S.P.; Sabanegh, E. Functional proteomic analysis of seminal plasma proteins in men with various semen parameters. *Reprod. Biol. Endocrinol.* **2013**, *11*, 38. [CrossRef]
46. Martins, A.D.; Panner Selvam, M.K.; Agarwal, A.; Alves, M.G.; Baskaran, S. Alterations in seminal plasma proteomic profile in men with primary and secondary infertility. *Sci. Rep.* **2020**, *10*, 7539. [CrossRef]
47. Vickram, S.; Rohini, K.; Anbarasu, K.; Dey, N.; Jeyanthi, P.; Thanigaivel, S.; Issac, P.K.; Arockiaraj, J. Semenogelin, a coagulum macromolecule monitoring factor involved in the first step of fertilization: A prospective review. *Int. J. Biol. Macromol.* **2022**, *209*, 951–962. [CrossRef]
48. Yamasaki, K.; Yoshida, K.; Yoshiike, M.; Shimada, K.; Nishiyama, H.; Takamizawa, S.; Yanagida, K.; Iwamoto, T. Relationship between Semenogelins bound to human sperm and other semen parameters and pregnancy outcomes. *Basic Clin. Androl.* **2017**, *27*, 15. [CrossRef]
49. Mitra, A.; Richardson, R.T.; O'Rand, M.G. Analysis of recombinant human semenogelin as an inhibitor of human sperm motility. *Biol. Reprod.* **2010**, *82*, 489–496. [CrossRef]
50. Lombardo, N.; Preianò, M.; Maggisano, G.; Murfuni, M.S.; Messina, L.; Pelaia, G.; Savino, R.; Terracciano, R. A rapid differential display analysis of nasal swab fingerprints to distinguish allergic from non-allergic rhinitis subjects by mesoporous silica particles and MALDI-TOF mass spectrometry. *Proteomics* **2017**, *17*, 1600215. [CrossRef]

MDPI
St. Alban-Anlage 66
4052 Basel
Switzerland
www.mdpi.com

International Journal of Molecular Sciences Editorial Office
E-mail: ijms@mdpi.com
www.mdpi.com/journal/ijms

www.ingramcontent.com/pod-product-compliance
Lightning Source LLC
Chambersburg PA
CBHW061445210926
51850CB00008B/1618
* 9 7 8 3 7 2 5 8 0 6 3 5 5 *